Springer Series in Chemical Physics

Volume 110

The purpose of this series is to provide comprehensive up-to-date monographs in both well established disciplines and emerging research areas within the broad fields of chemical physics and physical chemistry. The books deal with both fundamental science and applications, and may have either a theoretical or an experimental emphasis. They are aimed primarily at researchers and graduate students in chemical physics and related fields.

More information about this series at http://www.springer.com/series/676

Geir Humborstad Sørland

Dynamic Pulsed-Field-Gradient NMR

Springer

Geir Humborstad Sørland
Anvendt Teknologi AS
Trondheim
Norway

ISSN 0172-6218
ISBN 978-3-662-44499-3 ISBN 978-3-662-44500-6 (eBook)
DOI 10.1007/978-3-662-44500-6

Library of Congress Control Number: 2014947707

Springer Heidelberg New York Dordrecht London

Printed on acid-free paper

Springer is part of Springer Science+Business Media (www.springer.com)

Preface

This book summarizes 23 years of work in Pulsed Field Gradient Nuclear Magnetic Resonance (PFG NMR). When staring up as a Ph.D. student in 1991, I was working on a 4.7 Tesla Bruker DMX system with a homemade diffusion probe (except for the actively shielded gradient coil built by Prof. Paul Callaghan's group at Massey University, New Zealand). This Ph.D. work, including the development of the unequal bipolar Pulsed Field Gradient Stimulated Echo method, ended up in a defense in 1998 where I had the pleasure of having Prof. Ken Packer as the opponent. Since then I have worked on various tasks where PFG NMR have been extensively used, as in quantification of fat content in biological systems, characterization of rock core plugs saturated with different fluids, and characterization of emulsions, being either stable or unstable. Besides being a post doc in work with the PFG NMR method, a company called Anvendt Teknologi AS was founded in 1998, aiming at producing and implementing PFG NMR applications for commercial purposes. In this latter work, I have come to realize that there does not exist a better referee than a demanding quality control manager who wants a method and instrumentation to work properly. The bottom line is that if you do not solve an issue that may occur with respect to, for example, an implemented PFG NMR quality control method and/or an instrumentation that does not work as prescribed, you lose goodwill, which is essential for running a business. Fortunately, the company has been up and running since 1998, and I have made a living out of it the past 15 years. The applications presented in this book are mainly a consequence of the research and development work done in Anvendt Teknologi AS, either as internal development work or contract work for the industry, universities, and research institutions.

During the past 15 years I have also been cooperating with different companies working in the field of NMR, such as Bruker GmbH, Resonance Instruments Ltd.(no longer in operation), and Oxford Instruments. The past 5 years I have had the pleasure of cooperating with Advanced Magnetic Resonance Ltd. and its founder Dr. Tim Benson. He has played a crucial role in the construction of the low field permanent magnet system for the characterization of emulsions, comprising shaped pulsed field gradient as well as the latest within RF-technology and spectrometer utilities. He has also provided valuable feedback in the production of this book, for which I am very grateful. In particular, Chap. 4 in this book is his piece of work entirely.

I have also had the pleasure of working with Dr. Klaus Zick at Bruker GmbH on various issues, mostly related to PFG NMR sequence development and implementation on high-resolution systems. He also kindly supplied me with the gradient field plots from the actively shielded gradients (Chap. 3).

Without the dedicated work done by Prof. Johan Sjøblom and his co-workers at the Ugelstad Laboratory, NTNU (University of Trondheim), there would not have been a chapter on emulsion characterization. In addition, the spoiler recovery approach was developed due to their requirements for a robust PFG NMR method for characterizing unstable emulsions. In particular, I would like to thank Sebastien Simon for helping me out on emulsion sample preparation, and for his efforts in explaining to me what emulsion science is all about.

Dr. Hege C. Widerøe at Statoil Research Centre in Trondheim has made a significant contribution to Chap. 8, and kindly let me use the data we have acquired together on rock core plugs over the past 13 years.

I would also like to thank Frank Lundby at Nofima (former Matforsk) for his dedicated work within quantification of lipid content in biological systems using our PFG NMR methods, and consequently his faith in me.

Finally, thanks to Prof. Jostein Krane, who contributed to the founding of Anvendt Teknologi AS, also by initiating contact with hardware manufacturers of PFG NMR instrumentation.

Regarding the content of this book, it is divided into nine chapters as follows:

Chapter 1: *Pulsed Field Gradient—NMR Sequences*: Here the expected signal attenuation from a variety of PFG sequences combined with or without relaxation time measurements are evaluated. In addition, the concept of coherence transfer pathways is discussed.

Chapter 2: *Observation Time-Dependent Diffusion Measurements in Heterogeneous Media by PFG NMR*: Here the Fick's law is solved for the case of short observation times in restricting geometries. In particular, it is shown that the short observation time expansion of the diffusion propagator resolves the surface to volume ratio from the surface relaxivity, an important finding that enables the possibility of measuring physical properties as pore sizes. Comments for the diffusion propagator at longer timescales are also given.

Chapter 3: *Experimental, Pitfalls and Suggested Solutions*: The most important distortions that may affect the PFG NMR measurements in a destructive way are discussed in this chapter; unwanted coherence transfer pathways, eddy current transients, internal magnetic field gradients, convection, vibration, and the effect of finite gradient pulse lengths. Ways to circumvent or supress the effect from the above-mentioned artifacts will also be discussed, as the application of phase sequences, crusher gradients, preemphasis adjustment, bipolar pulsed field gradients, convection compensated PFG NMR sequences, the second cumulant approximation, and corrected effective diffusion times.

Chapter 4: *PFG NMR Spectrometer*: This chapter provides an overview of the hardware required in order to conduct PFG NMR applications. Author: Dr. Tim Benson.

Chapter 5: *Analysis of Dynamic NMR Data*: An analysis of one- and two-dimensional datasets, both real and synthetic, using the Laplace transforms (ILT) and a discrete approach called Anahess is presented. Dr. Åsmund Ukkelberg developed the Anahess approach due to the need for a more robust and reliable method for analysis for PFG NMR data, and consequently he is an important contributor to this chapter.

Chapter 6: *The Spoiler Recovery Approach (SR)*: The spoiler recovery approach is presented with examples of applications, both at low and high external magnetic fields. In addition, a method for fast Fourier-Transformed-Inverse-Laplace-Transform is presented, which was developed by Dr. Henrik W. Anthonsen.

Chapter 7: *Emulsion Characterization*: Fast and accurate methods for emulsion characterization are presented. Especially, the characterization of unstable emulsions is innovative.

Chapter 8: *SCAL (Special Core AnaLysis)*: A variety of applications dedicated to the characterization of rock core plugs at different fluid saturations is described with examples. Most of the results are made on core material supplied by Statoil, but some of the data were acquired together with Center of Integrated Petroleum Research at the University of Bergen.

Chapter 9: *PFG NMR Applied to Biological Systems*: A variety of applications dedicated to the characterization of biological systems is described with examples.

Trondheim Geir Humborstad Sørland

Contents

Chapter 1
Pulsed Field Gradient—NMR Sequences

Abstract The mechanism via which an NMR signal is generated and how diffusion can be measured from solving the equation of motion of the nuclear spins is described. A set of basic pulsed field gradient (PFG) NMR sequences are provided, along with their corresponding echo attenuations. The concept of spin echoes is given along with a description of coherence transfer pathways, a tool for acquiring the wanted spin echo signal. Two Mathematica programs are provided, which describe a calculation necessary in order to produce diffusion coefficients from PFG NMR experiments using quadratic and sinusoidal gradients.

The pulsed field gradient (PFG) nuclear magnetic resonance (NMR) method is a well-established technique for studying molecular motion without disturbing the system under investigation [1–17]. It could be described as a non-invasive approach, as the only perturbation performed on the system is to alter the phase and/or the direction of the nuclear magnetic moments of the diffusing molecules in order to get a detectable signal [18–20]. A large variety of diffusion weighted sequences or techniques have been proposed that are optimised for different tasks, such as diffusion measurements in the presence of internal magnetic field gradients induced due to magnetic susceptibility changes through the sample, convection within the sample, or large eddy current field transients arising from any conducting metal surrounding the sample. This variety of PFG sequences not only reflects the increasing interest in using the PFG NMR technique in diffusion studies, but also shows that it is not necessarily a trivial task to extract the true diffusion coefficient from a PFG NMR experiment. In the following section, the mechanism via which an NMR signal is generated and how diffusion can be measured from solving the equation of motion of the nuclear spins will be described. Then a set of basic PFG sequences will be provided, along with their corresponding echo attenuations. In the appendix to this section, the concept of spin echoes is discussed along with a description of coherence transfer pathways, a tool for acquiring the wanted spin echo signal. Two Mathematica programs can also be found in the appendix to this chapter, which describe a calculation necessary in order to produce diffusion coefficients from PFG NMR experiments using

G.H. Sørland, *Dynamic Pulsed-Field-Gradient NMR*,
Springer Series in Chemical Physics 110, DOI 10.1007/978-3-662-44500-6_1

quadratic and sinusoidal gradients. An elaborate treatment of the NMR-theory can be found in the book by Slichter [21]

When placing hydrogen or another NMR active nuclei in an external magnetic field, the nuclear magnetic moment will align along the direction of this field. The Hamiltonian for non-interacting nuclear magnetic spins in an external magnetic field can be written [21]

$$\mathbf{H}(t) = -\gamma \hbar \vec{I} \cdot \vec{H}(t) \tag{1.1}$$

γ = gyromagnetic ratio (which varies depending on the nucleus under investigation), $\hbar$ = Planck's constant, I = spin operator and $\vec{H}(t)$ = external magnetic field. The time dependency of $\vec{H}(t)$ is included in order to make (1.1) valid when the system is influenced by an oscillating magnetic field (RF-field) and magnetic field gradients (g). When the magnetic field, $\vec{H}(t)$, is constant and homogeneous (= H_0), the eigenvalues, the energy levels, of a nucleus with I = 1/2 (as in the case of the hydrogen) may be written

$$E = \pm\frac{1}{2}\gamma\, \hbar H_0 = \pm\frac{1}{2}\, \hbar\omega_0 \tag{1.2}$$

The difference between the two energy levels is written

$$\Delta E = \hbar\omega_0 \tag{1.3}$$

In thermal equilibrium a difference in population between upper and lower level is given by the Boltzmann factor

$$\frac{n_{upper}}{n_{lower}} = \mathrm{e}^{-\frac{\hbar\omega}{kT}} \tag{1.4}$$

T = absolute temperature and k = Boltzmann's constant. The difference in population will generate a net nuclear magnetic moment, which has an amplitude directly proportional to the number of nuclear spins present in the sample. In thermal equilibrium, the moment will be aligned with the external magnetic field H_0. By imposing an oscillating magnetic field, RF-field, transverse to the external magnetic field, transitions between the energy levels will occur. The direction of the net nuclear magnetic moment will then move away from alignment with the external field, i.e. thermal equilibrium. When the RF-field is switched off, the system will again tend to align with the external field, the direction of H_0. The rate back to thermal equilibrium given by the characteristic relaxation times T_1 (longitudinal relaxation) and T_2 (transverse relaxation) represent the rate of loss of coherence in the plane transverse to the external magnetic field. The return to thermal equilibrium in combination with an oscillating net nuclear magnetic moment transverse to H_0 will cause changes in the magnetic flux, which can be recorded with the same RF-coil as was used to excite the system. The concept of generating an NMR spin echo and a gradient spin echo signal is described in Appendix.

1.1 The Pulsed Field Gradient (PFG) Diffusion Experiment

A basic PFG stimulated echo sequence shown in Fig. 1.2 consists of 3 intervals, a prepare, a store, and a read interval. In the prepare interval the molecules are labelled with a phase proportional to the integral of the effective gradient g (t). In Fig. 1.1 the anti-helmholz gradient coil is shown, and with proper dimensions on this coil, a linear gradient is generated (Chap. 3)

$$\vec{H}(t) = \vec{H_0} + g(t) \cdot z \tag{1.5}$$

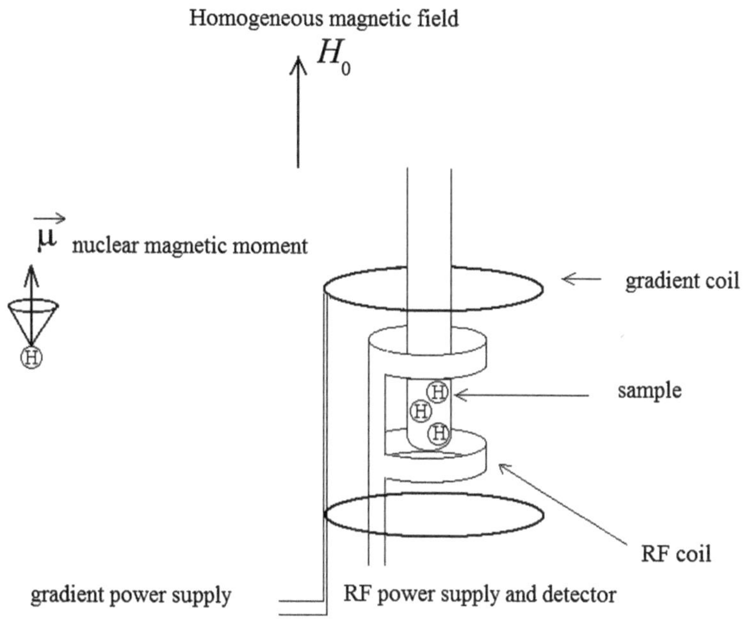

Fig. 1.1 A schematic drawing of the basics needed for performing a PFG-NMR experiment

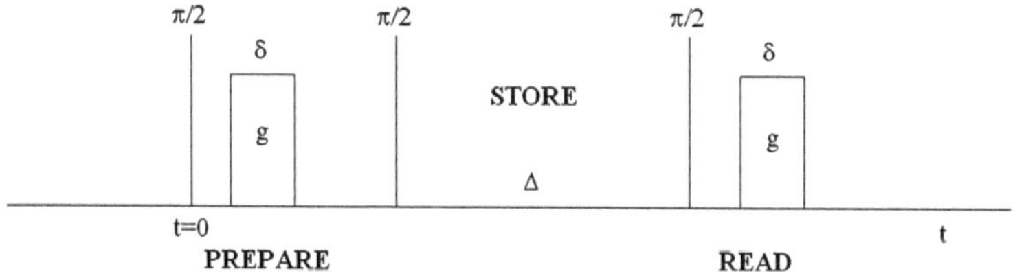

Fig. 1.2 The PFGSTE sequence where the PREPARE interval labels the phase of the spins, the STORE interval allow the spins to diffuse, and the READ interval labels the spins with a phase of opposite polarity

z is the position of the molecule along the direction of the magnetic field gradient, which again is collinear with H_0 or the longitudinal direction. The mobility of the hydrogen may be recorded by making use of this magnetic field gradient g, which imposes a position dependent frequency on the system, and with which the nuclear magnetic moment of the proton is oscillating in a plane transverse to H_0

$$\omega = \gamma H_0 + \gamma gz \tag{1.6}$$

After the application of the second pulse, the net magnetic moment is stored in the longitudinal direction, and is thus unaffected by the presence of any transverse relaxation. In the read interval the nuclear spins are labelled with a phase of opposite polarity to the prepare interval. If the molecules have travelled a distance during the PFG-sequence, there is a dephasing of the net magnetic moment given by

$$\Delta\phi = \gamma g(z_2 - z_1) \tag{1.7}$$

$(z_2 - z_1)$ is the distance travelled by the spins in one dimension, during the store interval, when assuming infinitely short gradient pulses. The induced current in the RF coil will be attenuated because of the dephasing. This will be apparent in the pulsed field gradient stimulated echo attenuation expressed as a function of the diffusion time and the gradient strength. In order to calculate the echo attenuation arising from an arbitrary PFG NMR experiment the Bloch equations [20] for a system of diffusing and interacting spins needs to be solved. By imposing a radio frequency field on a proton system situated in an external magnetic field, one may induce transitions between the energy levels, i.e. bringing the system out of thermal equilibrium. Assuming that the spins are interacting and diffusing in the presence of an external magnetic field the classical equation of motion of the spin system is known as the Bloch-Torrey equations in [20, 22]

$$\frac{d}{dt}\vec{M} = \gamma \vec{M} \times \vec{H} + D\nabla^2\vec{M} \tag{1.8}$$

Equation (1.8) may be written as

$$\begin{aligned}
\frac{d}{dt}M_z &= -\gamma M_y H_1 + \frac{M_0 - M_z}{T_1} + D\nabla^2 M_z \\
\frac{d}{dt}M_x &= \gamma M_y h_0 - \frac{M_x}{T_2} + D\nabla^2 M_x \\
\frac{d}{dt}M_y &= \gamma (M_z H_1 - M_x h_0) - \frac{M_y}{T_2} + D\nabla^2 M_y
\end{aligned} \tag{1.9}$$

H_1 is the rotating magnetic field with a frequency ω, $h_0 = H_0 + \omega/\gamma$, T_1 is the longitudinal or spin lattice relaxation time, T_2 is the transverse or spin-spin relaxation time, and D is the diffusion coefficient. The acquisition of the NMR signal is performed by sampling the induced magnetic moment in the transverse plane to the external field, and the phases for the transverse components are affected by diffusion in the presence of a magnetic field gradient. During the z-storage of the net magnetisation, there is no effect from diffusion in the presence of magnetic field gradients on the net magnetisation. When assuming a Gaussian diffusion propagator [23] and a mono-exponential attenuation of the NMR signal due to relaxation processes, it can be shown that the solution to the transverse magnetisation ($M^+ = Mx + i\,My$) for an arbitrary pulsed field gradient sequence is written [1]

$$M^+ = M_0^+ e^{-\frac{t_1}{T_2}} e^{-\frac{t_2}{T_1}} e^{-\gamma^2 D \int_0^t \left(\int_0^{t'} g(t'')dt''\right)^2 dt'} \quad (1.10)$$

M_+^0 is the integration constant depending on the initial conditions. The applied gradients in (1.10) are the effective gradients. That is, depending on the coherence transfer pathway, the polarity of the gradient may have to be changed after application of a 180° RF-pulse or two 90° RF-pulses (see Appendix or [1]). In the following, M^+ and M_0^+ will be denoted I and I_0 respectively. The induced current in the RF coil, the NMR signal, will be attenuated because of the dephasing. This will be apparent in the pulsed field gradient stimulated echo attenuation expressed as a function of the diffusion time and the gradient strength.

$$I = I_0 e^{-\frac{t_1}{T_2}} e^{-\frac{t_2}{T_1}} e^{-\gamma^2 D \int_0^t \left(\int_0^{t'} g(t'')dt''\right)^2 dt'} \quad (1.11)$$

where

- t_1 duration during which the NMR-signal is influenced by transverse relaxation processes
- t_2 duration during which the NMR-signal is influenced by longitudinal relaxation processes
- $g(t'')$ total effective magnetic field gradient, external and internal
- D diffusion coefficient
- T_1 Characteristic longitudinal relaxation time
- T_2 Characteristic transverse relaxation time
- I_0 Initial intensity of the NMR-signal

The usual way to conduct the diffusion experiment is to fix the durations and vary the applied magnetic field gradients, so that the reduction of the NMR echo signal due to transverse and longitudinal relaxation processes will be constant during the experiment. A general expression for the amplitude of echo signal can then be written

$$I = I_0 e^{-\gamma^2 D \int_0^t \left(\int_0^{t'} g(t'')dt''\right)^2 dt'} \quad (1.12)$$

In the following discussion, focus will be on the PFG NMR sequences where the diffusion coefficient is measured by generating an attenuated NMR signal by incrementing the applied gradient strength, g. The two relaxation terms containing t_1 and t_2 become constant and can be incorporated to the initial intensity of the NMR signal, I_0, as shown in (1.12). When incrementing the gradient strength and keeping the durations fixed, the diffusing molecules will probe the restrictions of a heterogeneous sample to the same degree during the NMR experiment. The measured diffusion coefficient will then represent a stable system with the respect to the fraction of molecules experiencing the heterogeneity of the sample. This is important when measuring physical parameters such as tortuosity, surface to volume ratio or surface relaxation in heterogeneous media [24–31] (Chap. 2).

Many PFG NMR experiments now combine diffusion measurements with relaxation or spectral measurements [11, 32]. Before showing some basic two-dimensional sequences, the spoiler recovery sequence is introduced. This sequence is designed to circumvent the need for waiting five times the longitudinal relaxation time between each scan while maintaining quantitative diffusion information, and may reduce the experimental time by ~80 % without any significant loss in signal to noise ratio. This method also reduces the impact from radiation damping, so its applicability is for any operating magnetic field, at least in the region 0.05–14.1 T (Chap. 6).

1.2 The Ordinary PFGSE Sequence

In Fig. 1.3 the simplest form of the NMR diffusion experiment is shown, the pulsed field gradient spin echo method. This sequence was introduced by Stejskal and Tanner in 1965 [33], and consists of a 90° excitation RF pulse and two mono polar gradient pulses separated by a 180° refocusing RF-pulse. As can be seen in Fig. 1.3 the impact of the refocusing 180° RF-pulse is to change the polarity of the effective gradient. This is a consequence of the impact the RF-pulse has on the evolving

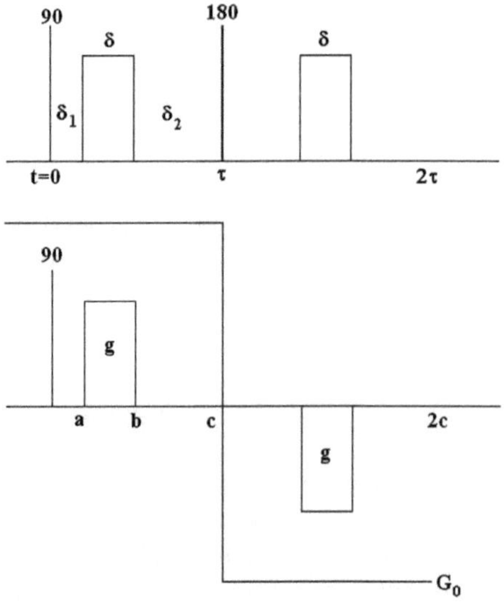

Fig. 1.3 The pulsed field gradient spin echo (*PFGSE*) with real (*upper sequence*) and effective gradients (*lower sequence*)

Hamiltonian when calculating the expectation value for the net nuclear magnetization (see Appendix). In order to evaluate the attenuation arising from the sequence it is taken into account that the system is subjected to transverse relaxation processes during the entire sequence. When assuming a Gaussian diffusion propagator, (1.12) can be applied to calculate the effective gradient seen by the nuclear spin system at each particular time interval as shown in column two in Table 1.1. Then the effective gradients for the successive time intervals are integrated, for which results are shown in column three in Table 1.1.

The next step is then to integrate the square of these terms, and the result of this rather tedious calculation even for the simplest PFG sequence is shown for each term in column four (Table 1.2).

Table 1.1 Column 1 is the time parameter with its corresponding effective gradients in column 2

Time	Effective gradient	$\int_0^{t'} g(t'')dt''$
(0, a)	$-G_0$	$-G_0 \cdot t'$
(a, b)	$-G_0 + g$	$-G_0 \cdot t' + g \cdot (t' - a)$
(b, c)	$-G_0$	$-G_0 \cdot t' + g \cdot (b - a)$
(c, c + a)	$+G_0$	$G_0 \cdot (t' - c) + g \cdot (b - a)$
(c + a, c + b)	$+G_0 - g$	$G_0 \cdot (t' - c) + g \cdot (b - a) - g \cdot (t' - (c + a))$
(c + b, 2c)	$+G_0$	$G_0 \cdot (t' - c)$

Column 3 yields the corresponding time integral of the effective gradient

Table 1.2 Column 1 is the time parameter with its corresponding time integral of the square of the time integral of the effective gradient in Column 2

Time	$\int_0^{t} (\int_0^{t'} g(t'')dt'')^2$
(0, a)	$\frac{a^3G_0^2}{3}$ or $\frac{\delta_1^3G_0^2}{3}$
(a, b)	$\frac{-a^3G_0^3+((b-a)(G-g)+aG)^3}{3(G_0-g)}$ or $\frac{-\delta_1^3G_0^3+(\delta(G-g)+\delta_1 G)^3}{3(G_0-g)}$
(b, c)	$\frac{(cG_0-(b-a)g)^3-((b-a)(G-g)+aG)^3}{3G_0}$ or $\frac{(\tau G_0-\delta g)^3-(\delta(G-g)+\delta_1 G)^3}{3G_0}$
(c, c + a)	$\frac{(cG_0-(b-a)g)^3+((b-a)g-(c-a)G)^3}{3G_0}$ or $\frac{(\tau G_0-\delta g)^3+(\delta g-(\delta+\delta_2)G)^3}{3G_0}$
(c + a, c + b)	$\frac{((c-b)G_0)^3+(-(b-a)g+(c-a)G)^3}{3(G_0-g)}$ or $\frac{(\delta_2 G_0)^3+(-\delta g+(\delta+\delta_2)G)^3}{3(G_0-g)}$
(c + b, 2c)	$\frac{(c-b)^3G_0^2}{3}$ or $\frac{\delta_2^3G_0^2}{3}$

The last time integral is either represented with the time parameters a, b, c or the more familiar gradient pulse length δ, dead times δ_1 and δ_2, and inter echo spacing 2τ

Finally, all the terms in column 2 are collected, simplified and the result is put into (1.12) to give the attenuation for the simplest PFG NMR sequence

$$I = I_0 e^{-\frac{2\tau}{T_2}} e^{-\gamma^2 g^2 \delta^2 D(\tau-\delta/3) - \gamma^2 G_0^2 D \frac{3}{2}\tau^3 - \frac{\gamma^2 D}{3}(-2\delta_1^3 + 3\delta_1^2\tau + 3\delta_1\tau^2 + (\delta_1+\delta)(2(\delta_1+\delta)^2 - 3(\delta_1+\delta)\tau - 3\tau^2))g \cdot G_0} \quad (1.13)$$

When neglecting the cross terms between the applied and internal magnetic field gradients, the final outcome is the well known PFGSE-attenuation

$$I = I_0 e^{-\frac{2\tau}{T_2}} e^{-\gamma^2 g^2 \delta^2 D(\tau-\delta/3)} \quad (1.14)$$

The observation time is half of the inter echo spacing, τ, minus a term that takes into account a finite duration of the gradient pulse, $\delta/3$. In Fig. 1.4, the natural logarithm of the attenuation from the spin echo due to incrementing the gradient strength is shown. Knowledge of the time parameters in the pulse sequence and the applied gradient strength used, which in this case goes up to 320 Gauss/cm for the strongest applied gradient, will provide the experimentally determined diffusion coefficient from the slope of the logarithm of the echo attenuation. The slope in this example is linear as the system measured on is bulk water, a single molecular liquid, which should result in a mono exponential decay, and thus a linear slope when taking the logarithm of the attenuation.

Other parameters within the experiment may be varied to achieve an attenuation of the NMR signal, including the inter echo spacing and the gradient pulse length. In bulk or homogeneous systems, where an observation time dependency is not expected on the diffusion coefficient, such an approach method would be

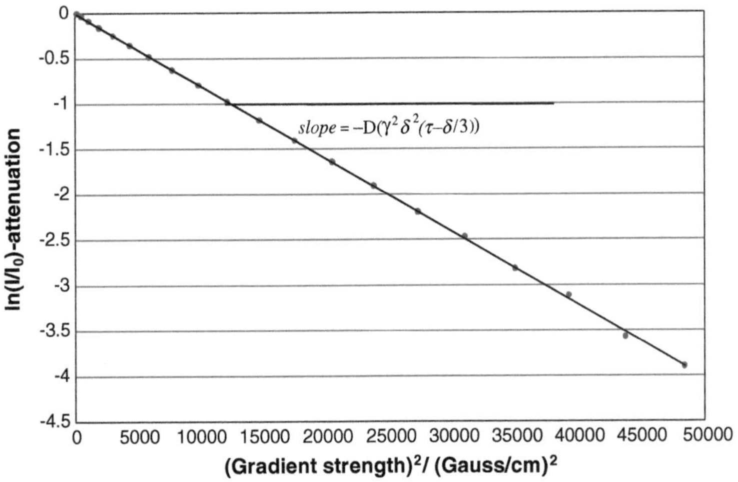

Fig. 1.4 The logarithm of the echo attenuation as a function of the square of the applied gradient strength

equally applicable. However, once there is any kind of restriction inside the system, be it a solid matrix or a different fluid component, a different apparent diffusion coefficient will be determined by the PFG experiment, which is also depend on the observation time (gradient pulse separation) and gradient pulse length applied. In these cases, the slope of the logarithm of echo attenuation will no longer be linear. In the case of restricted diffusion studies, alternative sequences are often preferred such as the stimulated echo sequence described in the next section, due to their ability to probe greater length scales.

1.3 The Ordinary PFGSTE Sequence

In many systems to be investigated, it is of interest to study the time evolution of the diffusion coefficient. Any spin experiencing restrictions will have their apparent mobility reduced when increasing the observation time, as the likelihood of encountering any obstacle increases as the observation time is increased. As the transverse relaxation time is significantly shorter than the longitudinal relaxation time in many heterogeneous systems, a z-storage delay can be introduced to the PFG sequence as shown in Fig. 1.5, producing the pulsed field gradient

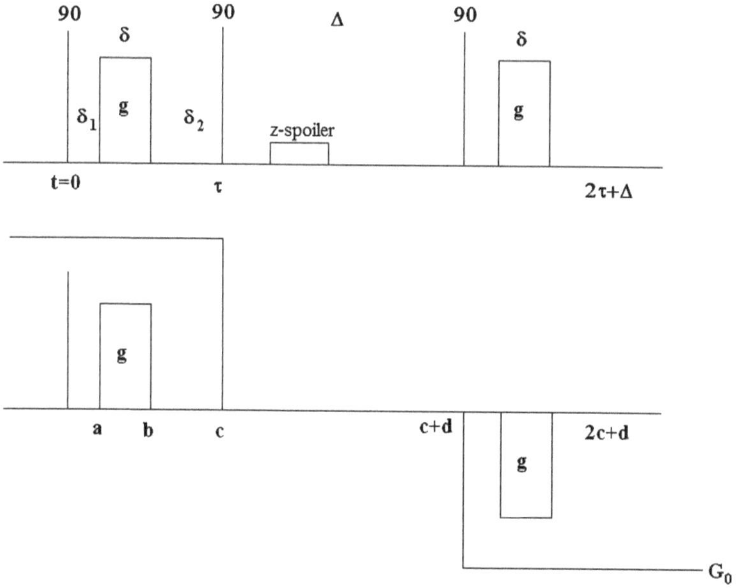

Fig. 1.5 The pulsed field gradient stimulated echo (*PFGSTE*) with real (*upper sequence*) and effective (*lower sequence*) gradients. In the z-storage interval a short spoiler gradient pulse is applied to remove unwanted transverse magnetization coherences. This is known as a spoiler technique and means that during the z-storage delay any transverse magnetization is subjected to a dephasing gradient pulse

stimulated echo sequence. Here the refocusing 180° RF-pulse is replaced by two 90° RF-pulses separated by a time interval Δ. In order to analyze the sequence, as before it is divided into different time intervals and the effective gradients are calculated as for the PFGSE sequence. However, now there is no refocusing RF-pulse that inverts the coupling to the Hamiltonian in the expectation value for the net nuclear magnetization to produce a spin echo. Instead, five echoes are generated in the stimulated echo sequence [34] (see Chap. 3 for further demonstration of the five echoes). As the information on diffusion is apparent in one of the stimulated echoes, the concept of coherence transfer pathways can be applied to follow the magnetization that contributes to the wanted stimulated echo signal. In order to collect this stimulated echo signal only, the phase sequence proposed by Fauth et al. [34], shown in Table 1.3 can be applied.

Following the correct coherence transfer pathway, the effective gradients and corresponding integrals are produced and shown in Table 1.4.

An integration of the square of the expressions in column 3 then provides the expressions that are collected and simplified. The expressions for the echo attenuation from the simplest PFGSTE is then written

$$I = I_0 e^{-\frac{2\tau}{T_2}} e^{-\gamma^2 g^2 \delta^2 D(\Delta+\tau-\delta/3) - \gamma^2 D \frac{\delta}{3}(2\delta_1^2 + 2(\delta_1+\delta)^2 + \delta_1(2(\delta_1+\delta)-3\tau) - 3(\delta_1+\delta)\tau - 3\tau(\tau+2\Delta))} \tag{1.15}$$

By neglecting the cross terms the more readable equation is found

$$I = I_0 e^{-\frac{2\tau}{T_2}} e^{-\gamma^2 g^2 \delta^2 D(\Delta+\tau-\delta/3)} \tag{1.16}$$

Table 1.3 The phase sequence that cancel all echoes except the stimulated echo in the PFGSTE sequence

Sequence #	π/2 Excitation	Second π/2	Third π/2	Receiver
1	+x or 0	+x or 0	+x or 0	−x or π
2	+x or 0	−x or π	+x or 0	+x or 0
3	+x or 0	+x or 0	−x or π	+x or 0
4	+x or 0	−x or π	−x or π	−x or π

Table 1.4 Column 1 is the time parameter with its corresponding effective gradients in column 2

time	effective gradient	$\int_0^{t'} g(t'')dt''$
(0, a)	$-G_0$	$-G_0 \cdot t'$
(a, b)	$-G_0 + g$	$-G_0 \cdot t' + g \cdot (t' - a)$
(b, c)	$-G_0$	$-G_0 \cdot t' + g \cdot (b - a)$
(c, c + d)	0	$-G_0 \cdot c + g \cdot (b - a)$
(c + d, c + d + a)	$+G_0$	$G_0 \cdot (t' - 2c - d) + g \cdot (b - a)$
(c + d + a, c + d + b)	$+G_0 - g$	$G_0 \cdot (t' - 2c - d) + g \cdot (b - a) - g \cdot (t' - (c + a + d))$
(c + d + b, 2c + d)	$+G_0$	$G_0 \cdot (t' - 2c - d)$

Column 3 yields the corresponding time integral of the effective gradient

Again, the slope of the logarithm of the attenuation as a function of the square of the applied gradient strength will return the diffusion coefficient provided all time parameters and gradient strengths are known. Now, the observation time has an extra term, the z-storage delay Δ, which is independent of the gradient pulse length. As the system is affected by T_1-relaxation during this period, the NMR signal persists longer than if it was subjected to T_2-relaxation only. Consequently, the observation time dependent diffusion coefficient can be studied up to the order of seconds. Figure 1.6 shows how the time dependency on the diffusion coefficient of water diffusing amongst compact mono sized plastic spheres can be studied. From the initial decay, the short observation times, the surface to volume ratio can be extracted [27], while for the longer observation times the tortuosity limit may be found. The latter is dependent on the NMR signal to persist long enough for the heterogeneity of the sample to be sufficiently probed. In the particular case shown in Fig. 1.6, the spheres have a diameter of 98 μm and then the root of the mean squared displacement assuming bulk diffusion is 150 μm for the longest observation time. Then it can be concluded that the experiment do measure the true tortuosity (Sect. 2.3).

As will be discussed in Chap. 3, there are some major drawbacks using the ordinary PFGSE and PFGSTE sequences. Because of the lack of a refocusing 180° RF pulse within the read and prepare intervals, when the magnetization is subjected to transverse relaxation, it produces a term that couples any background gradient within the sample to the observation time Δ (PFGSTE) or τ (PFGSE). Thus, the diffusion coefficient measured may be significantly dependent on an uncontrollable parameter, the so-called background gradient strength. The applied gradient strength is controlled by the geometry of the gradient coil and the electrical current being sent through. The background gradient may be an eddy current transient, i.e. varying with time, or simply an internal magnetic field gradient arising from differences in magnetic susceptibility within a

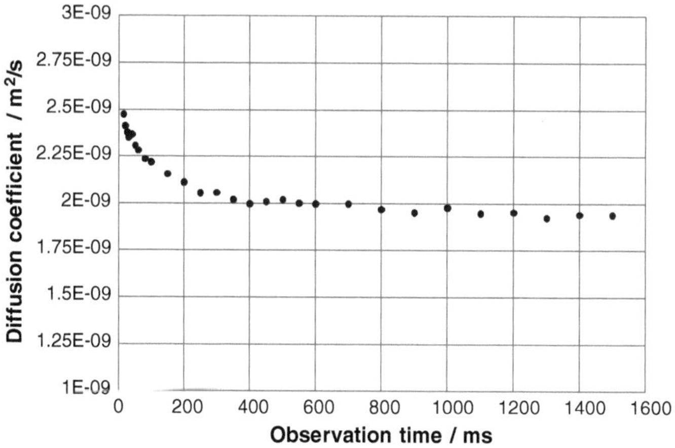

Fig. 1.6 Diffusion measurements amongst compact mono-sized spheres

heterogeneous sample [35]. The internal gradient may for some systems, such as porous rocks that may contain ferromagnetic impurities, exceed the values of the applied magnetic field gradient. Then the fitted diffusion coefficient from an experiment will be strongly dependent on the porous matrix, and not as much on the true molecular mobility of the fluid confined in the porous system. For many PFG NMR instrumentations (due to eddy current transients) and heterogeneous systems (due to internal magnetic field gradients), it is convenient to have the option of reducing the effect of background gradients, as when applying bipolar gradients in the PFG NMR sequence.

1.4 The 11 Interval Bipolar PFGSE

Due to the presence of internal magnetic field gradients, a pulsed field gradient technique that reduces the effect from internal gradients on the echo attenuation as a function of applied gradient strength improves the performance of the PFG NMR sequences. One intuitive approach is to apply such strong gradients that the cross terms that couples the internal and applied gradients are treated as insignificant in comparison [36]. Assuming that the available magnetic field gradient strength is not high enough to employ this technique, more complicated RF-pulse sequences can be employed together with gradients of both polarities in order to minimize the internal gradient effects. In 1973 Blinc et al. [7] proposed a pulsed field gradient sequence which made use of bipolar magnetic field gradients, but this experiment is more complex than the ones proposed by Karlicek et al. [1]. Blinc et al. employed a Waugh-type sequence [7, 21], which significantly reduces the term containing the dipolar interaction in a liquid crystal, together with a Carr Purcell train of 180° RF-pulses [37]. In addition to the removal of the term containing the dipolar interaction, R. Blinc established that their sequence, due to the application of bipolar gradient pulses, also averaged out the coupling to a magnetic field background gradient in the Hamiltonian. Later, Karlicek et al. [1] in 1980 and Cotts et al. [2] in 1989 proposed PFG NMR sequences which by applying bipolar gradient pulses was dedicated specifically to remove the coupling between the applied and internal magnetic field gradients.

The first sequence to be analyzed is the 11 interval bipolar PFGSE sequence [17]. It is the simplest sequence possible that applies bipolar pairs of magnetic field gradients. This sequence minimizes the coupling between the applied and the internal gradients and the observation time is kept at a minimum. It also employs the minimum amount of RF-pulses in the bipolar version of the PFG NMR experiment, from which it follows that signal arising from unwanted coherence transfer pathways will be at a minimum. An RF phase sequence for minimizing the signal arising from unwanted coherence transfer pathways is found in Table 1.5. This phase sequence is constructed specifically for the removal of unwanted coherence transfer pathways as $P = 0 \rightarrow +1 \rightarrow +1 \rightarrow +1$ and $P = 0 \rightarrow +1 \rightarrow 0 \rightarrow +1$, as will be shown in the following (the recipe is found in Appendix). For the first

Table 1.5 The phase sequence that cancel all echoes except wanted spin echo in the bipolar 11 interval PFGSE sequence

Sequence #	π/2 Excitation	First π	Second π	Receiver
1	+x or 0	+y or π/2	−y or 3π/2	+x or 0
2	+x or 0	+y or π/2	+y or π/2	+x or 0
3	+x or 0	+y or π/2	−x or π	−x or π
4	+x or 0	−x or π	+y or π/2	−x or π

coherence transfer pathway, ΔP_1 is +1 at the point of application of the first RF-pulse, then $\Delta P_{2,3}$ is zero for the rest of the pathway. Thus the resulting phase φ is 0 for all pathways, and as the receiver phase is 0, 0, π, and π, the signal from this pathway will be cancelled.

For the second pathway the calculation for the resulting phase for the 4 different phase sequences shown in Table 1.5 is

$$\varphi = \sum_{j=1}^{3} \Delta P_j \varphi_j = \begin{bmatrix} +1 \cdot 0 - 1 \cdot \pi/2 + 1 \cdot 3\pi/2 = \pi \\ +1 \cdot 0 - 1 \cdot \pi/2 + 1 \cdot \pi/2 = 0 \\ +1 \cdot 0 - 1 \cdot \pi/2 + 1 \cdot \pi = \pi/2 \\ +1 \cdot 0 - 1 \cdot \pi + 1 \cdot \pi/2 = -\pi/2 \end{bmatrix}, \quad \varphi_{RECEIVER} = \begin{bmatrix} 0 \\ 0 \\ \pi \\ \pi \end{bmatrix}$$

Thus the contribution from the pathway $P = 0 \rightarrow +1 \rightarrow 0 \rightarrow +1$ will also be cancelled. In addition unwanted coherence transfer pathways as $P = 0 \rightarrow -1 \rightarrow +1 \rightarrow +1$ will not contribute to the NMR echo signal as the sequence of effective gradient pulses does not result in a gradient echo (Fig. 1.7). The effective gradients and corresponding integrals are as shown in Table 1.6. The resulting equation from the calculation is

$$\mathrm{I} = \mathrm{I_0 e}^{-\frac{4\tau}{\mathrm{T}_2}} e^{-\gamma^2 f^2 \delta^2 \mathrm{D}(\tau-\delta/3) - \gamma^2 \mathrm{g}^2 \delta^2 \mathrm{D}(3\tau-\delta/3) - 2\gamma^2 fg\delta^2 \mathrm{D}(\tau) - \gamma^2 D\delta(\delta_1-\delta_2)(\tau)(\mathrm{g+f})\mathrm{G}_0 - \frac{4\tau^3 \mathrm{G}_0}{3}} \tag{1.17}$$

Denoting $g = g_{cal}(i)$ and $\mathrm{f} = g_{cal}(i - x)$, where g_{cal} is the gradient strength per arbitrary unit, and (i) and (i − x) are gradient amplitudes in arbitrary units, the equation is simplified to

$$\mathrm{I} = \mathrm{I_0 e}^{-\frac{4\tau}{\mathrm{T}_2}} \mathrm{e}^{-\gamma^2 \mathrm{g}^2 (2\delta)^2 \mathrm{D}\left[\frac{3\tau}{2} - \delta/6\right]\left[i - \frac{x}{2}\left(\frac{\tau-\delta/6}{\frac{3}{2}\tau-\delta/6}\right)\right]^2 - \gamma^2 D(2i-x)\delta(\delta_1-\delta_2)(\tau) \mathrm{g}_{cal} \mathrm{G}_0 - \frac{4\tau^3 \mathrm{G}_0}{3} + offset} \tag{1.18}$$

x provides the difference in applied gradient strength, and the offset is a small constant term due to this difference in amplitude of f and g, and is given by

$$offset = -\gamma^2 \delta^2 \mathrm{D}\left[\frac{(\tau - \delta/6)^2}{\frac{3}{2}\tau - \delta/6} - (\tau - \delta/3)\right] x^2 \tag{1.19}$$

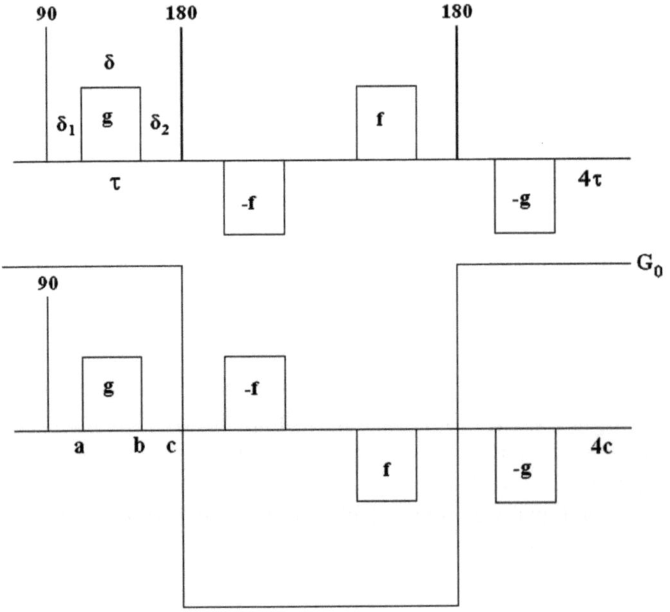

Fig. 1.7 The 11 interval bipolar PFGSE sequence with real (*upper sequence*) and effective gradients (*lower sequence*)

Table 1.6 Column 1 is the time parameter with its corresponding effective gradients in column 2

Time	Effective gradient	$\int_0^{t'} g(t'')dt''$
(0, a)	$+G_0$	$+G_0 \cdot t'$
(a, b)	$+G_0 + g$	$+G_0 \cdot t' + g \cdot (t' - a)$
(b, c)	$+G_0$	$+G_0 \cdot t' + g \cdot (b - a)$
(c, c + a)	$-G_0$	$-G_0 \cdot (t' - 2c) + g \cdot (b - a)$
(c + a, c + b)	$-G_0 + f$	$-G_0 \cdot (t' - 2c) + f \cdot (t' - (c + a)) + g \cdot (b - a)$
(c + b, 2c)	$-G_0$	$-G_0 \cdot (t' - 2c) + f \cdot (b - (c + a)) + g \cdot (b - a)$
(2c,2c + a)	$-G_0$	$-G_0 \cdot (t' - 2c) + f \cdot (b - a) + g \cdot (b - a)$
(2c + a,2c + b)	$-G_0 - f$	$-G_0 \cdot (t' - 2c) - f \cdot (t' - (2c + a)) + f \cdot (b - a) + g \cdot (b - a)$
(2c + b, 3c)	$-G_0$	$-G_0 \cdot (t' - 2c) + g \cdot (b - a)$
(3c, 3c + a)	$+G_0$	$+G_0 \cdot (t' - 4c) + g \cdot (b - a)$
(3c + a, 3c + b)	$+G_0 - g$	$+G_0 \cdot (t' - 4c) - g \cdot (t' - (3c + a)) + g \cdot (b - a)$
(3c + b, 4c)	$+G_0$	$+G_0 \cdot (t' - 4c)$

Column 3 yields the corresponding time integral of the effective gradient

When $x = 0$ there is no difference in amplitudes of f and g, and the offset term equals 0. Equations (1.17–1.19) are valid regardless of whether equal or unequal bipolar gradients are being used. A small cross term exists between the applied and the internal magnetic field gradient. This term is canceled provided

$\delta_1 = \delta_2$. It follows from the evaluation of the echo attenuation that the application of bipolar magnetic field gradients significantly reduces or cancels the cross term between the applied and the internal magnetic field gradients. But is there an intuitive explanation to this? Such an explanation can be arrived at by examination of the effective gradients in Table 1.1 for the ordinary PFGSE. It can be seen that the effective applied gradient changes polarity when the background gradient changes polarity. This is a consequence of the 180° RF pulse on the Hamiltonian (Appendix). As the square of the effective gradients generates cross terms between the applied and background gradients in both the prepare and read intervals, these terms will have the same polarity and will not cancel out. If the read and prepare intervals are extended by adding 180° RF-pulses in the PFGSE sequence and applying bipolar gradients in the prepare and read intervals, there will be two cross terms generated between the applied and background gradients in both intervals, but with opposite polarity (see Table 1.6). These cross terms are cancelled before taking the square of the effective gradients, leaving the square of the applied gradient and the square of the background gradient. As the latter is constant, the applied gradient strength may be incremented and the term containing the background gradient can be treated as an offset term.

1.5 The 13 Interval Bipolar PFGSTE

Even though the 11-interval sequence is the simplest and most effective in cancelling out the impact from internal magnetic field gradients in the PFG-NMR experiment, there is often a need for measuring the diffusion coefficient as a function of observation time. To do so, it is again convenient to proceed to the stimulated echo sequence where there is access to a z-storage delay, which can be used for the study of observation time dependent diffusivity. The 13 interval bipolar PFGSTE sequence, shown in Fig. 1.8, is a stimulated echo sequence but with refocussing 180° RF pulses in the prepare and read intervals [2, 5]. On each side of the 180° RF pulse, applied pulsed gradients of different polarity and different strength, f and g, are placed.

The appropriate phase sequence is found in Table 1.7. When using unequal bipolar gradients g and f it is sufficient to use the first four (#1–#4), as the unequal pairs cancel out the pathway $p \rightarrow 0 \rightarrow +1 \rightarrow +1$ in the read interval. The second part of the phase sequence must be included if equal bipolar gradients are applied (g = f). This phase sequence was developed by Wu et al. [38] and is designed to cancel the coherence transfer pathway $p \rightarrow 0 \rightarrow +1 \rightarrow +1$ from the read interval of the pulse sequence.

Following the correct coherence transfer pathway, the effective gradients and corresponding integrals are found as shown in Table 1.8.

The resulting equation from the calculation is

$$I = I_0 e^{-\frac{4\tau}{T_2}} e^{-\gamma^2 f^2 \delta^2 D(\Delta+\tau-\delta/3) - \gamma^2 g^2 \delta^2 D(\Delta+3\tau-\delta/3) - 2\gamma^2 fg\delta^2 D(\Delta+\tau) - \gamma^2 D\delta(\delta_1-\delta_2)(\tau)(g+f)G_0 - \frac{4\tau^3 G_0}{3}} \tag{1.20}$$

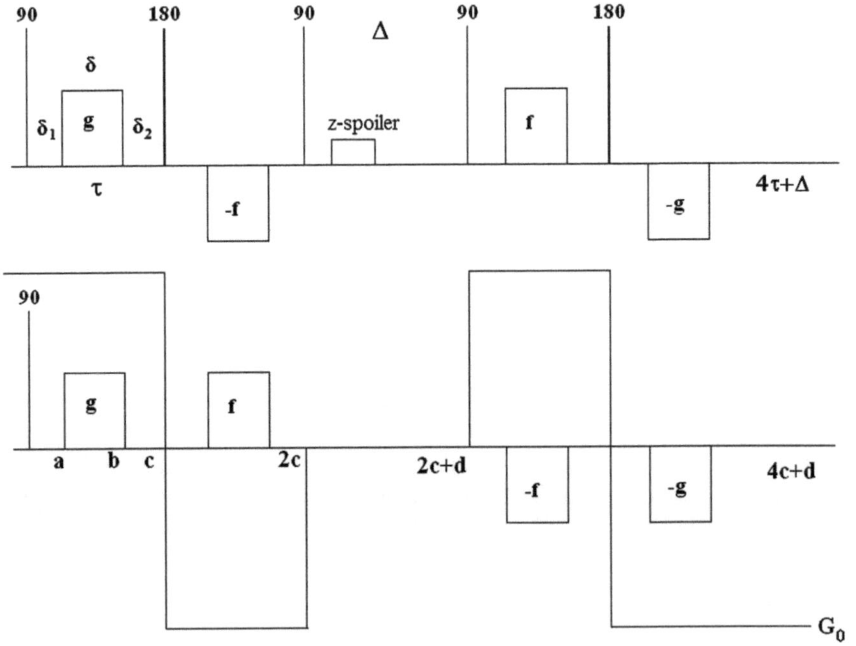

Fig. 1.8 The 13 interval bipolar PFGSTE sequence with real (*upper sequence*) and effective (*lower sequence*) gradients. In the z-storage interval a short spoiler gradient pulse is applied to spoil unwanted signal

Table 1.7 The phase sequences that cancel all echoes except wanted stimulated echo in the bipolar 13 interval PFGSTE sequence

Sequence #	π/2 Excitation	Second π/2	Third π/2	Receiver
1	+x or 0	+x or 0	+x or 0	−x or π
2	+x or 0	−x or π	+x or 0	+x or 0
3	+x or 0	+x or 0	−x or π	+x or 0
4	+x or 0	−x or π	−x or π	−x or π
5	+x or 0	+x or 0	+y or π/2	+y or π/2
6	+x or 0	−x or π	+y or π/2	−y or 3π/2
7	+x or 0	+x or 0	−y or 3π/2	−y or 3π/2
8	+x or 0	−x or π	−y or 3π/2	+y or π/2

Denoting $g = g_{cal}(i)$ and $f = g_{cal}(i - x)$, where g_{cal} is the calibrated gradient strength per arbitrary unit, and (i) and (i − x) are the gradient amplitudes in arbitrary units, the equation is simplified to

$$I = I_0 e^{-\frac{4\tau}{T_2}} e^{-\gamma^2 g^2 (2\delta)^2 D\left[\Delta+\frac{3\tau}{2}-\delta/6\right]\left[i-\frac{x}{2}\left(\frac{\Delta+\tau-\delta/6}{\Delta+\frac{3}{2}\tau-\delta/6}\right)\right]^2 - \gamma^2 D(2i-x)\delta(\delta_1-\delta_2)(\tau) g_{cal} G_0 - \frac{4\tau^3 G_0}{3} + offset} \quad (1.21)$$

Table 1.8 Column 1 is the time parameter with its corresponding effective gradients in column 2

Time	Effective gradient	$\int_0^{t'} g(t'')dt''$
(0, a)	$+G_0$	$+G_0 \cdot t'$
(a, b)	$+G_0 + g$	$+G_0 \cdot t' + g \cdot (t' - a)$
(b, c)	$+G_0$	$+G_0 \cdot t' + g \cdot (b - a)$
(c, c + a)	$-G_0$	$-G_0 \cdot (t' - 2c) + g \cdot (b - a)$
(c + a, c + b)	$-G_0 + f$	$-G_0 \cdot (t' - 2c) + f \cdot (t' - (c + a)) + g \cdot (b - a)$
(c + b, 2c)	$-G_0$	$-G_0 \cdot (t' - 2c) + f \cdot (b - (c + a)) + g \cdot (b - a)$
(2c, 2c + d)	0	$f \cdot (b - a) + g \cdot (b - a)$
(2c + d, 2c + d + a)	$+G_0$	$+G_0 \cdot (t' - 2c + d) + f \cdot (b - a) + g \cdot (b - a)$
(2c + d + a, 2c + d + b)	$+G_0 - f$	$+G_0 \cdot (t' - 2c + d) - f \cdot (t' - (2c + d + a)) + f \cdot (b - a) + g \cdot (b - a)$
(2c + d + b, 3c + d)	$+G_0$	$+G_0 \cdot (t' - 2c + d) + g \cdot (b - a)$
(3c + d, 3c + d + a)	$-G_0$	$-G_0 \cdot (t' - 4c + d) + g \cdot (b - a)$
(3c + d + a, 3c + d + b)	$-G_0 - g$	$-G_0 \cdot (t' - 4c + d) - g \cdot (t' - (3c + a)) + g \cdot (b - a)$
(3c + d + b, 4c + d)	$-G_0$	$-G_0 \cdot (t' - 4c + d)$

Column 3 yields the corresponding time integral of the effective gradient

x provides the difference in applied gradient strength, and the offset is a small constant term due to this difference in amplitude of f and g.

In work with observation time dependent apparent diffusion measurements it may be of interest to measure the initial slope of the time dependent diffusion coefficient, to produce the surface to volume ratio independent of surface relaxivity [27]. Due to the finite duration of the z-storage delay, there is a limit to how short the observation time can be for the 13 interval PFGSTE sequence. It may therefore be convenient to combine the 11 interval PFGSE and the 13 interval PFGSTE as shown in Fig. 1.9. Here a combination of the two sequences is shown, and it is quite evident that new information is gathered by adding the data points at the shortest observation times acquired with the 11 interval PFGSE [17].

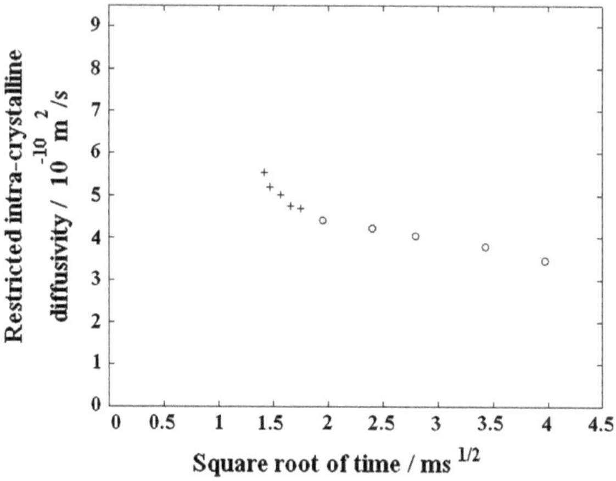

Fig. 1.9 The observation time dependent intra-crystallite diffusivity of ethane in H-ZSM5. "+", data acquired with the 11 interval PFGSE sequence; and "O", data acquired with the 13 interval PFGSTE sequence

1.6 The Convection Compensated PFG NMR Sequences

When performing diffusion experiments in solutions or in low viscosity systems, it is important to have a homogeneous heating of the sample and an accurate temperature control of the system. Otherwise, flow effects due to convection currents will disrupt the diffusion measurement. In addition, the effects of convection can be minimized by the use of so-called convection compensated PFG NMR sequences that serve to suppress effects caused by convection/laminar flow. It can be shown [8] that the consequence of convection within the sample on the expectation value for the magnetization from an arbitrary pulsed field gradient sequence in the presence of a velocity term, v, is

$$M^{+} = M_0^{+} e^{-\frac{t_1}{T_2}} e^{-\frac{t_2}{T_1}} e^{-\gamma^2 D \int_0^t \left(\int_0^{t'} g(t'')dt''\right)^2 dt'} e^{i v \gamma \int_0^t \left(\int_0^{t'} g(t'')dt''\right) dt'} \tag{1.22}$$

Already in 1954 Carr and Purcell [39] noted that even echoes in a multi spin-echo experiment (CPMG train) [37] were convection compensated. Thus, when assuming a constant velocity term experienced by the diffusing molecules, a double pulsed field gradient spin or stimulated echo sequence would suppress the effect from convection in the diffusion experiment. In principle, monopolar pulsed field gradient experiments (two times the ordinary PFGSE or PFGSTE) could be used, but this leads to substantial eddy current transients that would lead to increased line width and baseline distortions in high resolution Fourier transformed spectra. In order to simultaneously compensate for convection and simultaneously minimize the effect from eddy current transients, instead double versions of the 11 interval bipolar PFGSE or 13 interval bipolar PFGSTE [8] can be used. However, one major drawback using the double bipolar version is the lengthened duration of the sequence. If the molecules undergo fast transverse relaxation, the achievable to signal to noise ratio may become an issue. Another challenge might be the number of scans required to perform a proper phase sequence to eliminate signal from unwanted coherence transfer pathways [8]. This number could be much larger than what is required for a system to achieve a satisfactorily signal to noise ratio in the echo signal. This would unnecessarily extend the experiment time.

Realizing that the bipolar gradient pairs are not required for suppressing the coupling between applied and internal magnetic field gradients in liquid systems, the gradients of opposite polarity are placed as shown in Fig. 1.10 for the ordinary

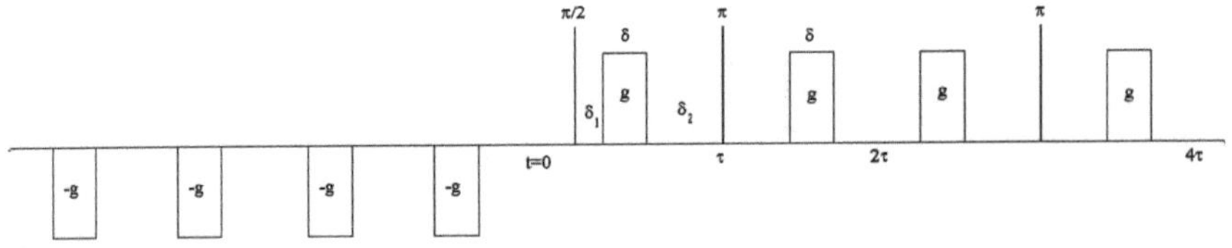

Fig. 1.10 A double PFGSE with preparatory gradients of opposite polarity

double PFGSE sequence [40]. Here the gradient pulses of opposite polarity are placed before the 90° RF excitation pulse, in order to reduce the eddy current transients following the PFGSE sequence. Conclusively, the eddy current transients will be greatly reduced while the double spin echo will be the simplest possible for achieving convection compensation. The echo attenuation for the sequence is written

$$I = I_0 e^{-\frac{2\tau}{T_2}} e^{-2\gamma^2 g^2 \delta^2 D(\tau - \delta/3)} \tag{1.23}$$

If it is crucial to include a storage period of the NMR signal between motional encoding and decoding, for example in order to have a proper attenuation of a slowly moving macro molecule in a solution, a double version of the pulsed field gradient stimulated echo sequence should be employed. The gradient pulses of opposite polarity should be placed as shown in Fig. 1.11. This arrangement allows the shortest duration between the bipolar pairs. Thus, the first gradient of opposite polarity is placed early in the first z-storage interval to compensate for the eddy current transients arising from first applied gradient pulse. In order to optimize the positioning of the preparatory gradients on high field systems, the spin lock signal can be monitored (if available). The lock signal is very sensitive to eddy current transients, and Fig. 1.12 shows how the application of preparatory gradient pulses

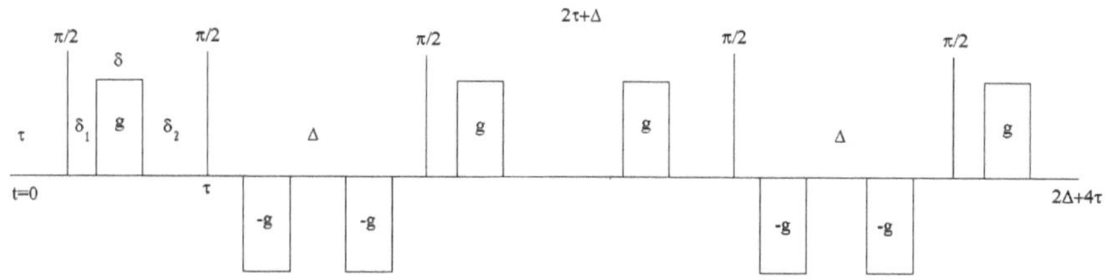

Fig. 1.11 A double PFGSTE with preparatory gradients of opposite polarity

Fig. 1.12 Recording of the lock deuterium signal during a mono-polar double PFGSTE sequence (*dashed line*) and the double PFGSTE with preparatory gradient pulses (*solid line*) [40]

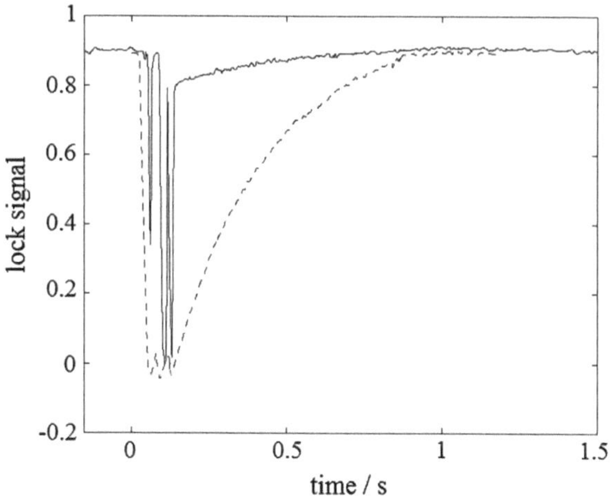

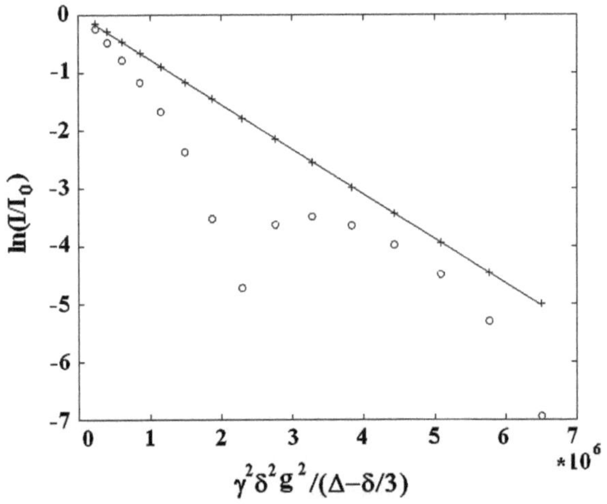

Fig. 1.13 Comparing the diffusion with ("+") or without ("O") convection compensation, using the double PFGSE sequence on acetonitrile at 15 °C [40]

significantly reduces the strength of eddy current transients after having optimized their positions. The impact of this reduction in eddy current field, is that both the phase and baseline error in the Fourier-transformed spectra are significantly reduced [40].

$$I = I_0 e^{-\frac{2\tau}{T_2}} e^{-2\gamma^2 g^2 \delta^2 D(\Delta + \tau - \delta/3)} \tag{1.24}$$

In Fig. 1.13, the effect of introducing a double pulsed field gradient spin echo sequence as compared to a single pulsed field gradient spin echo sequence in the presence of convection can be seen. The system studied was a solution of acetonitrile at 15 °C [6]. While the double version produces a linear decay of the logarithm of the echo attenuation as a function of the square of the applied gradient strength, the single spin echo approach yields a non-linear decay due to a non-zero convection term.

1.7 The Spoiler Recover Pulse Sequence

A long acquisition time may be an important limiting factor when performing a PFG NMR experiment. Samples, being biological tissues, emulsions or brine saturated rock core plugs, may not be stable for as long as the conventional PFG NMR experiment lasts. In conventional diffusion and two-dimensional experiments, it is usual to wait 5 times the longitudinal relaxation time between each scan in order to ensure that the system returns to its equilibrium state. Consequently, a 2-dimensional T_1-T_2 experiment may last for several hours depending on the number of scans and variable z-storage delays. Under these conditions, biological tissue may start to deteriorate, emulsions may start to separate, or the brine from the core plug may evaporate during the experiment, consequently leading to unstable conditions and unreliable results. In addition, in dynamic systems instability is often a key factor to be studied, such as the study of separation processes in water in oil emulsions by adding a demulsifier. This separation process may take from seconds up to days, depending on the experimental settings on the system under investigation (see Chap. 7). A PFG NMR method that bypassed the required 5 times T_1 between the scans while maintaining quantitative diffusion information enhances the range of systems that PFG NMR can be used to study. This requirement initiated the development of the spoiler recovery sequence at low resolution conditions (0.05 T permanent magnet) [41]. The development of the spoiler recovery sequence makes it possible to perform rapid measurements of diffusion characteristics and T_2 relaxation. In emulsion characterization that would implicate a possibility of measuring droplet size distribution and/or S/V profiles on unstable emulsions which have instability on a short timescale [42]. Further investigations established that the method can also be used at much higher field, where the effect from radiation damping [43] (Chap. 6) may complicate the diffusion measurements. Using the spoiler recovery approach radiation damping artefacts were removed at 11.7 T as well as the need for a delay of 5 times T_1 between the scans [44].

A common technique used in order to reduce the measurement time of experiments is the option of small flip angels for the initial 90° RF-pulse and/or RF-trains to saturate the artefacts arising from short wait time between the scans, but this approach is not well suited for low resolution NMR [45–47]. The reduction in signal intensity requires more scans to be accumulated, and thus increases the measurement time. Instead, a sequence that produces a situation where the net nuclear magnetic moment of the nuclear spins is zero as an outcome is a preferred solution, i.e. launching the spoiler recovery sequence as shown in Fig. 1.14. The first 90° RF-pulse flips the magnetization initially along the z-direction into the transverse (xy) plane where it is subjected to a relatively long gradient pulse of enough strength to change the phase of the spins by a relatively large value. After the application of the first gradient pulse, GS1, the net nuclear magnetization is zero and the nuclear spin vectors have dephased to cover the whole xy-plane. Then another 90° RF-pulse is applied to rotate the plane of nuclear spin vectors to a plane transverse to the xy-plane, followed by a new gradient pulse of opposite polarity to the first one and with a different gradient strength and duration. The

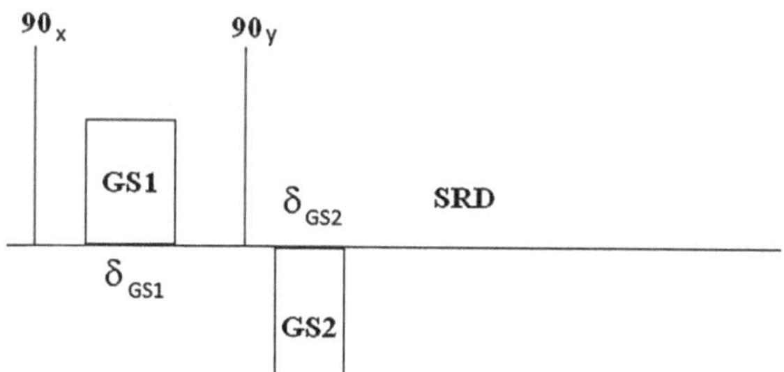

Fig. 1.14 The spoiler recovery sequence, applies two 90° RF-pulses and a bipolar pair of gradients of different duration and strength. Then a spoiler recovery delay (*SRD*) is set up for recovery of magnetization along the direction of the external field

idea is then to make the nuclear spin vectors to dephase uniformly in space, where again the net nuclear magnetization is zero after the application of the second gradient pulse.

The basic requirement for avoiding gradient echoes after the application of the second gradient pulse is the following relation

$$\gamma GS1 \int_0^{\delta_{GS1}} z_1(t)dt \neq \gamma GS2 \int_0^{\delta_{GS2}} z_2(t)\,dt \tag{1.25}$$

δ_{G1} and δ_{G2} are the durations of the spoiler gradient pulses, $Z_{1,2}$ (t) are the molecular positions at the time of the two gradient pulses, and γ is the gyromagnetic ratio. Then the following 3 coherence transfer pathways are not refocused, $p \to 0 \to 1 \to 1$, $p \to 0 \to -1 \to 1$, and $p \to 0 \to 0 \to 1$. To acquire noise only during the spoiler recovery delay (SRD) it is therefore important to choose the GS1 and GS2 values as a bipolar pair of different duration and strength and in accordance with (1.25). As will be shown in Chap. 6, the spoiler recovery approach applies equally well at an external magnetic field of 0.05 T as well as 14.1 T. With the introduction of the spoiler recovery (SR) sequence, two-dimensional acquisition schemes such as a two-dimensional T_1-T_2 experiment become more practical. These experiments will be introduced in the next section.

1.8 Combined Two Dimensional Diffusion, T_1 and/or T_2 Pulse Sequences

In the following section, the pulse sequence/experiment and its corresponding attenuation for a homogeneous system will be provided for three different two-dimensional PFG NMR experiments. These will serve as examples of sequences that may be constructed from the basic one-dimensional PFG NMR experiments, as there are numerous variants that can be constructed, some of which are presented in Chap. 6.

Figure 1.15 shows the combined SR-11-interval PFGSE-CPMG sequence. While fixing the spoiler recovery delay to a constant value, the applied gradient

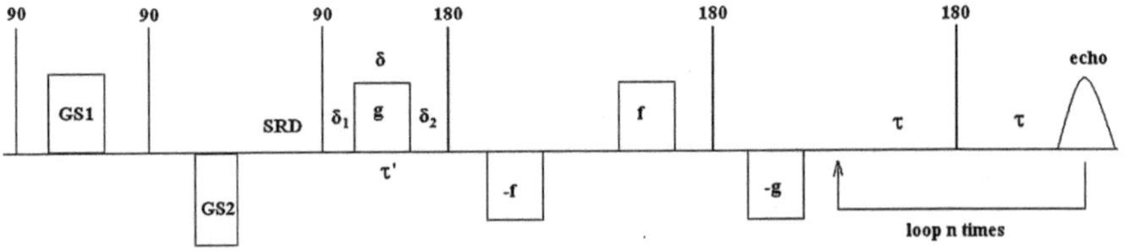

Fig. 1.15 The combined SR-Diffusion-T_2 experiment

strength may be incremented and a CPMG-echo train as a function of applied gradient strength can be recorded. The echo attenuation for this sequence is written as follows

$$I = I_0 \sum_n e^{-\frac{4\tau'}{T_2} - \gamma^2 g^2 (2\delta)^2 D\left[\Delta + \frac{3\tau'}{2} - \delta/6\right]\left[i - \frac{x}{2}\left(\frac{\Delta+\tau'-\delta/6}{\Delta+\frac{3}{2}\tau'-\delta/6}\right)\right]^2} e^{-\frac{2n\tau}{T_2}} \tag{1.26}$$

The experiment illustrated in Fig. 1.15 is referred to as a Diffusion-T_2 experiment.

Figure 1.16 shows a combined SR-11-interval PFGSE sequence. For different durations of SRD the applied pulsed field gradient strength is incremented. The attenuation for this sequence is written as follows

$$I = I_0(1 - e^{-\frac{SRD}{T_1}}) e^{-\frac{4\tau'}{T_2} - \gamma^2 g^2 (2\delta)^2 D\left[\Delta + \frac{3\tau'}{2} - \delta/6\right]\left[i - \frac{x}{2}\left(\frac{\Delta+\tau'-\delta/6}{\Delta+\frac{3}{2}\tau'-\delta/6}\right)\right]^2} \tag{1.27}$$

The experiment illustrated in Fig. 1.16 is referred to as a Diffusion-T_1 experiment, i.e. a diffusion experiment is conducted for a set of SRD's.

Figure 1.17 shows the combined SR - CPMG sequence. By varying the SRD a CPMG-train may be recorded as a function of the SRD. The attenuation for this sequence is written as follows

$$I = I_0(1 - e^{\frac{SRD}{T_1}}) \sum_n e^{-\frac{2n\tau}{T_2}} \tag{1.28}$$

The experiment illustrated in Fig. 1.17 is referred to as a T_1-T_2 experiment.

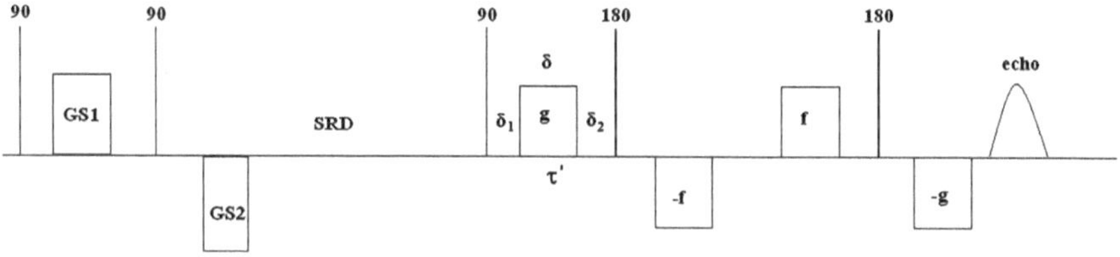

Fig. 1.16 The combined SR-Diffusion-T_1 experiment

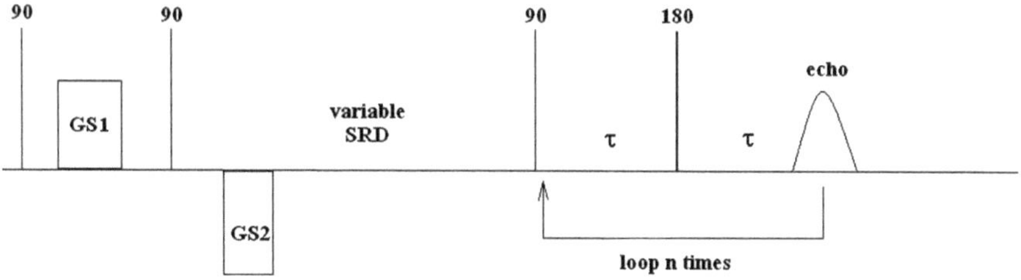

Fig. 1.17 The combined SR-T_1-T_2 experiment

There are numerous other variations of the two-dimensional experiments, but the experiments presented here are the basic ones that apply the spoiler recovery sequence to measure T_1, T_2 and/or Diffusion, and to minimize the experimental time. These sequences or variations of them are exploited in a number of applications which are presented in Chaps. 6–9.

Appendix

The Expectation Value for the Net Nuclear Magnetization in the PFGSE

The concept of a density matrix is convenient when working with practically non-interacting spins, as is the case for particles in motion. However, there is a requirement to calculate the magnetization that is detectable from an NMR pulse sequence containing RF-pulses and magnetic field gradient pulses. A differential equation reflecting the spin dynamics of the system must be found and solved. The solution of the equation will be dependent on the interaction between the spin system and the time dependent external field $\vec{H}(t)$, and will provide information on the evolution of the spin system when being subjected to RF pulses and magnetic field gradients. The derivation below is motivated by the need for a way to calculate the magnetization that is detectable from an NMR pulse sequence containing RF-pulses and magnetic field gradient pulses.

Assuming that the probability function of an ensemble of spins is the sum of a set of orthogonal eigen functions, it can be written as follows [21]

$$\psi = \sum_n c_n U_n \tag{1.29}$$

Here c_n is the time dependent part while U_n is the time independent part. The expectation value for the magnetization of the ensemble then is

$$\langle M_x \rangle = \sum_{m,n} c_m * c_n \langle m|M_x|n\rangle = \sum_{m,n} \langle n|P|m\rangle \langle m|M_x|n\rangle = \sum_n \langle n|PM_x|n\rangle = Tr\{PM_x\} \tag{1.30}$$

where P is defined as $\langle n|P|m\rangle = c_n c_m^*$. When looking at the ensemble average, the average over all possible time dependent wave functions c_n, the definition of the density matrix ρ_d may be produced.

$$\langle n|\rho_d|m\rangle = \overline{\langle n|P|m\rangle} = \overline{c_n c_m^*} \tag{1.31}$$

The average expectation value of the magnetization of the spin ensemble is then written

$$\overline{\langle M_x \rangle} = Tr\{\rho M_x\} \tag{1.32}$$

The physical interpretation of ρ_d becomes apparent when considering $\langle n|\rho_d|n\rangle = \overline{c_n c_n^*}$. This is the probability of finding the system in state n, and therefore the trace of ρM_x ($Tr\{\rho M_x\}$) is the sum of the expectation values for the magnetization of all states, each multiplied by the probability of finding that state occupied. Now consider the Schrödinger equation

$$-\frac{\hbar}{i}\frac{\partial \psi}{\partial t} = \mathbf{H}\psi \tag{1.33}$$

Expanding ψ in the orthogonal set of eigen functions, and multiplying it with one of the functions $c_k|k\rangle$, and perform a volume integration one finds [21]

$$-\frac{\hbar}{i}\frac{\partial c_k}{\partial t} = \sum_n c_n\langle k|\mathbf{H}|n\rangle \tag{1.34}$$

Regarding the ensemble average a differential equation for the density matrix ρ_d can now be found as

$$\frac{\partial \rho_d}{\partial t} = \frac{\partial}{\partial t}(c_k c_m^*) = c_k\frac{\partial c_m^*}{\partial t} + \frac{\partial c_k}{\partial t}c_m^* = \frac{i}{\hbar}[\rho_d, \mathbf{H}] \tag{1.35}$$

In the rotating frame of reference, the solution for a time independent Hamiltonian can be written [21]

$$\rho_R(t) = e^{-\frac{i}{\hbar}\mathbf{H}_{\mathbf{eff}}t}\rho_R(0)e^{\frac{i}{\hbar}\mathbf{H}_{\mathbf{eff}}t} \tag{1.36}$$

where $\mathbf{H}_{\mathbf{eff}} = -\hbar(\gamma H_0 - \omega)I_z\hat{k} - \gamma\hbar H_1(t)I_x\hat{i}$, and $\rho_R = e^{-iwtI_z}\rho_d e^{iwtI_z}$. This is the starting point when evaluating the influence of the magnetic field from an RF-coil has on the evolution of the magnetization of the nuclear spins. Before evaluating this influence, knowledge on the density matrix at thermal equilibrium must be obtained. For the system of non-interacting spins, $\rho_R(0)$ is given by

$$\rho_R(0) = \frac{1}{Z}e^{-\gamma\hbar H_0 I_z/kT} \tag{1.37}$$

Z is the partition function, k is the Boltzmann constant and T is the absolute temperature [21]

The Response to a 90° RF-Pulse

In the previous section the way of achieving the expectation value for the net nuclear magnetization was outlined, and now this can be used to evaluate the effect of applying a rotating H_1-field with frequency $\omega = \gamma H_0$ in the x-direction of the rotating frame of reference at $t = 0$, and with a duration t_1 such that $\gamma H_1 t_1 = \frac{\pi}{2}$. In this situation $\rho_R(t_1)$ is written [21]

$$\rho_R(t_1) = e^{\frac{i}{\hbar}\frac{\pi}{2}I_x}\rho_R(0)e^{-\frac{i}{\hbar}\frac{\pi}{2}I_x} = e^{\frac{i}{\hbar}\frac{\pi}{2}I_x}\frac{e^{-\gamma\hbar H_0 I_z/kT}}{Z}e^{-\frac{i}{\hbar}\frac{\pi}{2}I_x} \tag{1.38}$$

Employing the high temperature approximation for the partition function

$$\rho_R(0) = \frac{1}{(2I+1)}(1 - \frac{\gamma\hbar H_0}{kT}I_z) \tag{1.39}$$

Expanding the other exponentials in (1.38) and using the commutation relations, cyclic permutations from $[I_x, I_y] = I_x I_y - I_y I_x = iI_z$, it can be shown that the expectation value for the magnetization in the y-direction of the rotating frame of reference can be written [21]

$$\langle M_y(t_1)\rangle = \gamma\hbar Tr\{\rho_R(t_1)I_y\} = \frac{\gamma^2\hbar^2 H_0}{(2I+1)kT}Tr\{I_y^2\} = \gamma^2\hbar^2 H_0\frac{NI(I+1)}{3kT} \tag{1.40}$$

The Response to a Spin Echo Experiment

The pulse sequence for a spin echo experiment is shown in Fig. 1.18. At t = 0 the net magnetization is in the z-direction. After applying the first 90° RF-pulse, $\gamma H_1 t_1 = \frac{\pi}{2}$, the magnetization has been flipped to be aligned with the negative y-axis in the rotating frame of reference. At t_2 the second RF-pulse is applied, $\gamma H_1(t_3 - t_2) = \pi$, and depending on the direction of this 180° RF pulse, the nuclear spins will be flipped 180° around the negative y-axis or flipped 180° to be aligned with the positive direction of the y-axis. In the following, a 180° RF-pulse in the same direction as the 90° pulse will be assumed (i.e. H_1 along the x-axis in the rotating frame of reference). Now the evolution of the magnetization after the application of the second RF-pulse can be evaluated. In order to do this an expression for the density matrix at $t > t_3$ needs to be found, and this can be used to calculate the expectation value for the magnetization given by (1.32). In the rotating frame of reference, the Hamiltonian is independent of time except when switching on and off the RF magnetic field. Even if there is a discontinuity of the Hamiltonian at $t = 0$, t_1, t_2 *and* t_3, the density matrix, which reflects the probability density of spins, must be continuous. This can be shown by integrating over the discontinuities letting the limits approach the time of the discontinuity [21]. The solution of the density matrix subjected to the spin echo pulse sequence can therefore be done stepwise, and at $t = 0^-$ thermal equilibrium may be assumed. The density matrix is then given by the Boltzmann factor

$$\rho_R(0^-) = \frac{e^{-\gamma\hbar H_0 I_z/kT}}{Z} \tag{1.41}$$

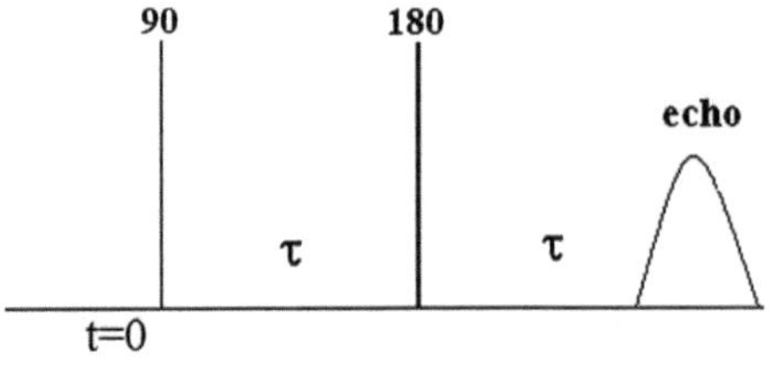

Fig. 1.18 The spin echo experiment [18]

Since the density matrix is continuous when the RF-field is switched on, $\rho_R(0^-) = \rho_R(0^+)$. During the RF-pulse $\rho_R(t)$ is evolving as

$$\rho_R(t) = e^{\frac{i}{\hbar}\gamma H_1 t I_x} \rho_R(0^-) e^{-\frac{i}{\hbar}\gamma H_1 t I_x} \tag{1.42}$$

Here it is assumed that the frequency of H_1 is so close to the resonance that the only contribution to H_{eff} is from the RF-field. In the previous section it was shown that when $\gamma H_1 t_1 = \frac{\pi}{2}$ the net magnetization is flipped into the xy-plane. After the application of the first RF-pulse the spin system will evolve according to the solution with a time independent Hamiltonian H_{eff} in the rotating frame of reference.

$$\rho_R(t) = e^{i(\gamma H_0 - \omega)(t - t_1) I_z} \rho_R(t_1) e^{-i(\gamma H_0 - \omega)(t - t_1) I_z} \tag{1.43}$$

Applying a RF-pulse at t = t_2 with a duration such that $\gamma H_1(t_3 - t_2) = \pi$ the evolution of the density matrix at $t > t_3$ is written

$$\rho_R(t) = e^{i\pi I_x - i(\gamma H_0 - \omega)(t - t_3) I_z} \rho_R(t_2) e^{i(\gamma H_0 - \omega)(t - t_3) I_z - i\pi I_x} \tag{1.44}$$

where $\rho_R(t_2) = e^{i(\gamma H_0 - \omega)(t_2 - t_1) I_z} \rho_R(t_1) e^{-i(\gamma H_0 - \omega)(t_2 - t_1) I_z}$. From (1.44) it can be seen that the effect of the last 180° RF-pulse is to change the sign in the exponents as compared with the evolution before the 180° RF-pulse. Thus, when assuming $t_3 - t_2 \ll t_2 - t_1$, the two parts will cancel at $t = 2t_2$ and result in an echo that is independent of $\gamma H_0 - \omega$

$$\rho_R(t = 2t_2) = e^{i\frac{3\pi}{2} I_x} \rho_R(0^-) e^{-i\frac{3\pi}{2} I_x} \tag{1.45}$$

Having found an expression for the density matrix an expression for the expectation value of the magnetization using the high temperature approximations (Z = 2I + 1 and $e^{-\gamma\hbar H_0 I_z/kT} \approx 1 - \frac{\gamma\hbar H_0 I_z}{kT}$) can be found

$$\langle M_y(2t_2)\rangle = \gamma\hbar Tr\{\rho_R(2t_2) I_y\} = -\gamma^2\hbar^2 H_0 \frac{NI(I+1)}{3kT} \tag{1.46}$$

Thus, all the initial magnetization following the first 90° RF-pulse has been refocused, i.e. a spin echo signal has been produced.

The Response to a Pulsed Field Gradient Spin Echo Experiment

In order to describe a pulsed field gradient spin echo experiment, it is necessary to include magnetic field gradients into the equations described in the previous section. It is convenient to start with a small volume, calculate the density matrix through all the durations in the pulse sequence (as the Hamiltonian varies depending on on/offset of RF- and gradient pulses) and thereafter conduct the integration over the volume of interest. The probability that a particle which is initially at a position z_0 at $t = 0$ moves to a position at time τ is given by the factor $p(z_0)P(z, z_0, \tau)$. This is a so called Markov process [48]; the experiment is insensitive to what happens between $t = 0$ and $t = \tau$. Only the probability $p(z_0)$ at $t = 0$ and the probability $P(z, z_0, \tau)$ at $t = \tau$ will affect the experimental outcome.

When the time evolution of the magnetization of the spins is taken into account, the probability factor should be multiplied by the expectation value for the magnetization of the spins. This is equivalent to evaluating the expectation value for the magnetization of the particles that move during the pulsed field gradient spin echo experiment

$$\left\langle M_{xy}(2\tau)\right\rangle = \int\int p(z_0)P(z, z_0, \tau)\gamma\hbar Tr\left\{\rho_R(2\tau)I_{xy}\right\}dzdz_0 \tag{1.47}$$

Using the result from the spin echo experiment and introducing a magnetic field gradient in the z-direction into the Hamiltonian, $H_z = H_0 + gz$, it may be shown that the density matrix at the time of the spin echo is written

$$\rho_R(2\tau) = -\mathrm{e}^{iI_z\gamma\delta g(z(\tau)-z_0)+i\frac{\pi}{2}I_x}\rho_R(0^-)\mathrm{e}^{-i\frac{\pi}{2}I_x-iI_z\gamma\delta g(z(\tau)-z_0)} \tag{1.48}$$

The major assumption used is that the spins do not move in the direction of the magnetic field gradient during the gradient pulse. The well known $\delta/3$ correction to the diffusion time τ, the duration between the gradient pulses, is then neglected. Again, using the commutator relations and the high temperature approximation, ρ_R can be written

$$\rho_R(2\tau) = Constant - \frac{\gamma\hbar H_0}{(2I+1)kT}(I_x\cos(\gamma\delta g(z(\tau)-z_0)) + I_y\sin(\gamma\delta g(z(\tau)-z_0)) \tag{1.49}$$

Letting the x-component of the magnetization be the real part and the y-component the imaginary part, the expectation value for the magnetization of moving spins at $t = 2\,\tau$ is then written

$$\langle M(2\tau)\rangle = M_0\int\int p(z_0)P(z, z_0, \tau)\mathrm{e}^{i\gamma\delta g(z(\tau)-z_0)}dzdz_0 \tag{1.50}$$

where

$$M_0 = -\gamma^2\hbar^2 H_0\frac{NI(I+1)}{3kT} \tag{1.51}$$

To solve the integrals in (1.50) knowledge about how the spins move must be used. That is, the solution of the differential equation called Fick's law must be found. For a homogeneous system it is shown in Chap. 2 that $p(z_0)P(z, z_0, \tau)$ is

Gaussian and (1.50) then yields the second order moment with respect to the magnetic field gradient

$$\langle M(2\tau)\rangle = M_0 e^{-\gamma^2\delta^2 g^2 D\tau} \tag{1.52}$$

D is the diffusion coefficient for the nuclear spins, and τ is the duration between the gradient pulses.

Coherence Transfer Pathways and Effective Gradients

When applying more than two RF-pulses in any pulse sequence, several echoes may be generated [18]. For example, in the ordinary pulsed field gradient stimulated echo experiment where three 90° RF-pulses are applied, five echoes will appear together with the FID's following each RF-pulse [34]. As the stimulated echo is the echo required in this sequence, the remaining echoes should be suppressed. A way to do this is to employ the concept of coherence transfer pathways aiming at constructing a phase sequence where the only surviving signal is the stimulated echo [49]. The coherence transfer pathway is given by

$$\varphi = \sum_{j=1}^{n} \Delta p_j \varphi_j \tag{1.53}$$

where φ is the accumulated phase change determined by the product of the change in coherence order Δp_j and the phase φ_j of the RF-pulse number j. n equals the number of RF-pulses applied in the experiment. The coherence order is dependent on the transitions possible within the spin system, and to the lowest order (non-interacting spin 1/2) one may define the coherence order $p = 0$ in the direction of the external field and $p = \pm 1$ transverse to it. A single spin initially at thermal equilibrium with the external field, will after the application of a 90° RF-pulse change the coherence level to $+1$ (rotating clockwise in the plane transverse to the external field), -1 (rotating counterclockwise), or it may be unaffected by the RF-pulse (i.e. coherence order 0). The ± 1 coherence levels are the two solutions for the transverse magnetization in the Bloch-Torrey equations. For the phase of the RF-pulses there are four possibilities; 0, $\frac{\pi}{2}$, π and $\frac{3\pi}{2}$. In the rotating coordinate system this implies that the RF-pulse is applied in the $\pm x$ or $\pm y$ direction.

As an example, the coherence transfer pathway from the spin echo sequence is to be considered. A phase sequence is to be constructed, such that the only surviving signal is the one following the coherence order pathway $p \rightarrow 0 \rightarrow -1 \rightarrow +1$, shown in Fig. 1.19 as the solid line.

For the first scan, let the phases of the RF-pulses equal 0. Then the resulting phases for the different pathways are written as follows

$$\varphi = -1 \cdot 0 + 2 \cdot 0 = 0 \quad \textit{and} \quad \varphi' = +1 \cdot 0 + 0 \cdot 0 = 0 \tag{1.54}$$

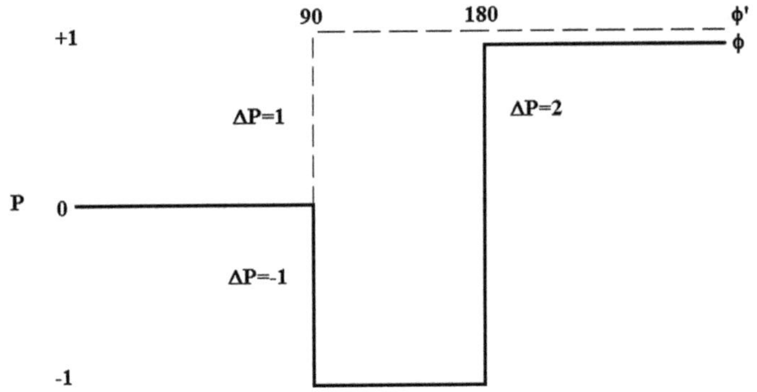

Fig. 1.19 Coherence order pathway for spin echo sequence (also equivalent to the read interval in the 13-interval PFGSTE sequence)

For the second scan let the phase of the 90° RF-pulse be shifted by $\frac{\pi}{2}$ while keeping the phase of the 180° RF-pulse unchanged. Then the resulting phases for the different pathways are written as follows

$$\varphi = -1 \cdot \frac{\pi}{2} + 2 \cdot 0 = -\frac{\pi}{2} \quad and \quad \varphi' = +1 \cdot \frac{\pi}{2} + 0 \cdot 0 = \frac{\pi}{2} \tag{1.55}$$

If the receiver phase is set to 0 and $-\frac{\pi}{2}$, and the two scans are added, the proposed phase sequence will result in a signal where the only contribution (ideally) is from the desired coherence transfer pathway $p \to 0 \to -1 \to +1$.

Finally, the concept of effective gradients is closely connected to the coherence transfer pathway. Karlicek et al. [1] showed using the Bloch-Torrey equations that the application of a 180° RF-pulse is equivalent to changing the polarity of the magnetic field gradient acting on the spin system. This can be seen using the concept of the density matrix: Consider the PFGSE in Fig. 1.3. The sequence yields an echo at $t = 2\tau$ with two positive, apparently dephasing magnetic field gradients. How is it then possible to acquire an echo at $t = 2\tau$? The mechanism for echo formation can be understood by examining the action of the 180° RF pulse on the transverse magnetization. The effect of this pulse is to change the coherence order from $+1$ to -1 (and vice versa), and (1.48) shows the resulting density matrix at $t = 2\tau$. When evaluating the resulting density matrix, it can be seen that the application of the 180° RF-pulse is equivalent to a change in the polarity of the effective magnetic field gradient. Thus, in the evaluation of the echo attenuation for more complicated PFGSTE sequences the 180° RF-pulse may be replaced by a change in sign of the magnetic field gradient following the RF-pulse.

Mathematica Program for Solving the Echo Attenuation from the Ordinary Monopolar PFGSTE Sequence with Rectangular Shaped Gradients

Mathematica 5.1 program for evaluation of attenuation from ordinary PFGSTE (Fig. 1.5);

```
h[x = −G*x
i[x] = −G*x+g*(x−a)
j[x] = −G*x+g*(b−a)
n[x] = g*(b−a)−G*c
o[x] = G*(x−(2*c+d))+g*(b−a)
p[x] = G*(x−(2*c+d))+g*(b−a)−g*(x−(c+d+a))
q[x] = G*(x−(2*c+d))
h1[x] = h[x]^2
i1[x] = i[x]^2
j1[x] = j[x]^2
n1[x] = n[x]^2
o1[x] = o[x]^2
p1[x] = p[x]^2
q1[x] = q[x]^2
h2[x] = Integrate[h1[x],]
i2[x] = Integrate[i1[x],]
j2[x] = Integrate[j1[x],]
n2[x] = Integrate[n1[x],{x,c,c+d}]
o2[x] = Integrate[o1[x],{x,c+d,c+d+a}]
p2[x] = Integrate[p1[x],{x,c+d+a,c+d+b}]
q2[x] = Integrate[q1[x],{x,c+d+b,2*c+d}]
sum1[x] = h2[x]+i2[x]+j2[x]
sum2[x] = n2[x]+o2[x]+p2[x]+q2[x]
sum5[x] = sum1[x]+sum2[x]
sum5[x] = Simplify[sum5[x]]
X1[x] = Coefficient[sum5[x],g^2]*g^2
sum5[x] = sum5[x]−X1[x]
sum5[x] = Simplify[sum5[x]]
X3[x] = Coefficient[sum5[x],G^2]*G^2
sum5[x] = sum5[x]−X3[x]
sum5[x] = Simplify[sum5[x]]
X1[x] = Simplify[X1[x]]
X3[x] = Simplify[X3[x]]
X4[x] = Coefficient[sum5[x],G,1]*G
Result[x]=X1[x]+X3[x]+X4[x]
```

Mathematica Programs for Solving the Echo Attenuation from the 13-Interval PFGSTE Sequence with Sinusoidal Gradients

Mathematica 5.1 program for evaluation of attenuation from the 13-PFGSTE (Fig. 1.20) *using f=g;*

```
h[x] = G*x
i[x] = G*x−g*((b−a)/Pi)*(Cos[Pi*(x−a)/(b−a)]−1)
j[x] = G*x+(2/Pi)*g*(b−a)
k[x] = −G*(x−2c)+(2/Pi)*g*(b−a)
l[x] = −G*(x−2c)+(2/Pi)*g*(b−a)−g*((b−a)/Pi)*(Cos[Pi*(x−(c+a))/(b−a)]−1)
m[x] = −G*(x−2c)+(4/Pi)*g*(b−a)
n[x] = (4/Pi)*g*(b−a)
o[x] = −G*(x−(2c+d))+(4/Pi)*g*(b−a)
p[x] = −G*(x−(2c+d))+(4/Pi)*g*(b−a)+g*((b−a)/Pi)*(Cos[Pi*(x−(2c+d+a))/
  (b−a)]−1)
q[x] = −G*(x−(2c+d))+(2/Pi)*g*(b−a)
r[x] = G*(x−(4c+d))+(2/Pi)*g*(b−a)
s[x]=G*(x−(4c+d))+(2/Pi)*g*(b−a)+(g*(b−a)/Pi)*(Cos[Pi*(x−(3c+d+a))/
  (b−a)]−1)
t[x] = G*(x−(4c+d))
```

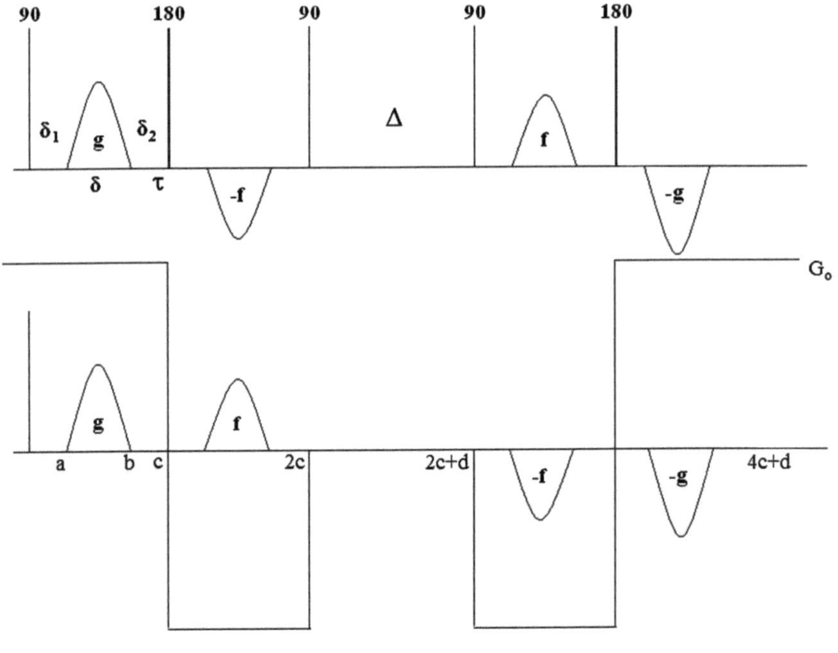

Fig. 1.20 The 13-interval PFGSTE sequence with sinusoidal shaped gradients

```
h1[x] = h[x]^2
i1[x] = i[x]^2
j1[x] = j[x]^2
k1[x] = k[x]^2
l1[x] = l[x]^2
m1[x] = m[x]^2
n1[x] = n[x]^2
o1[x] = o[x]^2
p1[x] = p[x]^2
q1[x] = q[x]^2
r1[x] = r[x]^2
s1[x] = s[x]^2
t1[x] = t[x]^2

h2[x] = Integrate[h1[x],{x,0,a}]
i2[x] = Integrate[i1[x],{x,a,b}]
j2[x] = Integrate[j1[x],{x,b,c}]
k2[x] = Integrate[k1[x],{x,c,c+a}]
l2[x] = Integrate[l1[x],{x,c+a,c+b}]
m2[x] = Integrate[m1[x],{x,c+b,2c}]
n2[x] = Integrate[n1[x],{x,2c,2c+d}]
o2[x] = Integrate[o1[x],{x,2c+d,2c+d+a}]
p2[x] = Integrate[p1[x],{x,2c+d+a,2c+d+b}]
q2[x] = Integrate[q1[x],{x,2c+d+b,3c+d}]
r2[x] = Integrate[r1[x],{x,3c+d,3c+d+a}]
s2[x] = Integrate[s1[x],{x,3c+d+a,3c+d+b}]
t2[x] = Integrate[t1[x],{x,3c+d+b,4c+d}]

sum1[x] = h2[x]+i2[x]+j2[x]+k2[x]+l2[x]+m2[x]
sum2[x] = n2[x]+o2[x]+p2[x]+q2[x]+r2[x]+s2[x]+t2[x]
sum5[x] = sum1[x]+sum2[x]
sum5[x] = Simplify[sum5[x]]

X1[x] = Coefficient[sum5[x],g^2]*g^2
sum5[x] = sum5[x]-X1[x]
sum5[x] = Simplify[sum5[x]]
X2[x] = Coefficient[sum5[x],G^2]*G^2
sum5[x] = sum5[x]-X2[x]
sum5[x] = Simplify[sum5[x]]
X3[x] = Coefficient[sum5[x],G,1]*G

X1[x] = Simplify[X1[x]]
X2[x] = Simplify[X2[x]]
X3[x] = Simplify[X3[x]]

Result[x] = X1[x]+X2[x]+X3[x]
```

The equation from the sequence with sinusoidal shaped gradients is

$$I = I_0 e^{-\frac{4\tau}{T_2}} e^{-\gamma^2 g^2 (\frac{4\delta}{\pi})^2 D \left[\Delta + \frac{3\tau}{2} - \delta/8\right] - \frac{\gamma^2 D 4\delta(\delta_1 - \delta_2)(\tau) g G_0}{\pi} - \frac{4\tau^3 G_0}{3}} \tag{1.56}$$

Even though the cross term vanishes only if $(\delta_1 - \delta_2) = 0$, it is significantly reduced as there is no Δ dependency.

References

1. R.F. Karlicek Jr., I.J. Lowe, A modified pulsed gradient technique for measuring diffusion in the presence of large background gradients. J. Magn. Reson. (1969) **37**(1), 75–91 (1980)
2. R.M. Cotts et al., Pulsed field gradient stimulated echo methods for improved NMR diffusion measurements in heterogeneous systems. J. Magn. Reson. (1969) **83**(2), 252–266 (1989)
3. A. Jerschow, N. Müller, 3D diffusion-ordered TOCSY for slowly diffusing molecules. J. Magn. Reson. A **123**(2), 222–225 (1996)
4. G.H. Sorland, D. Aksnes, L. Gjerdaker, A pulsed field gradient spin-echo method for diffusion measurements in the presence of internal gradients. J. Magn. Reson. **137**(2), 397–401 (1999)
5. G.H. Sørland, B. Hafskjold, O. Herstad, A stimulated-echo method for diffusion measurements in heterogeneous media using pulsed field gradients. J. Magn. Reson. **124**(1), 172–176 (1997)
6. G.H. Sørland et al., Improved convection compensating pulsed field gradient spin-echo and stimulated-echo methods. J. Magn. Reson. **142**(2), 323–325 (2000)
7. R. Blinc, J. Pirš, I. Zupančič, Measurement of self-diffusion in liquid crystals by a multiple-pulse NMR method. Phys. Rev. Lett. **30**(12), 546–549 (1973)
8. A. Jerschow, N. Müller, Convection compensation in gradient enhanced nuclear magnetic resonance spectroscopy. J. Magn. Reson. **132**(1), 13–18 (1998)
9. C.S. Johnson Jr, Diffusion ordered nuclear magnetic resonance spectroscopy: principles and applications. Prog. Nucl. Magn. Reson. Spectrosc. **34**(3–4), 203–256 (1999)
10. N.M. Loening, J. Keeler, G.A. Morris, One-dimensional DOSY. J. Magn. Reson. **153**(1), 103–112 (2001)
11. G.A. Morris, H. Barjat, Chapter 11 high resolution diffusion ordered spectroscopy, in *Analytical Spectroscopy Library*, ed. by K.E. Kévér, Gy. Batta, Cs. Széntay (Elsevier, 1997), pp. 209–226
12. M. Nilsson, G.A. Morris, Improving pulse sequences for 3D DOSY: convection compensation. J. Magn. Reson. **177**(2), 203–211 (2005)
13. A. Botana et al., J-modulation effects in DOSY experiments and their suppression: the Oneshot45 experiment. J. Magn. Reson. **208**(2), 270–278 (2011)
14. Y.Q. Song, M.D. Hürlimann, C. Flaum, A method for rapid characterization of diffusion. J. Magn. Reson. **161**(2), 222–233 (2003)
15. J.P. Stamps et al., Difftrain: a novel approach to a true spectroscopic single-scan diffusion measurement. J. Magn. Reson. **151**(1), 28–31 (2001)
16. S.J. Gibbs, C.S. Johnson Jr., A PFG NMR experiment for accurate diffusion and flow studies in the presence of eddy currents. J. Magn. Reson. (1969) **93**(2), 395–402 (1991)
17. G.H. Sørland, D. Aksnes, L. Gjerdåker, A pulsed field gradient spin-echo method for diffusion measurements in the presence of internal gradients. J. Magn. Reson. **137**(2), 397–401 (1999)
18. E.L. Hahn, Spin echoes. Phys. Rev. **80**(4), 580–594 (1950)
19. A. Abragam, *The Priciples of Nuclear Magnetism* (Oxford University Press, 1983), p. 618
20. F. Bloch, W.W. Hansen, M. Packard, Nuclear induction. Phys. Rev. **69**(3–4), 127 (1946)
21. C.P. Slichter, *Principles of Magnetic Resonance* (3rd enlarged and updated edition, Springer, 1990)
22. H.C. Torrey, Bloch equations with diffusion terms. Phys. Rev. **104**(3), 563–565 (1956)

23. J. Kärger, W. Heink, The propagator representation of molecular transport in microporous crystallites. J. Magn. Reson. (1969) **51**(1), 1–7 (1983)
24. L.L. Latour et al., Pore-size distributions and tortuosity in heterogeneous porous media. J. Magn. Reson. A **112**(1), 83–91 (1995)
25. L.L. Latour et al., Time-dependent diffusion coefficient of fluids in porous media as a probe of surface-to-volume ratio. J. Magn. Reson. A **101**(3), 342–346 (1993)
26. R.W. Mair et al., Probing porous media with gas diffusion NMR. Phys. Rev. Lett. **83**(16), 3324–3327 (1999)
27. P.P. Mitra, P.N. Sen, L.M. Schwartz, Short-time behavior of the diffusion coefficient as a geometrical probe of porous media. Phys. Rev. B **47**(14), 8565–8574 (1993)
28. P.P. Mitra, B.I. Halperin, Effects of finite gradient-pulse widths in pulsed-field-gradient diffusion measurements. J. Magn. Reson. A **113**(1), 94–101 (1995)
29. P.P. Mitra et al., Diffusion propagator as a probe of the structure of porous media. Phys. Rev. Lett. **68**(24), 3555–3558 (1992)
30. K.E. Washburn, C.H. Arns, P.T. Callaghan, Pore characterization through propagator-resolved transverse relaxation exchange. Phys. Rev. E **77**(5), 051203 (2008)
31. R.L. Kleinberg, W.E. Kenyon, P.P. Mitra, Mechanism of NMR relaxation of fluids in rock. J. Magn. Reson. A **108**(2), 206–214 (1994)
32. S. Godefroy, P.T. Callaghan, 2D relaxation/diffusion correlations in porous media. Magn. Reson. Imaging **21**(3–4), 381–383 (2003)
33. E.O. Stejskal, J.E. Tanner, Spin diffusion measurements: spin echoes in the presence of a time-dependent field gradient. J. Chem. Phys. **42**(1), 288–292 (1965)
34. J.M. Fauth et al., Elimination of unwanted echoes and reduction of dead time in three-pulse electron spin-echo spectroscopy. J. Magn. Reson. (1969), **66**(1), 74–85 (1986)
35. L.E. Drain, The broadening of magnetic resonance lines due to field inhomogeneities in powdered samples. Proc. Phys. Soc. **80**(6), 1380 (1962)
36. W. Heink et al., PFG NMR Self-diffusion measurements with large field gradients. J. Magn. Reson. A **114**(1), 101–104 (1995)
37. S. Meiboom, D. Gill, Modified spin-echo method for measuring nuclear relaxation times. Rev. Sci. Instrum. **29**(8), 688–691 (1958)
38. D.H. Wu, A.D. Chen, C.S. Johnson, An improved diffusion-ordered spectroscopy experiment incorporating bipolar-gradient pulses. J. Magn. Reson. A **115**(2), 260–264 (1995)
39. H.Y. Carr, E.M. Purcell, Effects of diffusion on free precession in nuclear magnetic resonance experiments. Phys. Rev. **94**(3), 630–638 (1954)
40. G.H. Sorland et al., Improved convection compensating pulsed field gradient spin-echo and stimulated-echo methods. J. Magn. Reson. **142**(2), 323–325 (2000)
41. G.H. Sørland, H.W. Anthonsen, K. Zick, J. Sjoblom, A spoiler recovery method for rapid diffusion measurements. Diffus. Fundam. **15**, 9 (2011)
42. G.H. Sørland, Characterization of emulsions by PFG-NMR. AIP Conf. Proc. **1330**(1), 27–30 (2011)
43. V.V. Krishnan, N. Murali, Radiation damping in modern NMR experiments: progress and challenges. Prog. Nucl. Magn. Reson. Spectrosc. **68**, 41–57 (2013)
44. H.W. Anthonsen, G.H. Sørland, K. Zick, J. Sjoblom, Quantitative recovery ordered (Q-ROSY) and diffusion ordered spectroscopy using the spoiler recovery approach. Diffus. Fundam. **16**, 12 (2012)
45. T. Stait-Gardner, P.G. Anil, W.S. Price, Steady state effects in PGSE NMR diffusion experiments. Chem. Phys. Lett. **462**(4–6), 331–336 (2008)
46. J. Mitchell, M.D. Hürlimann, E.J. Fordham, A rapid measurement of T1/T2: the DECPMG sequence. J. Magn. Reson. **200**(2), 198–206 (2009)
47. L. Venturi, K. Wright, B. Hills, Ultrafast T1–T2 relaxometry using FLOP sequences. J. Magn. Reson. **205**(2), 224–234 (2010)
48. R. Durret, *Probability Theory and Examples* (4th ed, Cambridge University Press, 2010), p. 438
49. A.D. Bain, Coherence levels and coherence pathways in NMR. A simple way to design phase cycling procedures. J. Magn. Reson. (1969) **56**(3), 418–427 (1984)

Chapter 2
Observation Time Dependent Diffusion Measurements in Heterogeneous Media by PFG NMR

Abstract Fick's law for diffusion within a homogeneous system is solved, followed by taking into account the impact of restricting geometries. The diffusing particles may then either reflect at the boundary or vanish at the surface due to a surface sink term in the diffusion equation. Focus is on the short observation time dependency of the diffusion coefficient. It is shown that under certain conditions the diffusion measurement may reveal information that are directly related to physical parameters as surface to volume ratio, curvature and surface relaxivity. In particular, it is shown that the short observation time expansion of the diffusion propagator resolves the surface to volume ratio from the surface relaxivity. Comments for the diffusion propagator at longer time scales are also given.

"Can one hear the shape of a drum?"

This question was posed during a lecture by Professor Mark Kac sometime in 1966 [1]. During this lecture he solved the equation of motion for the drum's membrane, and in doing so proved that one really can "hear" whether a drum is circular or not. The problem statement posed by Professor Kac has close resemblance to the observation time dependent diffusivity of a component within confined geometries, which is analyzed in the extensive work by Mitra and his co-workers [2–6]. Here the short observation time expansion of the diffusion coefficient provided the possibility of finding the surface to volume ratio (S/V) simply by measuring the diffusion of the component in a volume V bounded by a surface S at various observation times. The slope of the deviation from unrestricted diffusion as a function of the square root of the observation time provides a value for the S/V [2]

$$\frac{D(t)}{D_0} = 1 - \frac{4}{9\sqrt{\pi}}\sqrt{D_0 t}\frac{S}{V} \tag{2.1}$$

D(t) is the observation time dependent apparent diffusivity, D_0 is the unrestricted (bulk) diffusion coefficient and t is the observation time. Consequently, by measuring diffusion at short observation times the restricted geometries can be observed, a finding that is worthwhile looking into more thoroughly. As shown in

G.H. Sørland, *Dynamic Pulsed-Field-Gradient NMR*,
Springer Series in Chemical Physics 110, DOI 10.1007/978-3-662-44500-6_2

Chap. 1 pulsed field gradient nuclear magnetic resonance (PFG NMR) can be used to measure the displacement of a diffusing molecule at various observation times. The diffusion propagator [7], the probability of moving from one site to another during a defined observation time, represents this displacement as

$$G_{(\vec{r},\vec{r}_0,t)} = P_{(\vec{r},\vec{r}_0,t)}\, p_{(\vec{r}_0,0)} \tag{2.2}$$

$P(r_0, _0)$ is the probability of finding a particle at position r_0 at $t = 0$ and $P(r, r_0, t)$ is the conditional probability of finding a particle at position r at time t given $p(r_0, t)$. $G(r, r_0, t)$ can be interpreted as a probability density distribution function; the probability of finding a particle at position r at time t, given the position r_0 at time $t = 0$.

The echo attenuation from the PFG NMR experiment is written as a double integral over all possible initial and final positions of the diffusing molecule [6, 7]

$$I(t) = I_0 \int\int G(\vec{r}, \vec{r}_0, t) e^{i\gamma\delta\vec{g}(\vec{r}(t)-\vec{r}_0)} d\vec{r}\, d\vec{r}_0 \tag{2.3}$$

I_0 is the initial magnetization, γ is the gyromagnetic ratio, δ is the gradient pulse length and g is the applied magnetic field gradient strength. The evaluation of the diffusion propagator is essential in order to interpret the measured observation time dependent displacement in restricting geometries using PFG NMR.

At the shortest observation times, $t \to 0$, the molecules may experience unrestricted or bulk diffusion. As the observation time increases, the molecules are experiencing the restrictions within the heterogeneous medium, and the measured apparent diffusion coefficient will be less than the bulk diffusivity. At infinitely long observation times, the molecules have probed the entire geometry of the system, which is equivalent to a quasi-homogeneous system and a measured apparent diffusion coefficient that does not change with observation time. This is also called the tortuosity limit [8].

In the following section, Fick's law for a homogeneous system will be solved. Thereafter the impact of restricting geometries will be taken into account. The diffusing particles may then either reflect at the boundary or vanish at the surface due to a surface sink term in the diffusion equation [2]. As seen by the PFG NMR experiment, the equation of motion of the molecules must include the surface sink term into the border conditions when solving Fick's law in heterogeneous media, due to the existence of surface relaxation. Focus will be on the short observation time dependency of the diffusion coefficient. It can be shown that under certain conditions the diffusion measurement may reveal information that are directly related to physical parameters as surface to volume ratio, curvature and surface relaxivity. In particular, it turns out that the short observation time expansion of the diffusion propagator developed by Mitra et al. [2] resolves the surface to volume ratio from the surface relaxivity. This is a very important feature, because it provides the possibility of developing PFG NMR methods in combination with relaxation time measurements for the determination of absolute pore or droplet size distributions.

A few comments for the diffusion propagator at longer time scales will also be given. For a more general and broader treatment of diffusion in heterogeneous media there are comprehensive text books on this subject [9, 10], also in connection with PFG NMR [11, 12]. The selected theory presented here is primarily a consequence of the applications developed and used in Chaps. 6–9.

2.1 Solving Fick's Law for a Homogeneous System

The equation of motion for particles diffusing within cavities in a porous medium, may be described by Fick's law [9]

$$\frac{\partial G_{(\vec{r},\vec{r}_0,t)}}{\partial t} = D_0 \nabla^2 G_{(\vec{r},\vec{r}_0,t)} \tag{2.4}$$

The diffusion propagator, the conditional probability, is defined as [7]

$$G_{(\vec{r},\vec{r}_0,t)} = P_{(\vec{r},\vec{r}_0,t)}\, p_{(\vec{r}_0,0)} \tag{2.5}$$

Again $P(r_0, 0)$ is the probability of finding a particle at position r_0 at $t = 0$, and $P(r, r_0, t)$ is the conditional probability of finding a particle at position r at time t given $p(r_0, t)$. $G(r, r_0, t)$ can be interpreted as a probability density distribution function; the probability of finding a particle at position r at time t, given the position r_0 at time $t = 0$. The initial conditions required for a solution of (2.4), is written

$$G_{(\vec{r},\vec{r}_0,0)} = \delta_{(\vec{r},\vec{r}_0)} \tag{2.6}$$

2.1.1 The Gaussian Solution

When assuming an infinitely homogeneous or quasi-homogeneous system, a Fourier transform of (2.4) and solving it with respect to time yields

$$\widehat{G}_{(t)} = C_{(k)} e^{-Dk^2 t} \tag{2.7}$$

G(t) denotes the Fourier transformed propagator, and k is the Fourier variable. Using the initial conditions, the following equation can be produced

$$C_{(k)} = \left(\frac{1}{2\pi}\right)^{3/2} \int_{-\infty}^{\infty} \delta_{(\vec{\xi}-\vec{r}_0)} e^{i\vec{k}\vec{r}} d\vec{\xi} \tag{2.8}$$

Performing the inverse Fourier transform yields

$$G_{(\vec{r},\vec{r}_0,t)} = \left(\frac{1}{2\pi}\right)^{3} \int_{-\infty}^{\infty} \delta_{(\vec{\xi}-\vec{r}_0)} \left[\int_{-\infty}^{\infty} e^{ik(\vec{r}-\vec{\xi})-Dk^2 t} dk\right] d\vec{\xi} \tag{2.9}$$

The integral in the [...] brackets in (2.9) is not trivial but can be found in Rottmann's Matematische Formelsamlung [13] or may be verified by differentiation of the solution below. The diffusion propagator in a homogeneous system is then written as

$$G_{(\vec{r},\vec{r}_0,t)} = \left(\frac{1}{4\pi Dt}\right)^{3/2} \mathrm{e}^{-\frac{(\vec{r}-\vec{r}_0)^2}{4Dt}} \tag{2.10}$$

This is the Gaussian diffusion propagator, from which the expected mean squared displacement by integration can be derived

$$\left\langle R^2(t)\right\rangle = \int r^2 \left(\frac{1}{4\pi Dt}\right)^{3/2} \mathrm{e}^{-\frac{(\vec{r}-\vec{r}_0)^2}{4Dt}} d\vec{r} = 6Dt \tag{2.11}$$

This relation reflects the expectation value for the square of the distance travelled by molecules from their starting position. This is the unrestricted case where the molecules experience Brownian motion and no confining boundaries. The validity of the Gaussian approximation depends on the condition that the diameters of any cavities in the heterogeneous system are much larger than the root of the mean squared displacement. At the other end of the scale, the approximation is also valid for a heterogeneous system such as zeolites when the root of the mean squared displacement is considerably less than the dimension of the crystallites, such a system may be referred to as "quasi-homogeneous". In Fig. 2.1 it is shown how the diffusion propagator evolves as the observation time increases (D is fixed), when all particles initially (t = 0) are located in the center (x = 0).

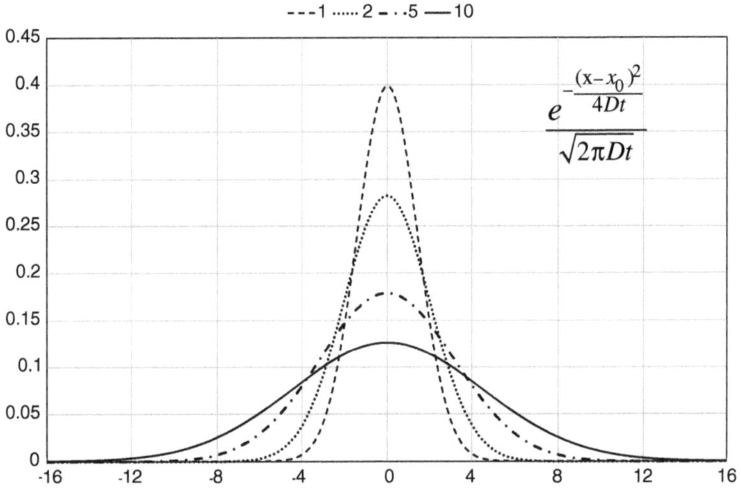

Fig. 2.1 The diffusion propagator in one dimension for various values of **Dt**; 1, 2, 5 and 10

2.2 Restricted Diffusion at Short Observation Times

The initial and boundary conditions required for a solution of (2.4) are written

$$G_{(\vec{r},\vec{r}_0,0)} = \delta_{(\vec{r},\vec{r}_0)} \tag{2.12}$$

$$D_0\,\vec{n}\cdot\nabla G_{r\in\Sigma} + \rho G_{r\in\Sigma} = 0 \tag{2.13}$$

$\vec{n}$ is the normalized surface vector, ρ is the surface relaxivity, δ is the Kronecker δ-function, and Σ is the surface surrounding the cavities (Fig. 2.2). The boundary condition in (2.13) merely states that the surface may act a sink for the NMR signal of the diffusing particles while in the physical picture the particles collide with the surface and bounce back into the cavity. If loss of coherence/relaxation takes place when the particle is in contact with the surface, its contribution to the NMR signal is lost, and thus the diffusing system is observed to behave differently from an NMR perspective. This is why it may be a challenge to relate the NMR diffusion experiments to physical properties as described by Fick's law without the surface relaxation term in the boundary condition. The existence of this term implies that particles at the boundary will have a larger probability of losing the coherence of the nuclear spins. Consequently, the possibility of contributing to the NMR-echo signal will be less than for particles still confined within the homogeneous region.

Before solving equation for short observation times in heterogeneous media, i.e. first order correction term to the Gaussian solution, a dimensional argument may be used to find the size of the exponents for the surface relaxivity, ρ, and the curvature $1/R_0$ of the cavities in the first order correction terms.

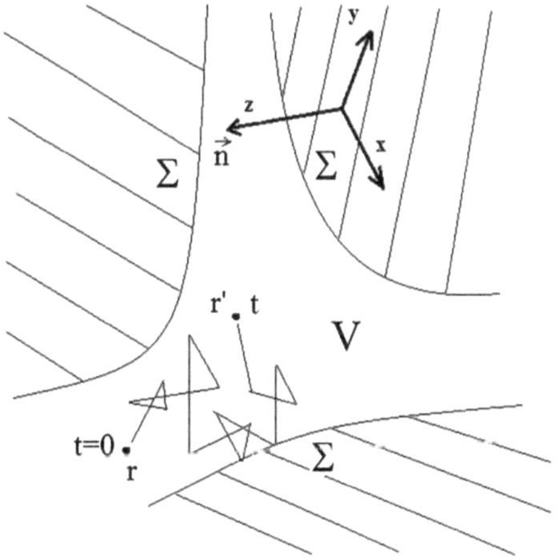

Fig. 2.2 The trajectory of a particle from $t = 0$, to $t = $ t within a volume V and restricted by a surface Σ

2.2.1 A Dimensional Perspective

The most important physical parameter, which affects the diffusivity within cavities at short observation times, is the fraction of particles experiencing the surface. Employing the Gaussian solution, particles within a distance $(6D_0\ t)^{1/2}$ from the surface are likely to sense the restrictions of the surface. This produces a volume fraction of particles, $(S\cdot(6D_0\ t)^{1/2}/V)$, which must be taken into account in a dimensional argument. Other physical parameters that will affect the diffusive process are

D_0 (m^2s^{-1})	bulk diffusion coefficient
t (s^1)	observation time
S (m^2)	surface of the restricting geometries
V (m^3)	cavity volume
ρ (m^1s^{-1})	surface relaxation strength
R_0^{-1} (m^{-1})	mean curvature of the restricting geometries

For bulk diffusion the diffusivity is D_0, and dividing the early observation time departure of the diffusion coefficient, D(t), by D_0, the correction term to the Gaussian solution will be dimensionless. The smallest dimensionless term including the fraction of particles sensing the border, $\approx(S/V)\cdot(D_0\ t)^{1/2}$ (S being the sum of all surfaces Σ in Fig. 2.3), is

$$A\frac{S}{V}\sqrt{D_0 t} \tag{2.14}$$

The smallest dimensionless term including the number of particles experiencing the boundary, $(D_0\ t)^{1/2}$ the surface relaxivity ρ, but not the curvature is

$$B\left(\sqrt{D_0 t}\frac{S}{V}\right)\rho\left(\sqrt{D_0 t}\frac{1}{D_0}\right) = B\rho\frac{S}{V}t \tag{2.15}$$

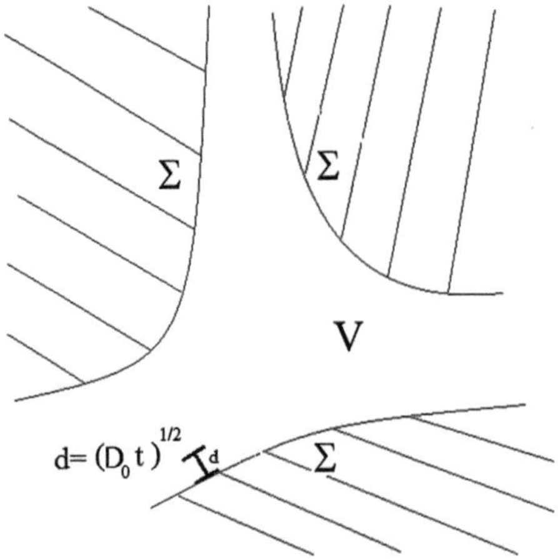

Fig. 2.3 Visualization of the region from the surface Σ to the distance d out from Σ, where the molecules are likely to sense the restrictions by Σ during the observation time t

For the curvature part, the smallest dimensionless term is

$$C\left(\sqrt{D_0 t}\frac{S}{V}\right)\frac{1}{R_0}\left(\sqrt{D_0 t}\right) = C\frac{1}{R_0}\frac{S}{V}D_0 t \tag{2.16}$$

From the product between curvature and surface relaxivity, the smallest dimensionless term is

$$E\left(\sqrt{D_0 t}\frac{S}{V}\right)\frac{\rho}{R_0}\left(\sqrt{D_0 t}\frac{S}{V}\right)^{-1} t = E\frac{\rho}{R_0}t \tag{2.17}$$

Using the dimensional argument, s^1 is the dimension for the time in the lowest order correction terms to $D(t)/D_0$ when ρ and R_0 are involved. This is also true when ρ and R_0 are present simultaneously. The smallest dimensionless term that can be derived from the parameters is equivalent to the fraction of particles sensing the surface. The lowest order correction terms are thus written

$$\frac{D}{D_0} = 1 - A\frac{S}{V}\sqrt{D_0 t} - B\rho\frac{S}{V}t - C\frac{1}{R_0}\frac{S}{V}D_0 t - E\frac{\rho t}{R_0} \tag{2.18}$$

From (2.18) describing the early observation time departure of the measured diffusion coefficient in a porous medium, it can be seen that a square root of time dependency on the lowest order correction term to the bulk diffusion coefficient exists. Most important is that this correction term does not involve the product of ρ and S/V, only the surface to volume ratio. This can be used to find absolute pore size distributions and droplet sizes by PFG NMR (Chaps. 6–9).

2.2.2 *Solving the Diffusion Equation at Short Observation Times*

For particles confined in an arbitrary region as seen by the PFG NMR experiment, the boundary conditions used to find the Gaussian behavior of the diffusion propagator do not apply. When looking at the differential equation in (2.4), the boundary condition in (2.13) makes the equation difficult to solve in a straightforward manner. It is more convenient to start with the Laplace transformation of (2.4) [2]. For the boundary conditions, the geometry of the cavities must somehow be taken into account and the fact that the surface relaxivity makes particle "disappear" at the surface, as shown in (2.13). In evaluating the short-time expansion of $G(r, r_0, t)$ first a simplified solution is used followed by a perturbation expansion. The simplified Gaussian propagator G_0 used here will the one with reflecting boundaries at a smooth surface. The perturbation expansion can then be performed as follows [2];

- Start with the Laplace transform of (2.4) for both $\tilde{G}_0(r', r'', s)$ and $\tilde{G}(r, r', s)$ multiply them respectively with $\tilde{G}(r, r', s)$ and $\tilde{G}_0(r', r'', s)$. Both propagators are assumed to be functions that satisfy the diffusion equation within the restricted geometries.

- Integrate the two equations over r′ and subtract.
- Convert the volume integrals to surface integrals and make use of the boundary conditions.

The calculations following the recipe above are performed in detail in Sect. 2.5.1, and the resulting propagator is as follows

$$\tilde{G}(\mathbf{r},\mathbf{r}'',s) = \tilde{G}_0(\mathbf{r},\mathbf{r}'',s) - D_0 \iint \tilde{G}_0(\mathbf{r},\mathbf{r}',s)[\hat{n}\nabla' + \frac{\rho}{D_0}]\tilde{G}_0(\mathbf{r}',\mathbf{r}'',s)d\Sigma \tag{2.19}$$

When finding an expression for the short time expansion of diffusion in restricted geometries, it is convenient to start with a known and simple solution, the Gaussian diffusion propagator for reflecting boundary conditions at a smooth surface. Doing this, the surface relaxation term in (2.13) is neglected. It has already been argued from dimensional analysis in the previous section that the dimension of the time where ρ is involved must be s^1. Thus, it will not affect the lowest order correction term, as the dimension s^1 is larger than the expected square root of time behavior ($s^{1/2}$) of the lowest order correction term. The following reflecting boundary conditions can then be applied

$$\left(D_0\vec{n}\cdot\nabla\tilde{G}_0\right)_{r\in\Sigma} = 0 \tag{2.20}$$

The mean squared displacement is then written as

$$R^2(t) = (1/V)\int\int drdr'(r-r')^2 G(r,r',t) \tag{2.21}$$

From the time derivative of (2.21), it can be found that

$$\begin{aligned}\frac{\partial}{\partial t}R^2(t) &= (1/V)\int\int drdr'(r-r')^2\frac{\partial}{\partial t}G(r,r',t)\\ &= (D_0/V)\int\int drdr'(r-r')^2\nabla^2 G(r,r',t)\end{aligned} \tag{2.22}$$

The Laplace transform of (2.22) will simplify the derivation and is written

$$s\tilde{R}^2(s) = (D_0/V)\int\int drdr'\,\overbrace{(r-r')^2}^{u}\,\overbrace{\nabla^2\tilde{G}(r,r',s)}^{v'} \tag{2.23}$$

Using a 2-step partial integration, and remembering that $\nabla\tilde{G}(r,r',s)$ vanishes at the surface

$$\left\{\iiint u'\cdot v dr = \iiint \nabla(u\cdot v)dr - \iiint u\cdot v' dr = \iint (u\cdot v)d\sigma - \iiint u\cdot v' dr\right\} \tag{2.24}$$

$$s\tilde{R}^2(s) = -\frac{2D_0}{V}\int\int dr'd\sigma\,\hat{n}\cdot(r-r')\tilde{G}(r,r',s) + \frac{2D_0}{V}\int\int drdr'\tilde{G}(r,r',s) \tag{2.25}$$

The last term in (2.25) will produce the relation for unrestricted Brownian movement, as it is an integral over a normalized density distribution function:

$$\frac{2D_0}{V}\int\int drdr'\tilde{G}(r,r',s) = \frac{2D_0}{V}\int\int\int drdr'dtG(r,r',t)\mathrm{e}^{-st} = 6D_0\int_0^\infty \mathrm{e}^{-st}dt = \frac{6D_0}{s} \tag{2.26}$$

Conducting the inverse Laplace transform produces a relation in three dimensions. Equation 2.26 is then written

$$R^2 = 6D_0 t \tag{2.27}$$

Putting this relation into (2.25) provides the following relation

$$s\tilde{R}^2(s) = \frac{6D_0}{s} - \frac{2D_0}{V}\int\int dr'd\sigma\,\hat{n}\cdot(r-r')\tilde{G}(r,r',s) \tag{2.28}$$

The second term of (2.28) consists of a surface integral over the point r and a volume integral over r′. Now, the approximation that disregards curvature of the surface can then be incorporated: At the shortest observation times, the surface may be approximated by a plane transverse to z, i.e. the tangent plane at r. Then the initial propagator becomes the Gaussian diffusion propagator with reflecting boundary conditions

$$\begin{aligned}
\tilde{G}_0(\mathbf{r}'',\mathbf{r}',s) &= \frac{1}{4\pi D_0}\left[\frac{\mathrm{e}^{-(\mathrm{s/D_0})^{1/2}\mathrm{r_1}}}{r_1} + \frac{\mathrm{e}^{-(\mathrm{s/D_0})^{1/2}\mathrm{r_2}}}{r_2}\right]\\
r_1 &= \sqrt{(x'-x'')^2+(y'-y'')^2+(z'-z'')^2}\\
r_2 &= \sqrt{(x'-x'')^2+(y'-y'')^2+(z'+z'')^2}\\
z'' &= \frac{1}{2}\left[\frac{x''^2+y''^2}{R_0}\right]
\end{aligned} \tag{2.29}$$

Next, the perturbation expansion for the propagator (2.19) can be used and incorporated into (2.28)

$$\begin{aligned}
s\tilde{R}^2(s) &= \frac{6D_0}{s} - \frac{2D_0}{V}\int\int dr'd\sigma\,\hat{n}\cdot(r-r')\tilde{G}(r,r',s)\\
&= \frac{6D_0}{s} - \frac{2D_0}{V}\int\int dr'd\sigma(r-r')\\
&\quad\left(\begin{array}{l}\tilde{G}_0(\mathbf{r},\mathbf{r}',s)-\\ D_0\int\int\tilde{G}_0(\mathbf{r},\mathbf{r}'',s)\,\hat{n}''\nabla''\,\tilde{G}_0(\mathbf{r}'',\mathbf{r}',s)d\sigma''\end{array}\right)
\end{aligned} \tag{2.30}$$

The first order correction term to the unrestricted diffusion in (2.30) is

$$\frac{2D_0}{V}\int\int dr'd\sigma(r-r')\tilde{G}_0(\mathbf{r},\mathbf{r}',s) \tag{2.31}$$

By assuming piecewise smooth and flat surfaces and that only a small fraction of the particles are sensing the restricting geometries, it can be shown (see Sect. 2.5.2 for details) that the restricted diffusion coefficient can be written as

$$\frac{D(t)}{D_0} \approx 1 - \frac{4}{9\sqrt{\pi}}\sqrt{D_0 t}\frac{S}{V} + \phi(\rho, R, t) \tag{2.32}$$

$D(t)$ is the time dependent diffusion coefficient, D_0 is the unrestricted diffusion coefficient, in bulk fluid, and t is the observation time. The higher order terms in t, $\varphi(\rho, R, t)$ contain the contribution due to finite surface relaxivity and curvature ($1/R_0$) of the surfaces. At the shortest observation times, these terms may be neglected such that the deviation from bulk diffusion depends on the surface to volume ratio alone.

2.2.3 Second Order Corrections

The second term in equation contain a curvature dependency on the expansion at short observation times. However, as can be seen from the dimensional arguments, the curvature dependent correction terms must depend linearly on time. Mitra et al. [2] states that the second correction term involving curvature dependency is written

$$-\frac{D_0 t}{6R_0}\frac{S}{V} \tag{2.33}$$

where the curvature dependency on z is defined as

$$z = \frac{1}{2R_0}\left[x^2 + y^2\right] \tag{2.34}$$

Regarding the possible impact from the surface relaxivity on the expansion at short observation times, the reflecting boundary conditions must be replaced with the possible "loss" of molecules at the surface, This provides the sinks at the boundaries

$$\begin{aligned} \tilde{G}(\mathbf{r}, \mathbf{r}'', s) &= \tilde{G}_0(\mathbf{r}, \mathbf{r}'', s) \\ &- D_0 \iint \tilde{G}_0(\mathbf{r}, \mathbf{r}', s)[\hat{n}\nabla' + \frac{\rho}{D_0}]\,\tilde{G}_0(\mathbf{r}', \mathbf{r}', s)dS \quad + \cdots \end{aligned} \tag{2.35}$$

As anticipated from dimensional analysis this correction term is also linear with time and as shown by Mitra et al. [2] to be

$$\frac{\rho t}{6}\frac{S}{V} \tag{2.36}$$

Finally, the restricted diffusion at short observation times is written

$$\frac{D(t)}{D_0} \approx 1 - \frac{4}{9\sqrt{\pi}}\sqrt{D_0 t}\frac{S}{V} - \frac{D_0 t}{6R_0}\frac{S}{V} + \frac{\rho t}{6}\frac{S}{V} \tag{2.37}$$

The sign of the different terms can be interpreted as follows: The term that is proportional to the square root of time reduces the measured diffusivity, as it represents the reflecting boundaries for a flat surface without any "loss" of molecules at the border. The term that involves the curvature is negative when the curvature R_0 is positive, i.e. a concave surface, and the concavity with positive R_0 follows from the functionality of z in (2.29) and the positive z axis defined by Mitra et al. to be along the inward surface normal [2]. For a concave surface the pore space is reduced as compared to flat surfaces, and thus a larger fraction of molecules are experiencing the restrictions, leading to a further reduction of the measured diffusivity. For a convex surface, R_0 is negative, the sign of the term changes and the pore space available increases as compared to the flat surface, leading to smaller reduction of the measured diffusivity. Finally, the third term involving a finite and positive surface relaxivity will weigh the PFG NMR towards the molecules that have not experienced the surface yet. Thus, it contributes to a smaller reduction of the measured diffusivity.

2.2.4 On the Use of the Short Time Expansion of the Diffusion Propagator

In Fig. 2.4 the observation time dependent diffusivity of water amongst compact mono sized spheres (Duke Scientific Ltd.) is displayed [14]. Knowing the porosity of the system, ϕ, the average diameter of the spheres can be established from the measured surface to volume ratio obtained from the PFG-NMR experiment

$$d_{sphere} = \frac{6\left(\frac{1}{\phi} - 1\right)}{\frac{S}{V}} \tag{2.38}$$

Using the corrected effective diffusion times, which take into account restricted diffusion during the motional encoding and decoding [15] (see Chap. 3 for further discussions on corrected effective diffusion times), the diameter of the spheres is found to be 100.6 ± 4 μm. The value reported from Duke Scientific is 98.7 ± 4.9 μm [14]. Thus, it is possible to establish the diameter of the compact spheres by combining experimental data from PFG NMR with the short observation time expansion.

The result of the short time expansion of the diffusion propagator is important in the sense that it disconnects the surface to volume ratio from the surface relaxivity. Consequently, it is possible to measure absolute values of the surface to volume ratio even in the presence of a finite surface relaxivity, which produces a

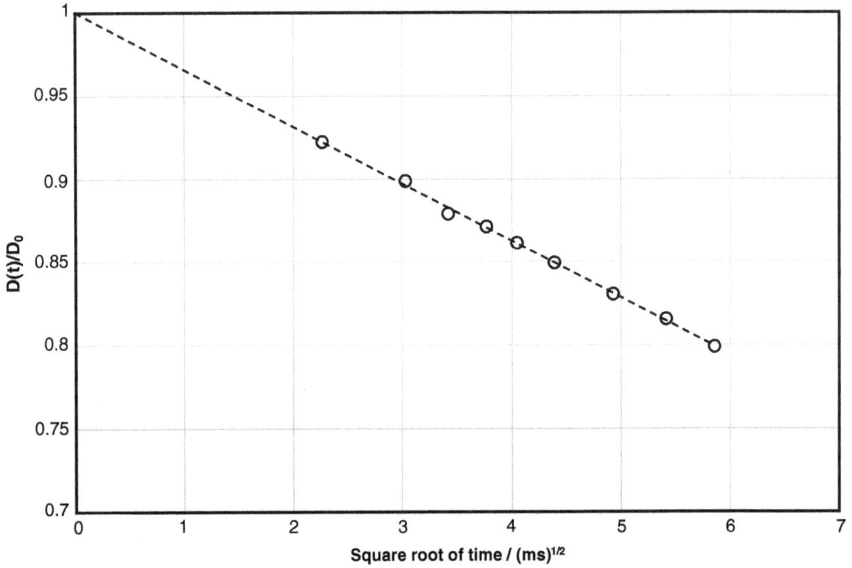

Fig. 2.4 Normalized diffusion coefficient as a function of the square root of observation times [14]. The system investigated is water amongst compact mono sized spheres from Duke Scientific

sink term in the diffusion equation. This can be combined with the relaxation time measurements in heterogeneous media in the fast diffusion limit [16–18]

$$\frac{1}{T_2} \approx \rho \frac{S}{V} \tag{2.39}$$

Then, an absolute pore size or droplet size distribution may be acquired (see Chaps. 6–9). However, this approach has limitations, which can be found in the assumptions used to derive the short time expansion. Most important ones are the assumption of piecewise smooth surfaces, i.e. neglecting the curvature of the surfaces and that the root of the mean squared displacement is much smaller than the typical size of the cavities.

2.3 Diffusion in Heterogeneous Media at Longer Observation Times

In this section the heterogeneous system will be divided into two classes, one where the diffusing particles are allowed to diffuse throughout the medium (Sect. 2.3.1), as in a rock core plug with permeability >0. The second class is where the diffusing particles are confined within a region which is much less than the sample dimension (Sect. 2.3.2), such as water droplets trapped within a continuous oil phase (water in oil emulsion).

2.3.1 Tortuosity and Micro Connectivity

Measuring diffusion or the mean squared displacement due to Brownian motion in heterogeneous media by PFG NMR is a truly challenging task. The existence of a surface relaxation term means that the pore space as seen by the NMR experiment may not reflect the true geometry (see Sect. 2.4). In addition, internal magnetic field gradients may influence the measured diffusivity significantly [19–24].

The tortuosity is a parameter that describes the connectivity of the pore space. Assuming that the steady state diffusion coefficient is measured, i.e. when the diffusing molecules have probed the heterogeneity of the pore space completely, then an asymptotic value for the diffusion denoted D_∞ is measured, the tortuosity limit. The tortuosity factor T, is defined as the ratio between D_0 and D_∞ [8]

$$\frac{1}{T} = \frac{D_\infty}{D_0} \tag{2.40}$$

However, due to the finite lifetime of the NMR signal (the order of seconds), it will not always be possible to probe the entire heterogeneity of the system. For example, in a porous rock saturated with water, under fortuitous circumstances, the lifetime of the water signal may be so long that length scales up to ~100 μm can be probed. Previous studies have used hyperpolarized Xenon gas in order to circumvent this problem [25–27]. Under these conditions, the PFG NMR experiment can be used to probe length scales of the order of millimeters. In Fig. 2.5, the measured diffusivity from water amongst compact monosized plastic spheres of diameter 45 μm can be seen. For these plastic spheres, the surface relaxation is weak, so the lifetime of the NMR signal is long enough for the molecules to probe the heterogeneity of the sample. For the longest observation time, the root of the mean squared displacement is 135 μm. This is almost three times the diameter of the spheres, so it can be concluded that the heterogeneity of the system has been probed in its entirety. The measurements also confirm that an asymptotic value for the diffusion coefficient has been reached, as there is no significant observation time dependence on the measured diffusivity from 800 ms and onwards.

When analyzing porous systems, such as brine saturated rock core plugs, it is due to the finite life time of the NMR signal more the exception than the rule that the asymptotic limit for the measured apparent diffusivity is reached [8]. Therefore, the measured diffusivity cannot be related to a physical parameter such as tortuosity or permeability. However, by combining the short observation time diffusion measurements with the longest possible observation time diffusion measurements, information on the micro connectivity of the porous system may be obtained. The diffusion measurements at short observation times provide a value for the surface to volume ratio (S/V). This value is a physical parameter that is

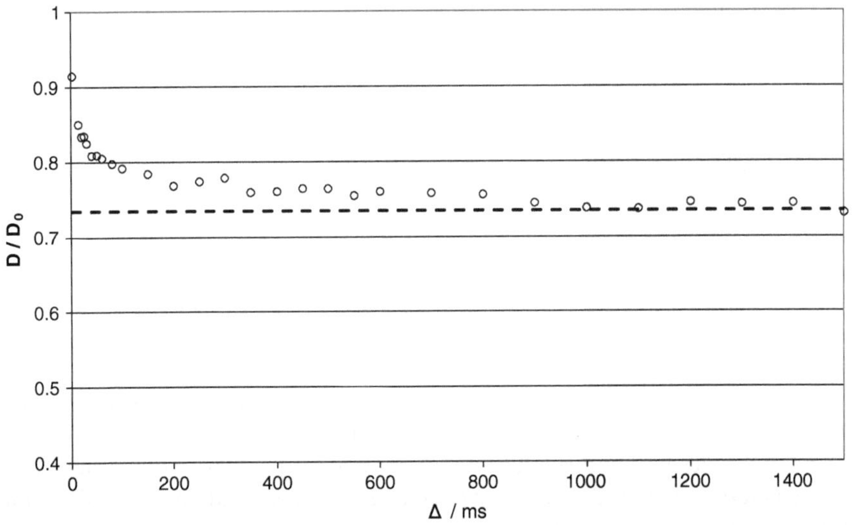

Fig. 2.5 The measured diffusivity as a function of observation time (Δ) of water amongst mono-sized plastic spheres (Duke Scientific Ltd.), and normalized against the bulk diffusion. Tortuosity limit, $1/T$, was fitted to ~0.74

characteristic of the porous system. If instead of plotting the measured diffusion coefficient as a function of time, it is plotted against the product of the root of the mean squared displacement for bulk diffusion and (S/V), i.e. a plot against a dimensionless variable x defined as

$$x = \sqrt{D_0 t} \cdot \left(\frac{S}{V}\right) \tag{2.41}$$

When the value of this parameter is 1, the diffusing molecules have probed the cavity that is characteristic for the system being investigated. If x is >1 the diffusing molecule is probing heterogeneities larger than the typical size of the cavity within the system, represented by $(S/V)^{-1}$. In other words, by plotting the measured diffusivity against the dimensionless variable, the attenuation provides a measure for the connectivity on a micro scale (up to the order of 100 μm depending on the lifetime of the NMR signal). Furthermore, if there are systems having the same micro-connectivity, but having different scale on the cavities, the plots of the measured diffusivity against the dimensionless variable should follow the same slope. In Fig. 2.6 the measured diffusion coefficient as a function of observation time for two systems can be seen, one with water diffusing amongst compact mono sized spheres of 45 μm and one with water diffusing amongst compact mono sized spheres of 98 μm. As expected, the measured diffusivity reaches the tortuosity limit faster for the 45 μm (at ~800 ms), while it takes more than a second before the 98 μm system reaches that limit. In Fig. 2.7, the attenuations are plotted against the dimensionless variable. It can be seen that the two systems

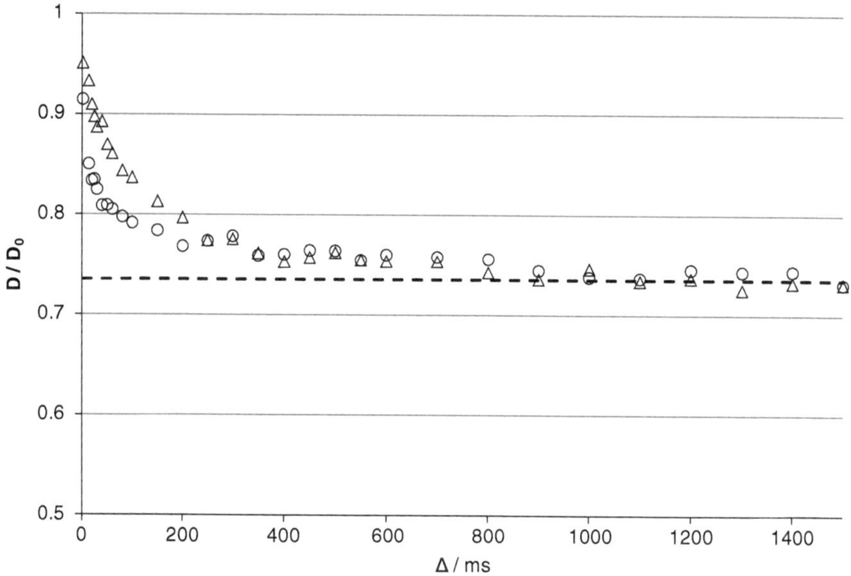

Fig. 2.6 The measured normalized diffusion coefficients as a function of the observation time Δ. "o" represents diffusion data for the water amongst 45 μm polymer microspheres while "Δ" originates from water amongst 98 μm polymer microspheres

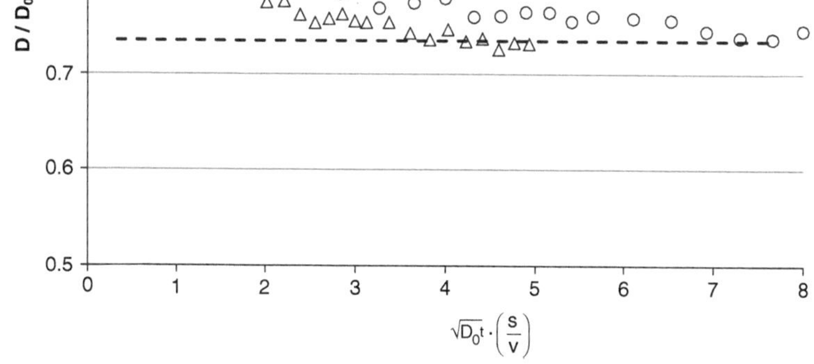

Fig. 2.7 The measured normalized diffusion coefficients as a function of the dimensionless variable x in (2.41) "o" represents diffusion data for the water amongst 45 μm polymer microspheres while "Δ" originates from water amongst 98 μm polymer microspheres

are more or less identical (the small discrepancy can be explained from a different random packing of the two sets of compact mono sized spheres). Therefore, in the absence of the ability to probe the heterogeneity on a larger length scale, at least the micro connectivity may provide insight into how cavities are connected on a micro scale. In Chap. 8, this procedure is applied on brine saturated rock core plugs and compared against permeability measurements.

2.3.2 Diffusion Within Closed Spherical Geometries at Longer Observation Times

There exist exact solutions to the Fick's diffusion equation for well-defined cavities, for example between planes, within cylinders and closed spheres [9, 11, 28]. Assuming a surface relaxation sink, ρ, in the boundary condition (2.13) the diffusion propagator for spherical cavities is [12]

$$G(\mathbf{r}_0, \mathbf{r}, t) = \sum_{n=0}^{\infty}\sum_{m=0}^{\infty}\left(\frac{j_n(\alpha_{nm} r_0/R) P_n(cos\theta_0) j_n(\alpha_{nm} r/R) P(\cos\theta)}{j_n^2(\alpha_{nm}) - j_{n-1}(\alpha_{nm}) j_{n+1}(\alpha_{nm})}\right)$$
$$\left(\frac{2n+1}{2\pi R^3}\right)\exp^{\left(-\frac{\alpha_{nm}^2 Dt}{R^2}\right)} \quad (2.42)$$

R is the spherical cavity radius, $P_n(x)$ is a Legendre polynomial, j_n are the spherical Bessel functions, θ and θ_0 are the angles between r and the direction of the measurement (i.e. the direction of the applied magnetic field gradient in PFG NMR experiment), α_{nm} is the eigenvalue determined by

$$\alpha_{nm}\frac{\frac{n}{\alpha_{nm}} j_n(\alpha_{nm}) - j_{n+1}(\alpha_{nm})}{j_n(\alpha_{nm})} = \alpha_{nm}\frac{j_n'(\alpha_{nm})}{j_n(\alpha_{nm})} = -\frac{\rho R}{D} \quad (2.43)$$

Which defines $j_n'(a_{nm})$.

Obviously, the diffusion propagator for diffusion within spherical geometries results in a PFG NMR attenuation function that is not easy to interpret and one cannot obtain an algebraic form of the distribution function, if indeed such a function exists at all. Packer and Rees set in 1971 the standard for how to achieve droplet size distributions from PFG NMR experiments, and their method is still, four decades later, extensively used in the food industry for determination of droplet size distributions [29]. In addition to the theoretical framework done by others [30–32], they introduced a log-normal distribution function for the droplet sizes, which was an appropriate approximation for their particular system under investigation

$$\mathrm{P(r)} = \frac{1}{2\mathrm{r}\sigma\sqrt{2\pi}}\exp^{-\left[\frac{(\ln(2\mathrm{r})-\ln(\xi))^2}{2\sigma^2}\right]} \quad (2.44)$$

ξ is the diameter median, σ is the standard deviation of the distribution and r is the droplet radius. Then, by measuring the time dependent diffusion coefficient in a region where

$$\Delta \sim \left[\frac{r^2}{2D}\right] \tag{2.45}$$

the data were fitted assuming the following relation for the attenuation of the PFG-NMR signal

$$I^{total}_{(\Delta,g,\delta)} = \frac{\int_0^\infty r^3 P_{(r)} I_{(\Delta,g,\delta,D,r)} dr}{\int_0^\infty r^3 P_{(r)} dr} \tag{2.46}$$

where the echo attenuation is written [29]

$$I_{(\Delta,g,\delta,D,r)} = \exp^{-\left[2\gamma^2 g^2 \sum_{m=1}^{\infty} \frac{1}{\alpha_m^2(\alpha_m^2 r^2 - 2)}\left(\frac{2\delta}{\alpha_m^2 D} - \frac{2 + \exp^{(-\alpha_m^2 D(\Delta-\delta)} - 2\exp^{-\alpha_m^2 D\Delta} - 2\exp^{(\alpha_m^2 D\delta)} + \exp^{-\alpha_m^2 D(\Delta+\delta)}}{(\alpha_m^2 D)^2}\right)\right]} \tag{2.47}$$

and α_m is the mth positive root of the Bessel function equation

$$\frac{J_{3/2}(\alpha r)}{\alpha r} = J_{5/2}(\alpha r) \tag{2.48}$$

This approach will not be investigated here further, as it is well documented elsewhere [33–36]. However, attention should be paid to the statement found at the end of the article by Packer and Reese, where they speculate that another approach could be to measure at longer observation times where the echo attenuation is independent of the diffusion coefficient. This approach will be explored extensively in Chap. 7, where PFG NMR experiments will be applied on spherical systems as water droplets confined in an oil phase. Then emphasis will be on the asymptotic behaviour of the diffusion propagator in (2.42). This behaviour is reached when the following conditions are satisfied:

1. the root of the means squared unrestricted displacement is much larger than the droplet sizes ($R_{droplet}$)

$$\sqrt{6D_0\Delta} \gg R_{droplet} \tag{2.49}$$

2. the root of the means squared displacement during the gradient pulse is much less than the droplet size

$$\sqrt{6D_0\delta} \ll R_{droplet} \tag{2.50}$$

Under these conditions, it can be shown that the attenuation of the PFG-NMR signal is simplified to [30]

$$I = I_0 e^{-R_{droplet}^2 \gamma^2 g^2 \delta^2 / 5} \tag{2.51}$$

This equation was derived under the assumption of reflecting boundaries [31]. However, there is no reason for this not to be valid for finite surface relaxivity given that (2.49–2.50) are also satisfied. At long observation times, i.e. (2.49) is fulfilled; all molecules have experienced contact with the restricting surface numerous times. Thus, there is no distinction between the bulk and surface relaxivity, and the system can be interpreted as having reflecting boundaries, but with a relaxivity weighted between the bulk and surface values. Figure 2.8 shows two PFG NMR experiments performed on an emulsion containing water droplets within an oil phase. The observation time is set so long that the asymptotic limit is reached, in this case to 750 and 1,000 ms. As expected from (2.51) the attenuation is independent on the observation time and the slope should then be proportional to the droplet size of the water droplets. The attenuation is not entirely linear as a function of the square of the applied magnetic field gradient strength. In the upper right corner of Fig. 2.8 an understanding of why the attenuation is nonlinear can be found. A microscopy picture of the water droplets can be seen, and this shows a distribution of droplet sizes. Thus, it is expected that the attenuation of the echo signal will be multi exponential in nature due to the distribution. This raises the question of whether the conditions exist for assuming that (2.51) is correct for all droplet sizes? In Chap. 7 the validity of assuming these conditions will be discussed.

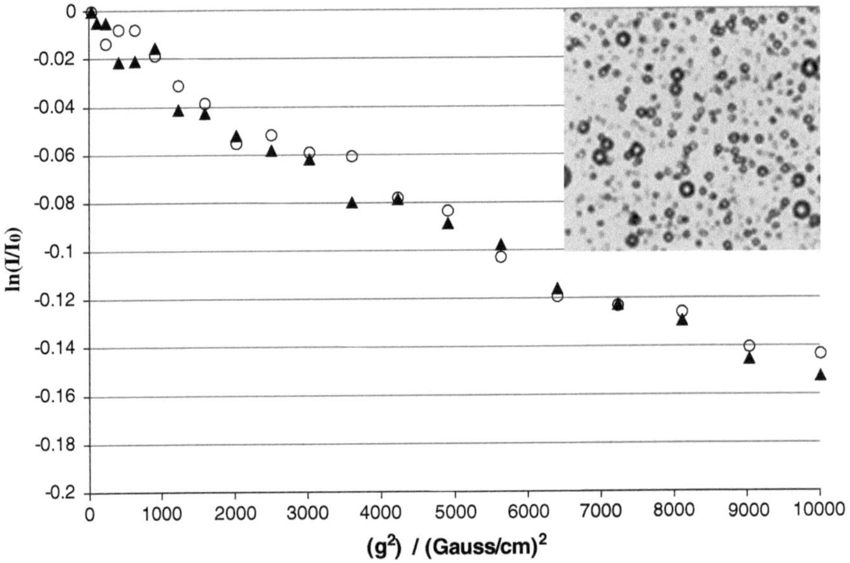

Fig. 2.8 The PFG-NMR attenuation of water in oil emulsion as a function of applied gradient strength and at two observation times, $\Delta = 750$ and 1,000 ms. In the *upper right* corner there is a microscopy photograph of the emulsion. The oil signal does not contribute as its T_1 relaxation time is 100 ms

The average droplet diameter can be extracted from the initial slope in Fig. 2.8. This can be combined with the relaxation time measurements in heterogeneous media in the fast diffusion limit [16–18]

$$\frac{1}{T_2} \approx \rho \frac{S}{V} \tag{2.52}$$

The absolute pore size or droplet size distribution may then be acquired in the so-called asymptotic limit approach (Chaps. 7–9).

2.4 The Impact of Surface Relaxivity on the Restricted Diffusion Measurements by PFG-NMR

The existence of the surface relaxation term makes it necessary to gain some knowledge on the strength of the surface relaxivity, as it can significantly alter the geometry as seen by the PFG NMR experiment compared to the real geometry. In Fig. 2.9 a set of cavities are shown that are connected with narrow channels. Imagine a strong surface relaxivity is present at the boundary's surface. During the PFG NMR experiment, the diffusion time is made long enough so the

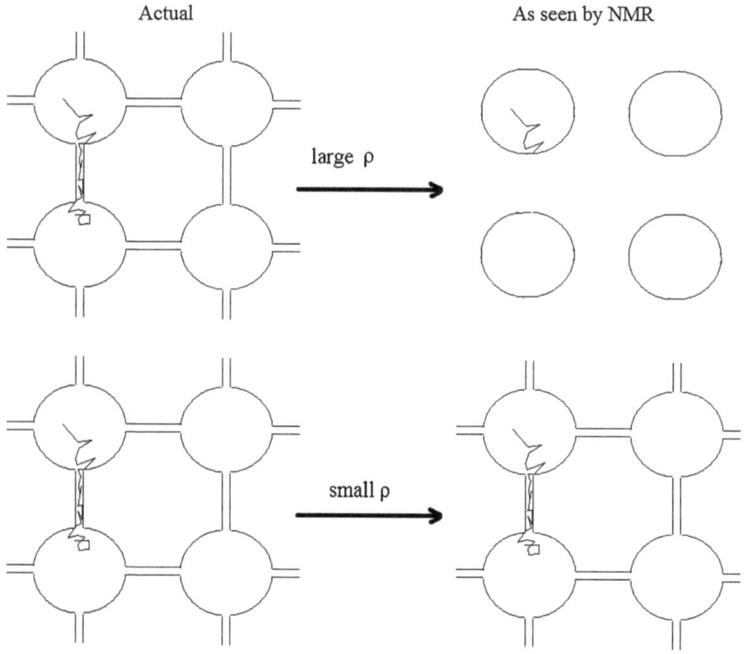

Fig. 2.9 Schematic drawings of porous systems either subjected to strong (*upper half*) or weak (*lower half*) surface relaxivity. The *left* part reflects the actual system while the *right* reflects how it might be interpreted from PFG-NMR measurements

particles will have time to diffuse from one cavity to another. Then a root of the mean squared displacement that exceeds the dimensions of the cavities should be measured. However, when the relaxation processes are strong at the surface of the channels, the spin coherence of the particles is lost by the time they reach the next cavity. Under these conditions, no constructive NMR-signal will result from this portion of particles and the PFG NMR experiment will not be sensitive to the existence of the channels connecting the cavities. The probability density distribution function will be biased towards those particles staying within the same cavity during the diffusion measurement, and the data could be wrongly interpreted as if the heterogeneous medium consisted of closed or isolated cavities. It should be noted that if a significant portion of the system consisted of dead end cavities, i.e. just one way in and out, it would be impossible to distinguish between that system and a well-connected system in the presence of large surface relaxivity value as shown in Fig. 2.9. Therefore, it must be noted that any geometry as seen by the PFG NMR experiment in the presence of strong surface relaxivity does not necessarily reflect the true geometry of the heterogeneous sample being studied [37].

If the strength of the surface relaxivity is weak enough for a significant portion of the diffusion particles to migrate from one cavity to another without losing their spin coherence, the results from the PFG NMR experiment have a greater tendency to reflect the actual confining geometries. This situation is also encountered when the dimensions of the cavities are much less than the root of the mean squared displacement, as in zeolites where pore sizes are of the order of Angstroms [7]. Then one may treat the system as quasi homogeneous [38]. A quasi-homogeneous system is defined as a heterogeneous system where dimension of the cavities are much less than the shortest root of mean squared displacement measurable by the PFG NMR experiment.

Appendix

Perturbation Expansion of the Diffusion Propagator

Consider the diffusion equation of the form

$$\frac{\partial G(\mathbf{r},\mathbf{r}',t)}{\partial t} = D\nabla^2 G(\mathbf{r},\mathbf{r}',t) \tag{2.53}$$

The partial time derivative may be removed by applying the Laplace transform

$$s\tilde{G}(\mathbf{r},\mathbf{r}',s) - \delta(r-r') = D\nabla^2\tilde{G}(\mathbf{r},\mathbf{r}',s) \tag{2.54}$$

Performing the Laplace transform with the boundary conditions produces:

$$D_0\hat{n}\nabla\tilde{G}(\mathbf{r},\mathbf{r}',s) + \rho\tilde{G}(\mathbf{r},\mathbf{r}',s)|_{r\in\Sigma} = 0 \tag{2.55}$$

Now, let $\tilde{G}_0(\mathbf{r}',\mathbf{r}'',s)$ be any other function that satisfies the diffusion equation in the cavities of the porous medium:

$$s\tilde{G}_0(\mathbf{r}', \mathbf{r}'', s) - \delta(r' - r'') = D\nabla'^2\tilde{G}_0(\mathbf{r}', \mathbf{r}'', s) \tag{2.56}$$

Multiplying (2.54) by $\tilde{G}_0(\mathbf{r}', \mathbf{r}'', s)$, (2.56) by $\tilde{G}(\mathbf{r}, \mathbf{r}', s)$, integrating over r$'$, gives us the two equations

$$\begin{aligned}\int \tilde{G}(\mathbf{r}, \mathbf{r}', s)\tilde{G}_0(\mathbf{r}', \mathbf{r}'', s)d\mathrm{r}' - \int \tilde{G}_0(\mathbf{r}', \mathbf{r}'', s)\delta(r - r')d\mathrm{r}' \\ = \int D_0\tilde{G}_0(\mathbf{r}', \mathbf{r}'', s)\nabla'^2\tilde{G}(\mathbf{r}, \mathbf{r}', s)d\mathrm{r}'\end{aligned} \tag{2.57}$$

$$\begin{aligned}\int \tilde{G}_0(\mathbf{r}', \mathbf{r}'', s)\tilde{G}(\mathbf{r}, \mathbf{r}', s)d\mathrm{r}' - \int \tilde{G}(\mathbf{r}, \mathbf{r}', s)\delta(r' - r'')d\mathrm{r}' \\ = \int D_0\tilde{G}(\mathbf{r}, \mathbf{r}', s)\nabla'^2\tilde{G}_0(\mathbf{r}', \mathbf{r}'', s)d\mathrm{r}'\end{aligned} \tag{2.58}$$

Subtraction of those two equations then yields

$$\begin{aligned}&\int \tilde{G}(\mathbf{r}, \mathbf{r}', s)\tilde{G}_0(\mathbf{r}', \mathbf{r}'', s)d\mathrm{r}' - \int \tilde{G}_0(\mathbf{r}', \mathbf{r}'', s)\delta(r - r')d\mathrm{r}' - \int \tilde{G}_0(\mathbf{r}', \mathbf{r}'', s)\tilde{G}(\mathbf{r}, \mathbf{r}', s)d\mathrm{r}' \\ &- \int \tilde{G}(\mathbf{r}, \mathbf{r}', s)\delta(r' - r'')d\mathrm{r}' = \int D_0\tilde{G}_0(\mathbf{r}', \mathbf{r}'', s)\nabla'^2\tilde{G}(\mathbf{r}, \mathbf{r}', s)d\mathrm{r}' \\ &\qquad - \int D_0\tilde{G}(\mathbf{r}, \mathbf{r}', s)\nabla'^2\tilde{G}_0(\mathbf{r}', \mathbf{r}'', s)d\mathrm{r}'\end{aligned} \tag{2.59}$$

$$\begin{aligned}\tilde{G}(\mathbf{r}, \mathbf{r}', s) = \tilde{G}_0(\mathbf{r}, \mathbf{r}'', s) + D_0\Big[\int \tilde{G}_0(\mathbf{r}', \mathbf{r}'', s)\nabla'^2\tilde{G}(\mathbf{r}, \mathbf{r}', s)d\mathrm{r}' \\ - \int \tilde{G}(\mathbf{r}, \mathbf{r}', s)\nabla'^2\tilde{G}_0(\mathbf{r}', \mathbf{r}'', s)d\mathrm{r}'\Big]\end{aligned} \tag{2.60}$$

In order to simplify this expression, the theorem of divergence can be applied [39]. Consider a function u defined as

$$u = G\nabla G_0 \tag{2.61}$$

Then

$$\nabla u = \nabla G\nabla G_0 + G\nabla^2 G_0 \tag{2.62}$$

Taking the volume integral and applying the theorem of divergence one finds

$$\iiint \nabla u\, dV = \iint u\, dS = \iiint \nabla G\nabla G_0 + G\nabla^2 G_0\, dV \tag{2.63}$$

which can be simplified to

$$\iiint \nabla G\nabla G_0 + G\nabla^2 G_0\; dV = \iint G\nabla G_0 dS \tag{2.64}$$

Now consider another function v defined as

$$v = G_0 \nabla G \tag{2.65}$$

Then

$$\nabla v = \nabla G_0 \nabla G + G_0 \nabla^2 G \tag{2.66}$$

Taking the volume integral and applying the theorem of divergence one finds

$$\iiint \nabla v \, dV = \iint v \, dS = \iiint \nabla G_0 \nabla G + G_0 \nabla^2 G \, dV \tag{2.67}$$

which can be simplified to

$$\iiint \nabla G_0 \nabla G + G_0 \nabla^2 G \, dV = \iint G_0 \nabla G dS \tag{2.68}$$

Finally, taking the divergence of the difference between u and v the following relation is produced

$$\iiint \nabla (u - v) \, dV = \iiint (\nabla G \nabla G_0 + G \nabla^2 G_0 - \nabla G_0 \nabla G - G_0 \nabla^2 G) \, dV \tag{2.69}$$

Thus, with the aid of 2.64 and 2.68 the final equation is

$$\iiint (G \nabla^2 G_0 - G_0 \, \nabla^2 G) dV = \iint (G \nabla G_0 - G_0 \nabla G) dS \tag{2.70}$$

Application of the boundary conditions then produces the first two terms in a series expansion when G has been substituted with G_0 on the right hand side:

$$\tilde{G}(\mathbf{r}, \mathbf{r}', s) = \tilde{G}_0(\mathbf{r}, \mathbf{r}', s) - D_0 \iint \tilde{G}_0(\mathbf{r}, \mathbf{r}', s)[\hat{n}\nabla' + \frac{\rho}{D_0}] \, \tilde{G}_0(\mathbf{r}', \mathbf{r}', s) dS \quad + \cdots \tag{2.71}$$

The First Part of the Second Term in (2.30)

The following integral needs to be evaluated

$$-\frac{2D_0}{V} \int \int dr' d\sigma (r - r') \tilde{G}_0(\mathbf{r}, \mathbf{r}', s) \tag{2.72}$$

However, before this, it is convenient to scale the r-variable with the aim of simplifying the expression. By choosing $\varepsilon = \sqrt{\frac{D_0}{s}}$, the exponent in the diffusion propagator will contain only the dimensionless variable $r_0 = \frac{r}{\varepsilon}$. By placing the coordinate system as shown in Fig. 2.2 with r in the center and assuming a piecewise flat surface (i.e. n = [0, 0, 1] and $z_0 = 0$) the diffusion propagator is written

$$\tilde{G}_0(\mathbf{r}, \mathbf{r}_0, s) = \frac{e^{-r_0}}{2\pi D_0 r_0} \tag{2.73}$$

where $r_0 = \frac{r'}{\sqrt{D_0/s}} = \frac{r'}{\varepsilon}$
Then (2.72) can be evaluated

$$
\begin{aligned}
&-\frac{2D_0}{V}\int\int dr'd\sigma\,\hat{n}\cdot(r-r')\tilde{G}_0(\mathbf{r},\mathbf{r}',s)\\
&=-\frac{2D_0}{V}\int \varepsilon dx_0\varepsilon dy_0\,\varepsilon dz_0\int_\Sigma d\sigma\, z_0\frac{\mathrm{e}^{-r_0}}{2\pi D_0 r_0}\\
&=-\frac{\epsilon^3}{\pi V}\int_\Sigma d\sigma\int_{-\infty}^{\infty} dx_0dy_0\int_{z_0(\approx 0)}^{\infty} dz_0z_0\frac{\mathrm{e}^{-r_0}}{r_0}\\
&=\frac{\varepsilon^3}{\pi V}\int_\Sigma d\sigma\int_{-\infty}^{\infty} dx_0dy_0\int_{z_0(\approx 0)}^{\infty} d\mathrm{e}^{-r_0}\\
&=-\frac{\varepsilon^3}{\pi V}\int_\Sigma d\sigma\int_{-\infty}^{\infty} dx_0dy_0\mathrm{e}^{-\sqrt{x_0^2+y_0^2}}\\
&=-\frac{\varepsilon^3}{\pi V}\int_\Sigma d\sigma\int_{0}^{\infty} 2\pi\,\mathrm{Re}^{-\mathrm{R}}d\mathrm{R}\\
&=-\frac{\varepsilon^3}{\pi V}\int_\Sigma d\sigma\left[-2\pi\,\mathrm{Re}^{-\mathrm{R}}\right]_0^\infty-\frac{\varepsilon^3}{\pi V}\int_\Sigma d\sigma\int_0^\infty 2\pi\,\mathrm{e}^{-\mathrm{R}}d\mathrm{R}\\
&=-\frac{2\varepsilon^3}{V}\int_\Sigma d\sigma=-\frac{2\varepsilon^3 S}{V}
\end{aligned}
\tag{2.74}
$$

where S is the total surface area. As a piecewise flat surface (i.e. n = [0, 0, 1] and $z_0 = 0$) is assumed, i.e. neglecting the curvature, the first part of the second term in (2.30) may be solved. This results in an expression that is proportional to S/V, and the mean squared displacement is written as

$$s\tilde{R}^2(s)=\frac{6D_0}{s}-\frac{2\varepsilon^3 S}{V} \tag{2.75}$$

When performing an inverse Laplace transform (L^{-1}) of both sides of (2.75), and denoting the mean squared displacement as 6D(t), the following relation is produced

$$
\begin{aligned}
6D(t)t &= \mathrm{L}^{-1}\left(\frac{6D_0}{s^2}-\frac{2D_0^{3/2}S}{s^{5/2}V}\right)\\
&=6D_0t-\frac{2D_0^{3/2}S}{V}\mathrm{L}^{-1}\left(\frac{1}{s^{5/2}}\right)\\
&=6D_0t-\frac{2D_0^{3/2}S}{V}\frac{t^{3/2}}{\Gamma(5/2)}
\end{aligned}
\tag{2.76}
$$

A value for the gamma function $\Gamma(5/2)$ can be found [13]

$$\Gamma(\frac{5}{2}) = \Gamma(\frac{3}{2}+1) = \frac{3}{2}\Gamma(\frac{3}{2}) = \frac{3}{2}\frac{1}{2}\Gamma(\frac{1}{2}) = \frac{3}{4}\sqrt{\pi} \tag{2.77}$$

Consequently

$$\begin{aligned} 6D(t)t &= 6D_0t - \frac{2D_0^{3/2}S}{V}\frac{t^{3/2}}{\frac{3}{4}\sqrt{\pi}} \\ &= 6D_0t\left(1 - \frac{4}{9\sqrt{\pi}}\frac{S}{V}\sqrt{D_0t}\right) \end{aligned} \tag{2.78}$$

References

1. M. Kac, Can one hear the shape of a drum? The Am. Monthly **73**(4), 23 (1966)
2. P.P. Mitra, P.N. Sen, L.M. Schwartz, Short-time behavior of the diffusion coefficient as a geometrical probe of porous media. Phys. Rev. B **47**(14), 8565–8574 (1993)
3. P.P. Mitra et al., Diffusion propagator as a probe of the structure of porous media. Phys. Rev. Lett. **68**(24), 3555–3558 (1992)
4. P.P. Mitra et al., Pulsed-field-gradient NMR measurements of restricted diffusion and the return-to-the-origin probability. J Magn. Reson. Ser. A **114**(1), 47–58 (1995)
5. L.L. Latour et al., Time-dependent diffusion coefficient of fluids in porous media as a probe of surface-to-volume ratio. J. Magn. Reson. Ser. A **101**(3), 342–346 (1993)
6. P.P. Mitra, P.N. Sen, Effects of microgeometry and surface relaxation on NMR pulsed-field-gradient experiments: simple pore geometries. Phys. Rev. B **45**(1), 143–156 (1992)
7. J.R. Kärger, M. Douglas, in *Diffusion in Zeolites and other Microporous Solids*. (Wiley-Interscience, 1992)
8. L.L. Latour et al., Pore-size distributions and tortuosity in heterogeneous porous media. J. Magn. Reson. Ser. A **112**(1), 83–91 (1995)
9. J. Crank, in *The Mathematics of Diffusion*. (Oxford University Press, 1975), 414
10. H.S. Carslaw, J.C Jaeger, in *Conduction of Heat in Solids*. (Oxford University Press, 1946), 510
11. W.S. Price, in *NMR Studies of Translational Motion Principles and Applications*. (Cambridge University Press, 2010)
12. P.T. Callaghan, in *Translational Dynamics & Magnetic Resonance*. (Oxford University Press, 2011), 547
13. K. Rottmann, in *Mathematische Formelsammlung*. (Bibliographisches Institut, 1984)
14. G.H. Sørland, Short-time PFGSTE diffusion measurements. J. Magn. Reson. **126**(1), 146–148 (1997)
15. E.J. Fordham, P.P. Mitra, L.L. Latour, Effective diffusion times in multiple-pulse PFG diffusion measurements in porous media. J. Magn. Reson. Ser. A **121**(2), 187–192 (1996)
16. M.H. Cohen, K.S. Mendelson, Nuclear magnetic relaxation and the internal geometry of sedimentary rocks. J. Appl. Phys. **53**(2), 1127–1135 (1982)
17. K.S. Mendelson, Nuclear magnetic resonance in sedimentary rocks: effect of proton desorption rate. J. Appl. Phys. **53**(9), 6465–6466 (1982)
18. A. Valfouskaya et al., Nuclear magnetic resonance diffusion with surface relaxation in porous media. J. Colloid Interface Sci. **295**(1), 188–201 (2006)
19. R.M. Cotts et al., Pulsed field gradient stimulated echo methods for improved NMR diffusion measurements in heterogeneous systems. J. Magn. Reson. **83**(2), 252–266 (1989). (1969)
20. R.F. Karlicek Jr, I.J. Lowe, A modified pulsed gradient technique for measuring diffusion in the presence of large background gradients. J. Magn. Reson. **37**(1), 75–91 (1980)

21. M.D. Hürlimann, Effective gradients in porous media due to susceptibility differences. J. Magn. Reson. **131**(2), 232–240 (1998)
22. L.L. Latour, L.M. Li, C.H. Sotak, Improved PFG stimulated-echo method for the measurement of diffusion in inhomogeneous fields. J. Magn. Reson. Ser. B **101**(1), 72–77 (1993)
23. G.H. Sorland, D. Aksnes, L. Gjerdaker, A pulsed field gradient spin-echo method for diffusion measurements in the presence of internal gradients. J. Magn. Reson. **137**(2), 397–401 (1999)
24. G.H. Sørland, B. Hafskjold, O. Herstad, A stimulated-echo method for diffusion measurements in heterogeneous media using pulsed field gradients. J. Magn. Reson. **124**(1), 172–176 (1997)
25. R.W. Mair et al., Tortuosity measurement and the effects of finite pulse widths on xenon gas diffusion NMR studies of porous media. Magn. Reson. Imaging **19**(3–4), 345–351 (2001)
26. R.W. Mair et al., The narrow pulse approximation and long length scale determination in xenon gas diffusion NMR studies of model porous media. J. Magn. Reson. **156**(2), 202–212 (2002)
27. R. Wang et al., Xenon NMR measurements of permeability and tortuosity in reservoir rocks. Magn. Reson. Imaging **23**(2), 329–331 (2005)
28. P.T. Callaghan, Pulsed-gradient spin-echo NMR for planar, cylindrical, and spherical pores under conditions of wall relaxation. J. Magn. Reson. Ser. A **113**(1), 53–59 (1995)
29. K.J. Packer, C. Rees, Pulsed NMR studies of restricted diffusion. I. Droplet size distributions in emulsions. J. Colloid Interface Sci. **40**(2), 206–218 (1972)
30. J.E. Tanner, E.O. Stejskal, Restricted self-diffusion of protons in colloidal systems by the pulsed-gradient, spin-echo method. J. Chem. Phys. **49**(4), 1768–1777 (1968)
31. C.H. Neuman, Spin echo of spins diffusing in a bounded medium. J. Chem. Phys. **60**(11), 4508–4511 (1974)
32. J.S. Murday, R.M. Cotts, Self-diffusion coefficient of liquid Lithium. J. Chem. Phys. **48**(11), 4938–4945 (1968)
33. M.A. Voda, J. van Duynhoven, Characterization of food emulsions by PFG NMR. Trends Food Sci. Technol. **20**(11–12), 533–543 (2009)
34. D. Gabriele et al., Characterisation of dairy emulsions by NMR and rheological techniques. Food Hydrocolloids **23**(3), 619–628 (2009)
35. K. Van Lent et al., Determination of water droplet size distribution in butter: pulsed field gradient NMR in comparison with confocal scanning laser microscopy. Int. Dairy J. **18**(1), 12–22 (2008)
36. B. Balinov et al., Characterization of water-in-crude oil emulsions by the NMR self-diffusion technique. Colloids Surf., A **82**(2), 173–181 (1994)
37. P.N. Sen et al., Surface relaxation and the long-time diffusion coefficient in porous media: periodic geometries. Phys. Rev. B **49**(1), 215–225 (1994)
38. L. Gjerdåker, G.H. Sørland, D.W. Aksnes, Application of the short diffusion time model to diffusion measurements by NMR in microporous crystallites. Microporous Mesoporous Mater. **32**(3), 305–310 (1999)
39. G.B. Arkfen, .H.J. Weber, in *Mathematical Methods for Physicists*, 6th edn. (Elsevier Academic Press, 2005)

Chapter 3
Experimental, Pitfalls and Suggested Solutions

Abstract The most important distortions that may affect the PFG NMR measurements in a destructive way will be discussed; unwanted coherence transfer pathways, eddy current transients, internal magnetic field gradients, convection, vibration and the effect of finite gradient pulse lengths. All the distortions discussed may be present on high field equipment (i.e. superconducting magnets) as well as on low resolution equipment (i.e. permanent magnets). Although, the magnitude and nature of the effects may vary depending on the type of equipment used. Ways to circumvent or suppress the effect from the abovementioned artefacts will also be discussed, as the application of phase sequences, crusher gradients, preemphasis adjustment, bipolar pulsed field gradients, convection compensated PFG NMR sequences, the second cumulant approximation and corrected effective diffusion/observation times.

Having access to a well behaving spectrometer, as described in Chap. 4, it can be applied for measuring the PFG NMR signal. This signal could for example be a free induction decay (FID) following a single radio frequency (RF) pulse, a spin echo (SE), or a stimulated echo (STE). In addition, magnetic field gradients may be on during the acquisition period, to provide spatial resolution of the recorded signal, as in a diffusion experiment. One of the major challenges when acquiring the NMR signal is the presence of distortions that have destructive influences on the NMR signal. There exist multiple origins of these distortions, and in this section, the most important distortions that may affect the NMR measurements in a destructive way will be discussed. This will include the generation of unwanted NMR signal in multi RF-pulsed NMR experiments, which may affect the echo attenuation significantly. In addition, the presence of eddy current field transients due to the application of pulsed magnetic field gradients that may lead to errors in interpretation of the PFG NMR data, as well as the existence of internal magnetic field gradients present in heterogeneous media will be discussed. In the formula describing the PFG NMR attenuation one must be aware of the impact of finite duration of the applied gradient pulse lengths and the existence of spatially varying internal magnetic field gradients. All the distortions discussed may be present on high field

G.H. Sørland, *Dynamic Pulsed-Field-Gradient NMR*,
Springer Series in Chemical Physics 110, DOI 10.1007/978-3-662-44500-6_3

equipment (i.e. superconducting magnets) as well as on low resolution equipment (i.e. permanent magnets). Although, the magnitude and nature of the effects may vary depending on the type of equipment used. Ways to circumvent or suppress the effect from the abovementioned artefacts will also be discussed.

3.1 Selection of the Wanted Signal in PFG-NMR Experiments

As discussed in Chap. 1 the concept of coherence transfer pathways may be used in order to design phase sequences where one after a set of scans in the pulse sequence ends up, in theory, with the desired signal only in a PFG NMR experiment [1]. In principle, the stimulated echo sequence shown in Fig. 3.1 generates 5 echoes which will interfere more or less with the wanted stimulated echo [2].

Assuming $\tau_2 > \tau_1$ the echoes appear at the following times [2]; $(2\tau_1)$, $(\tau_1 + \tau_2)$, $2(\tau_2 - \tau_1)$, $(2\tau_2 - \tau_1)$ and $2(\tau_2)$. The echo-signal being used for measuring diffusion, i.e. impose motional encoding/decoding on the echo signal, is the one that appears at $t = (\tau_1 + \tau_2)$. All other echoes and free induction decays (FID) following the RF-pulses will contribute to the unwanted signal. In Fig. 3.2, the acquired signal from the stimulated echo sequence is shown when the phases of all the RF pulses and the receiver are identical, producing both the wanted and unwanted echo signals in one scan.

The NMR signal appearing in Fig. 3.2 originates from the following:

- **FID1**: Free induction decay from the first $\pi/2$ RF pulse
- **FID2**: Free induction decay from the second $\pi/2$ RF pulse
- **FID3**: Free induction decay from the third $\pi/2$ RF pulse
- **ECHO_12**: Two pulse echo (Hahn-echo) arising from first and second $\pi/2$ RF-pulse
- **STE**: The stimulated spin echo arising from all three $\pi/2$ RF-pulses (the second $\pi/2$ stores the magnetization along the external field)
- **RE**: Refocusing of ECHO12 by the third $\pi/2$ RF pulse
- **ECHO_23**: Two pulse echo (Hahn-echo) arising from second and third $\pi/2$ RF-pulse

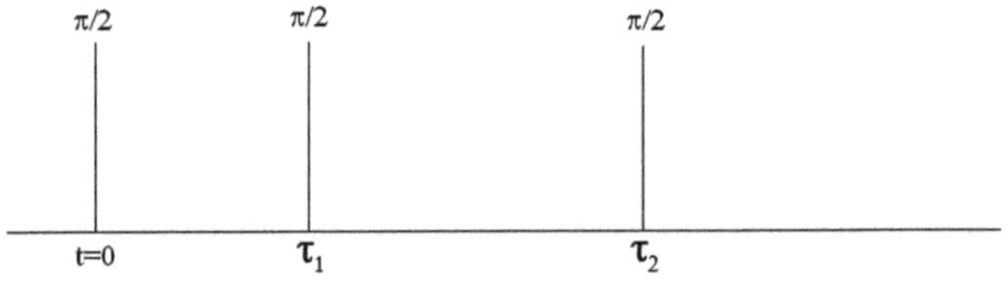

Fig. 3.1 The basic stimulated echo sequence applying three 90° ($\pi/2$) RF pulse. At $t = 0$ the first 90° pulse, while two other 90° RF pulses are initiated at $t = \tau_1$ and $t = \tau_2$ successively

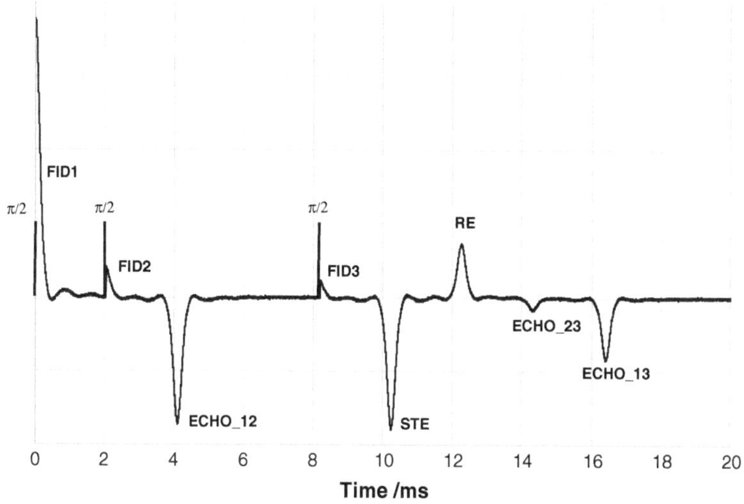

Fig. 3.2 The acquisition of the NMR signal throughout a stimulated echo sequence. The time of appearance of the three $\pi/2$ RF-pulses is indicated by bars

- **ECHO_13**: Two pulse echo (Hahn-echo) arising from first and third $\pi/2$ RF-pulse

Even though the stimulated spin echo signal looks well resolved from the other unwanted echoes and FID's, this is not necessarily the situation when the measurement parameters are optimized in a PFG NMR experiment [3]. In the following section, different approaches for acquiring the wanted stimulated spin echo signal only will be utilized, either alone or in a combination.

3.1.1 Suppressing Unwanted Coherences by RF-Phase Sequencing

The use of different phases on the RF pulses is an important tool for resolving the wanted echo in a pulse sequence containing three or more RF-pulses [1, 2]. As can be seen from Fig. 3.2 there is already one echo (RE) and the FID's having an amplitude of opposite sign to the other echoes even when the phase of all RF-pulses are set to the same value. This is a consequence of the coupling of the RF-field in the Hamiltonian (see Chap. 1). For all the experimental work presented in this book the stimulated echo sequences have the phase sequence proposed by Fauth et al. [2] as the basic sequence, which was developed using the concept of coherence transfer pathways [1]. The phase sequence that suppresses all echoes except for the stimulated one is given in Table 3.1.

Table 3.1 The phase sequences as proposed by Fauth et al. [2]

Sequence #	π/2 excitation	Second π/2	Third π/2	Receiver
1	+x or 0	+x or 0	+x or 0	−x or π
2	−x or π	+x or 0	+x or 0	+x or 0
3	+x or 0	−x or π	+x or 0	+x or 0
4	−x or π	−x or π	+x or 0	−x or π

Table 3.2 The minimum phase sequences needed when applying crusher gradients in the z-storage interval

Sequence #	π/2 excitation	Second π/2	Third π/2	Receiver
1	+x or 0	+x or 0	+x or 0	−x or π
2	+x or 0	+x or π	+x or 0	+x or 0

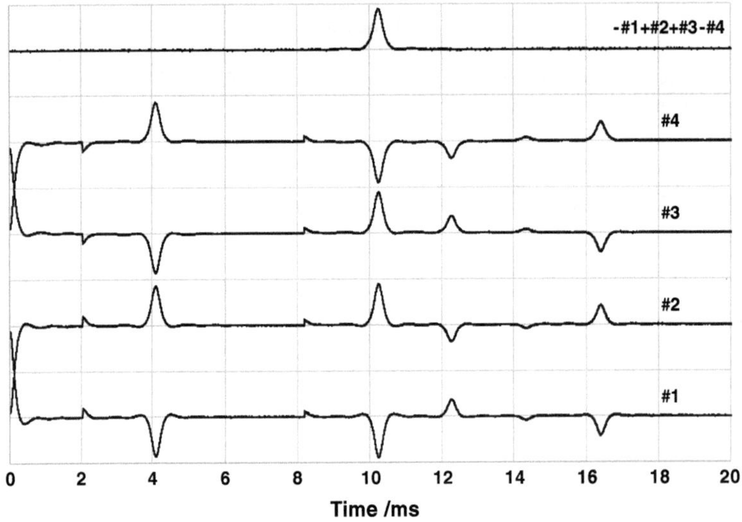

Fig. 3.3 The acquired NMR signal from the four parts in the phase sequence given in Table 3.1. The *upper curve* is the four parts added according to the receiver phase

To visualize the effect of the phase sequence shown in Table 3.1, the individual parts in phase sequence are shown in Fig. 3.3 together with the resulting signal when the scans are added up according to the phase sequence given by the receiver in Table 3.1. It is quite evident that the use of a well-defined set of phases on the RF-pulses and receiver, the stimulated echo can be recorded without any interference from other echoes and FID signals. However, a minimum of 4 scans is required to produce the phase sequence used to suppress the unwanted signals. If there is a requirement to reduce the number of scans, the concept of magnetic field crusher gradients must be applied to assist in suppression of the unwanted echoes. An interesting feature with the two pulse echoes (Hahn echoes [4, 5]) is that the polarity of the echo amplitude is insensitive to the phase of the second $\pi/2$ RF pulse that generates the echo. Comparing the phase sequences #1 and #3 for example, the first $\pi/2$ RF pulse has the same phase while the second $\pi/2$ RF pulse has opposite phases. Still the two pulse echo signal at $t = 2\tau_1$ appears with the same negative amplitude. As will be shown in the next section this characteristic can be used together with magnetic field crusher gradients to reduce the number of phase sequences to two only.

3.1.2 Suppressing Unwanted Coherences by Crusher Gradients

There is a simple way to suppress many of the unwanted echoes as shown in Fig. 3.2, without any too complicated set of phase sequences. This is achieved by adding a so-called magnetic field crusher gradient in the z-storage interval, i.e. between the second and third $\pi/2$ RF pulses (Fig. 3.4). Then a small gradient pulse is applied for a finite duration, which will dephase any coherent NMR-signal in the transverse plane to the external magnetic field. The use of crusher gradients is an effective tool for minimising the phase sequence needed for acquiring the wanted signal only, and is since the early 1990s has been widely used in high resolution for the "cleaning" of spectra [6]. However, care must be taken when choosing the strength and positioning of the crusher gradient in the pulse sequence. The strength and duration must be sufficient to dephase the unwanted echo signal while at the same time not affect the wanted stimulated echo signal. In Fig. 3.5 the acquired signal is shown for an increasing value on the crusher gradient which is initiated just after the second $\pi/2$ RF pulse and is on for 2 ms. Even at the lowest strength on the crusher gradient the unwanted echoes are significantly reduced, while at the intermediate strength they are hardly visible. Using the strongest crusher gradient, the unwanted echoes have vanished, but then the wanted echo is also significantly affected by the crusher gradient or the following eddy current transients, both in amplitude and of the symmetry of the echo signal. It can also be seen in Fig. 3.5 that the FID following the second $\pi/2$ RF pulse is more or less crushed due to the onset of the crusher gradient. Thus for the settings used there is a value of the crusher gradient that removes most of the unwanted echo signal and leaves the wanted echo signal undistorted, the intermediate value Gspoil = 0.8 G/cm.

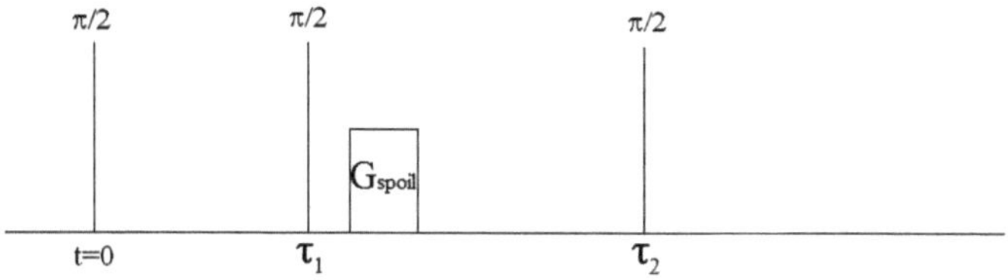

Fig. 3.4 The stimulated echo sequence with a crusher gradient in the z-storage interval

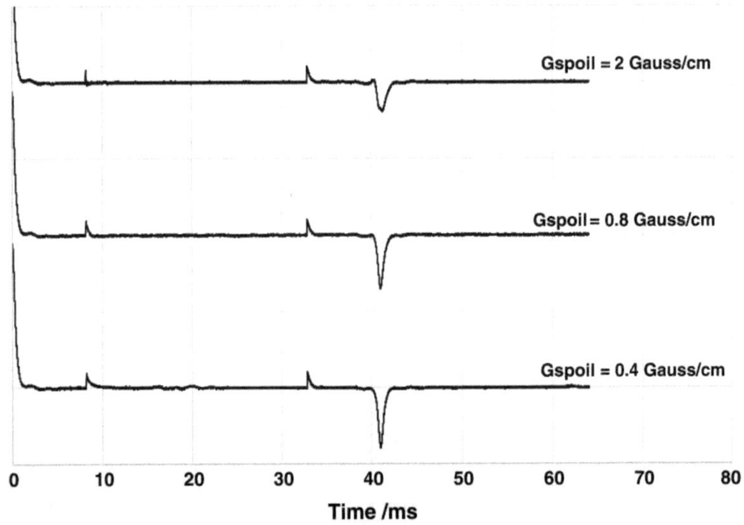

Fig. 3.5 Acquisition of the NMR-signal during the stimulated echo sequence for different values on the crusher gradient Gspoil

In order to acquire the wanted and undistorted echo signal only, it is convenient to apply the intermediate crusher strength in a combination with a two-step phase sequence where ECHO12, RE, and ECHO13 are suppressed. This is done by shifting the phase of the second $\pi/2$ RF pulse and leaving the phase of first $\pi/2$ RF pulse unaltered. Then the remaining signals are the wanted echo signal and the initial FID signal from the second $\pi/2$ RF pulse. The latter is spoiled during the onset of the crusher gradient pulse, but initially it produces a signal in the resulting acquisition (+#2 −#1) because it takes a finite time for the crusher gradient to dephase the FID signal. The reduction in experiment time can be used to study behaviour in systems, which are time unstable without introducing experimental artefacts from unwanted coherences (Fig. 3.6).

3.1.3 Suppressing Unwanted Coherences in Bipolar PFG-NMR Diffusion Experiments

The phase sequence and crusher gradients applied for the stimulated echo sequence are shown to be sufficient for acquiring an attenuation of the PFG NMR signal without any interference from unwanted signal. However, in heterogeneous media where internal magnetic field gradients exist, it becomes necessary to apply a bipolar pulsed gradient version of the stimulated echo [7–9], as for example the 13-interval PFG NMR sequence presented in Chap. 1 [3]. Thus, more RF-pulses are introduced into the sequence, which again will produce more unwanted coherence transfer

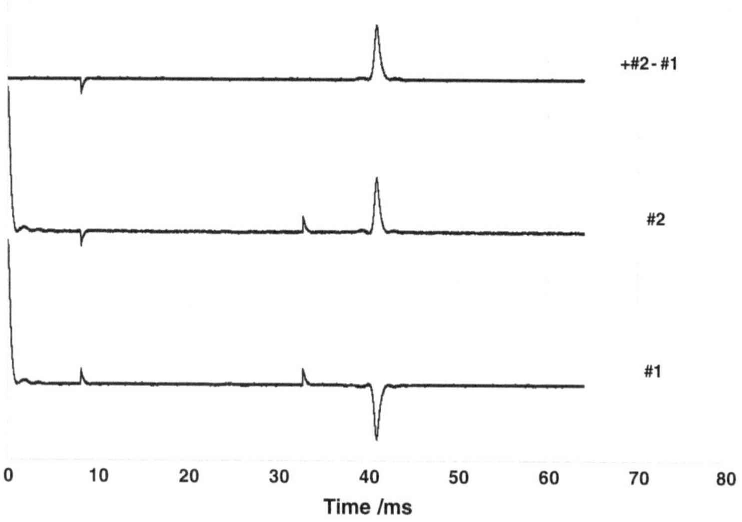

Fig. 3.6 The acquired NMR signal from the two-phase sequences given in Table 3.2. The *upper curve* is the two sequences added according to the receiver phase

pathways. Using the simple phase sequence from Table 3.1, and with crusher gradients in the z-storage intervals, there will no longer be a linear decay of the logarithm of the echo attenuation as a function of the square of the applied gradient strength [3]. The origin of this non-linear decay is due to the refocusing of coherence transfer pathways where the π RF pulses does not alter the coherence of the magnetisation. Thus using bipolar gradients of opposite polarity but of the same strength, the FID signal following the first $\pi/2$ pulse is restored at the time of the second $\pi/2$ RF-pulse. The same applies for the FID following the third $\pi/2$ RF-pulse, for which the FID signal is restored at the time when the wanted echo signal is recorded.

There are at least two solutions to this problem. If it is undesirable to increase the number of scans by applying a more complex phase cycle, crusher gradients may be used in the prepare and read intervals. This is done by adding an extra gradient strength to one of gradients in the gradient pairs, and has been denoted the unequal bipolar gradient method [3, 10]. However, for this method to work one must choose the difference in gradient strength between the pairs, such that the unwanted signal is properly crushed. This level of difference is dependent on the duration of the gradient pulses and the initial intensity of the FID. In Fig. 3.7 the attenuations of the echo signal from a 13 interval PFG NMR sequence are shown on distilled water. Here again the phase sequence by Fauth et al. [2] is used, with and without crusher gradients for comparison in the prepare and read intervals. The natural logarithm of the attenuation without the crusher gradient is linear only down to -1 while the same attenuation is observed to be linear down to where the NMR signal is of the same order of magnitude as the experimental noise when the crusher gradient is applied.

Fig. 3.7 The 13 interval PFGSTE experiment on distilled water using the same phase sequences [2], and the same applied bipolar gradient strengths. The 'o' signs show the logarithm of the echo attenuation without crusher gradients in the prepare and read intervals while the 'x' are with crusher gradients. The gradient coils produced a gradient strength of 4.6 G/cm A. The spectrometer used was a 200 MHz Bruker DMX-system

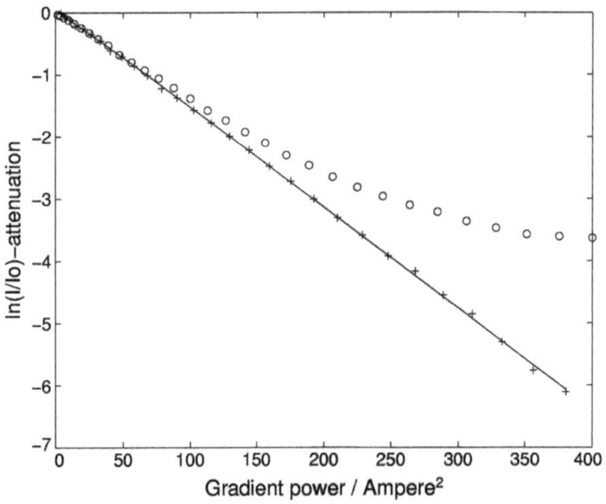

Another solution is to use a more complicated phase sequence than the one proposed by Fauth et al., and such a sequence was proposed by Wu et al. [11] in 1989. When neglecting the eddy current dead time interval in their proposed phase sequence, the minimum number of scans in the sequence increases to 8, i.e. an increase of the experimental time by factor of 2. The way this phase sequence works is to suppress the restoration of the FID arising from the third $\pi/2$ RF pulse from the wanted stimulated echo. While the stimulated echo follows the coherence order pathway $p \to 0 \to -1 \to +1$ in the read interval, the restoration of the FID signal follows the coherence order pathway $p \to 0 \to +1 \to +1$, and in Chap. 1 the phase sequence for accumulating the wanted signal while canceling the unwanted signal is shown.

When applying a new PFG sequence for the first time it is important to check the performance with respect to the possible interference from unwanted signal. In many instances, the signal fraction of the component of interest may be quite small, maybe down to the order of 0.001 (i.e. 0.1 % of the NMR signal following a $\pi/2$ RF-pulse). Then the appearance of just small portions of unwanted coherences from the remaining 99.9 % of NMR signal may interfere destructively with or completely dominate the wanted NMR echo signal. In order to ensure that the signal acquired is the desired one, it is convenient is to start with setting the z-storage interval to 5 times T_1. Then there should be an insignificant portion of wanted signal acquired at the time of the stimulated echo. If such is the case, the next step is to reduce the z-storage delay until a finite NMR signal is recorded at the time of the echo. By applying crusher gradients in the z-storage interval and an appropriate phase sequence, the logarithm of the echo attenuation from a homogeneous system as a function of the square of the applied gradient strength should then be linear, as shown in Fig. 3.7.

When introducing further RF pulses, as for the double stimulated echo approach for suppression of convection artefacts [12, 13], the number of scans has to be increased significantly in order to suppress the unwanted coherences. In low resolution experiments, where one is sampling the signal of echo peak only, it is easier to avoid the unwanted echo signals. Then the parameters of the pulse sequence can be set such that the unwanted echoes are located away from the wanted echo. In high resolution experiments, unwanted coherences may be superimposed on the FID and lead to significant artefacts in the Fourier transformed spectrum. In order to obtain acceptable results, where the FID has to be sampled in order to produce the Fourier transformed spectrum, an increased number of scans must be applied to select the wanted coherence transfer pathway only. However, as for the double stimulated echo using bipolar gradient in liquid samples, the gradients of opposite polarity may be placed in the z-storage interval instead of in the prepare and read intervals [13]. Here the bipolar pairs of gradients are used to suppress eddy current transients and not the coupling to internal magnetic field gradients, and thus there is no need for refocusing π RF-pulses in the prepare and read intervals. Consequently there is a reduction of RF-pulses by a number of 4 and gradient pulses placed in the prepare and read intervals act both as preparatory and crusher gradients (see Sect. 3.2.3 for a discussion on preparatory gradients).

A final and revealing test for the performance of the PFG NMR sequence with respect to measuring the wanted echo signal only, is to deliberately set the $\pi/2$ RF-pulses away from the correct ones. Then the amount of unwanted signal will increase, especially from the coherence order pathway $p \rightarrow 0 \rightarrow +1 \rightarrow +1$ in the read interval of the sequence. This pathway is the main contributor in producing the non-linear decay in Fig. 3.7. If there is a considerable amount of unwanted signal interfering in the echo signal, the slope of the attenuation would change as a function of the offset of the $\pi/2$ and π RF-pulses. Thus, the apparent diffusion coefficient measured will change and become a function of the duration of the RF-pulses. In Fig. 3.8 the measured apparent diffusivity using a sample of distilled water is shown as a function of the duration of the $\pi/2$ RF-pulse used in the 13 interval PFGSTE sequence, either with ('x') or without ('o') crusher gradients in the prepare and read intervals. The phase sequences used is the one proposed by Fauth et al. [2], i.e. 4 scans only. Evidently, there is a large difference in the response to the offset of the RF-pulses for the two approaches. While the measured diffusivity varies between (2.27–2.35) $\times$ 10^{-9} m^2/s when applying crusher gradients in the prepare and read intervals, it varies between (1.40–2.08) $\times$ 10^{-9} m^2/s without. The conclusion is that the amount of unwanted signal is removed or satisfactorily reduced using the crusher gradients. For the experiment without crusher gradients, the presence of the unwanted signal results in a nonlinear decay of the logarithm of the echo attenuation and also produces an error in the initial slope [3].

3.2 Reducing the Impact from Eddy Current Transients

When placing the sample to be investigated in the NMR spectrometer it will be subjected to a strong external field (H_0), a time dependent oscillating electromagnetic field (RF-field) and pulsed field magnetic field gradients. The last factor is

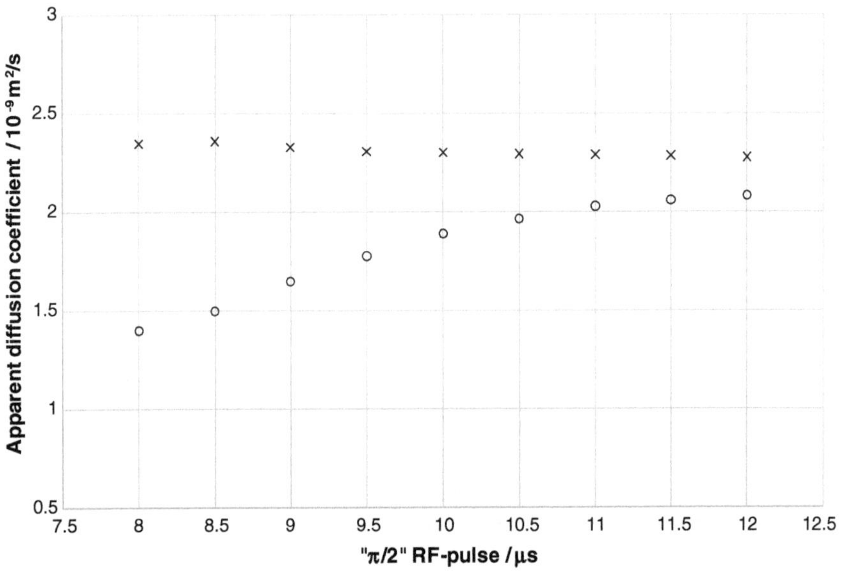

Fig. 3.8 Measurements of the apparent diffusion coefficient of distilled water at 25 °C as a function of the length of the $\pi/2$ RF-pulse, where the apparent diffusivities were extracted from the initial decay. The 'o'-symbols show the measured apparent diffusion coefficient for the 13 interval PFGSTE sequence without crusher gradients in the prepare and read intervals while the 'x'-symbols are with crusher gradients. The true $\pi/2$ RF-pulse length is 10.5 μs for the diffusion probe used. The spectrometer used was a 200 MHz Bruker DMX-system

the one that causes the biggest challenge in PFG NMR studies, as the onset and offset of the magnetic field gradients causes magnetic flux changes, not only experienced by the sample being investigation but also by conducting materials nearby. As any conducting metal experiencing magnetic flux changes will produce eddy currents to counteract the flux changes, this will set up a transient magnetic field that varies in space and is damped in time; eddy current field transients [14].

What effect will this have on the effective gradients as experienced by the sample subjected to the PFG NMR experiment? When applying squared pulsed field gradients, there are two incidents where the eddy currents appear, at the onset and offset of the gradient pulses. The effect of the eddy current transients can be visualized by replacing the sample with a Hall probe [15]. In short, this probe records the time dependent magnitude of the magnetic field gradient, and can be monitored on an oscilloscope, as shown in Fig. 3.9. Whilst the requested profile is square with short rise and fall times (lower part in Fig. 3.9), the rise and fall times of the magnetic field induced by the current through the gradient coils and measured by the Hall probe is seen to be substantially longer (upper part in Fig. 3.9). It differs significantly from the ideal requested waveform and therefore the waveforms used in all our theoretical calculations to date. This is the effective magnetic field that labels the nuclear spins with a position dependent phase as follows

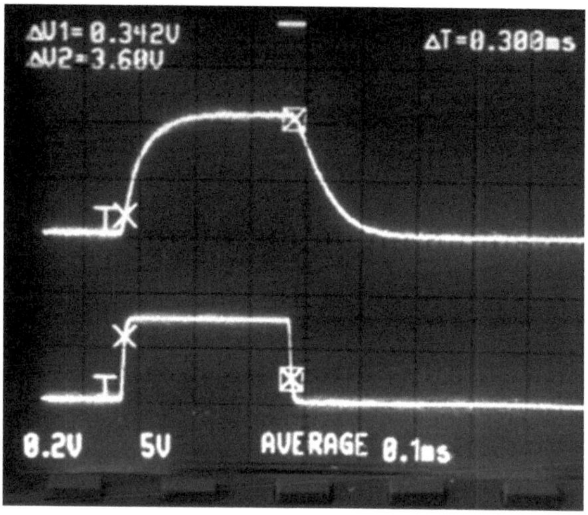

Fig. 3.9 The recording of response in a Hall probe [15] located where the sample is placed (*upper figure*) from a current pulse going through a gradient coil (*lower figure*)

$$\phi = \gamma \int_0^{\tau} g(t')\, z(t') dt' \tag{3.1}$$

However, the effective field is not necessary scaled linearly to the input current through the gradient coils, for the following reason: There is limited control on the transient field, as it may change with current input and consequently its relative contribution will change as a function of magnetic field gradient strength. Thus, there is necessarily no linear relationship between the logarithm of the echo attenuation and the square of the applied gradients. In addition, if the transient field has significant values after the first tau interval, the effective gradients in the PFG NMR will not be the same in the prepare and read intervals. In Fig. 3.10 the pulsed field gradient spin echo sequence is shown where the accumulated phase ϕ_1 in the prepare interval does not equal ϕ_2 in the read interval. Then the appearance of the expected spin echo at time $t = 2\tau$ will be shifted to longer times if $\phi_2 > \phi_1$. This is shown in the left part of Fig. 3.11, where an undistorted echo peak (solid line) is displayed together with a distorted apparent echo peak (dotted line). Both the shape and intensity of the distorted peak is altered, and this will be apparent

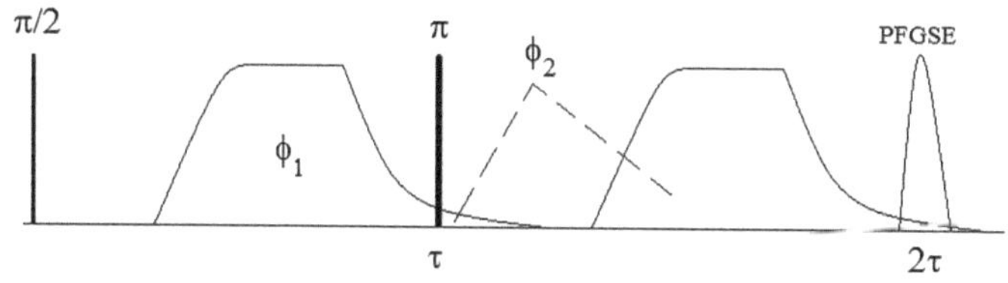

Fig. 3.10 A schematic drawing of the gradient as experienced by the sample during an ordinary PFGSE experiment

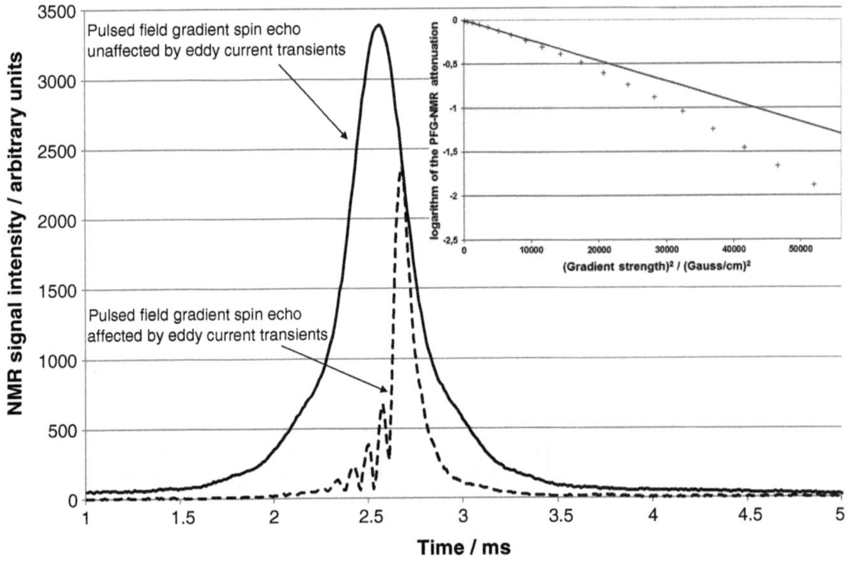

Fig. 3.11 The acquisition of two spin echoes (*left part*), one unaffected by eddy current transients (*solid*) line and one affected by eddy current transients (*dotted line*). The logarithm of the PFG-NMR attenuation in the *upper right* shows the deviation from a linear decay

in the logarithm of the PFG NMR attenuation as a function of the square of the applied gradient strength as shown in the upper right part of Fig. 3.11. Here the PFGSE experiment is applied on a bulk sample, where a linear decay is expected to appear. Instead there is a non linear decay with an increasing slope as a function of the square of the applied gradient strength. An increased absolute value of the slope is expected as the eddy current field imposes a stronger eddy current transient field as a function for current input to the gradient coils. This produces a larger mismatch between $\phi_2 > \phi_1$ and thus an increase in the decay.

The obvious approach for reducing the impact from the eddy current transients is to compensate for the higher phase decoding in the read interval of the PFG NMR sequence ($\phi_2 > \phi_1$) by increasing the applied gradient strength in the prepare interval (i.e. increase ϕ_1). This compensation of effective gradient mismatch, aiming at ϕ_1 to equal ϕ_2, will be strongly dependent on the gradient strength applied and its duration. To be able to conduct a diffusion measurement with incrementing gradients, there will be a need for a specific compensation for each applied gradient strength. Moreover, even if the echo intensity is restored at its maximum, the eddy current transient will still distort the shape of the FID following the echo. In other words this approach is at best only really applicable for low resolution diffusion experiments. Fortunately, there are several other approaches that can be applied in reducing the impact eddy current transients on the PFG NMR diffusion experiment, and these will be discussed in the following sections.

3.2.1 The Actively Shielded Gradient Coil Approach

The primary gradient coil shown in Fig. 3.12 is called an anti-Helmholz coil and produces a linear gradient provided the radius $\boldsymbol{a}$ of the coil and and distance $\boldsymbol{l}$ between the two coils with opposite current has the following relation [16] (also shown in Appendix)

$$l = \sqrt{3}a \tag{3.2}$$

Then the functionality of the magnetic field along the z-direction is given as (Appendix)

$$H_z = +\frac{4I}{7a}\sum_{n=1}^{\infty}\left(\frac{r}{a}\right)^{2n-1}\frac{4}{7}^{\frac{2n-1}{2}}P^1_{2n\sqrt{\frac{3}{7}}}\left[P_{2n\cos\theta}\cos\theta - \frac{1}{2n}P^1_{2n\cos\theta}\sin\theta\right] \tag{3.3}$$

The first order correction to H_z along the z-axis ($\theta = 0$) will go as $(z/a)^5$, and thus a gradient coil that generates a magnetic field gradients in a region near the centre of the gradient coil is produced, as indicated in Fig. 3.12. When switching the currents on and off with the primary coil present only, there are considerable magnetic flux changes outside the coil. In this region in an NMR spectrometer, there is conducting metal in which the unwanted eddy current transients or transient mirror currents are generated [16, 17].

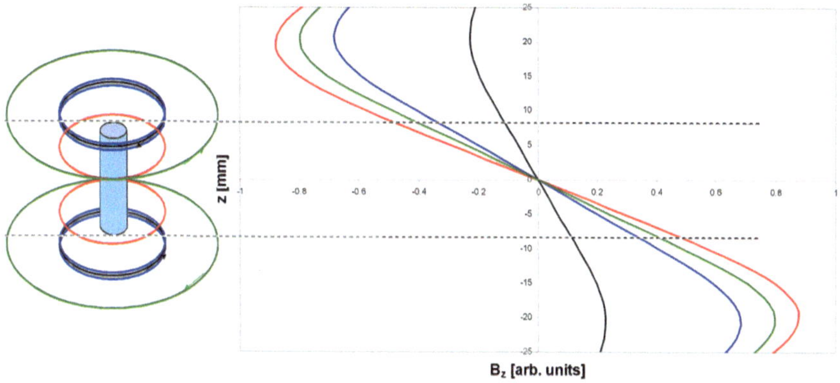

Fig. 3.12 A schematic drawing of the principle behind actively shielded gradient coil design from the target field approach and minimizing the inductance [18, 19]. The *blue coil in the left part* of the figure represents an ordinary anti-Helmholz coil, which produces the z-gradient shown as a *blue line in the right part* of the figure. The *red coil* is positioned to cancel out the $(z/a)^5$-correction term (where a is the primary coil radius), and together with the *blue coil* the configuration produces as a stronger z-gradient (*red line in the right part* of the figure). The *green coil* represents the shielding coil with a current direction opposite to the primary coils. Compared to the primary coils (*red line*) the total configuration produces a weaker z-gradient (*green line in the right part* of the figure). The *dotted line* indicates the 5 % limit for gradient field homogeneity along the z-axis (diagram courtesy of Bruker Biospin)

A shielding of the primary gradient coil can reduce the production of uncontrollable eddy current transients in the surrounding structures significantly [16, 17, 20]. This is done by adding the shielded coils to the configuration as shown in Fig. 3.12, also denoted the actively shielded gradient coil approach. The idea is then to have two gradient coil sets. One inner set of the primary coils creating a magnetic field gradient in the region where the sample is located and one outer set with the main target to minimize the magnetic field generated by the two sets in the region where the eddy current transients are generated. This configuration increases the volume in which gradient field around the centre of the coils is homogeneous within ±5 %. It also reduces the impedance of the gradient coil set, which improves the rise time when switching on the current passing through the coils. The application of shielding of the gradient coils has shown to reduce the eddy current transients' impact on the diffusion experiment significantly, typically reducing the eddy current dead time by one order of magnitude [21]. In addition the effect from base line distortions in high resolution spectra is reduced considerably [11]. The major drawback is that it requires more space for the gradient coils, the effective applied gradient strength is reduced compared to an unshielded gradient set with the same number of windings and applying the same current through the gradient coils, and the construction is more complex.

In Fig. 3.13 two sets of gradient coils are compared for demonstration of the improved performance using actively shielded gradients coils; a single anti-Helmholz gradient coil (to the right) and an actively shielded gradient coil set (to the left). The theoretical gradient strength from the two sets of coils are within the same order of magnitude (~5 G/cm A). While the performance of the actively shielded coil set was satisfactorily within the operating region (0–40 A), The anti-helmholz set returned a measured magnetic field gradient pulse that was affected both by strong eddy current transients and by a relatively long rise time of the current pulse due to the much higher impedance of the gradient coil. Although the eddy current transients are greatly reduced, it is quite evident from Fig. 3.13 that

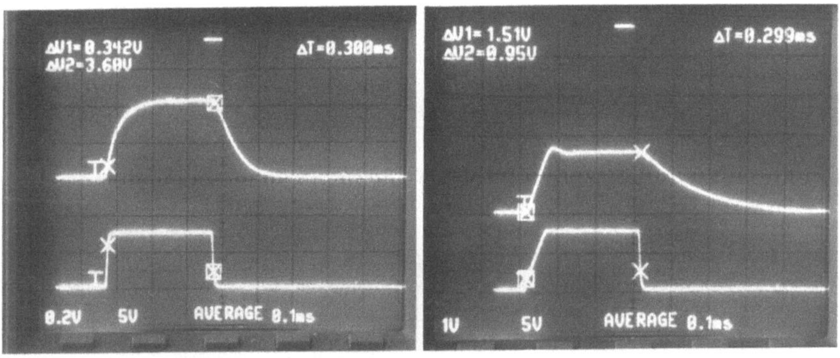

Fig. 3.13 The respond in a Hall probe to a current pulse going through an actively shielded gradient coil (*left picture*) and an unshielded ani-Helmholz coil (*right picture*)

they are still present in a significant amount, also from the shielded gradient coil. However, this can be compensated for by introducing preemphasis adjustment and/or preparatory gradients.

3.2.2 Preemphasis Adjustment

Another approach to reduce the impact from eddy current transients further is to apply so called preemphasis adjustment [22]. In Fig. 3.13, the eddy current generates a magnetic field that decays in time, and the following multi exponential behaviour represent the decay of the current transient

$$I_{transient} = \sum_i I^i e^{-\frac{time}{T^i_{transient}}} \tag{3.4}$$

where $\boldsymbol{I^i}$ is the transient component $\boldsymbol{i}$ with a damping factor (time constant) $\boldsymbol{T^i_{transient}}$. The idea is to mimic the effect of the eddy current transients on the effective magnetic field, and cancel the effect from the transients by applying the opposite polarity on a controlled and decaying current through the gradient coils.

An efficient way to document the eddy current transients' impact on the PFG NMR signal is to run a set of gradients as shown in Fig. 3.14, then after a variable delay apply a 90° RF pulse followed by and acquisition of the FID. In Fig. 3.15 the FID's recorded after three different gradient pulse configurations (maximum gradient strength 100 G/cm) are seen with a variable delay ranging from 0.1 to 500 ms. The largest impact from eddy current transients is found when applying a square gradient pulse. Then the magnetic flux change is at its largest when switching on/

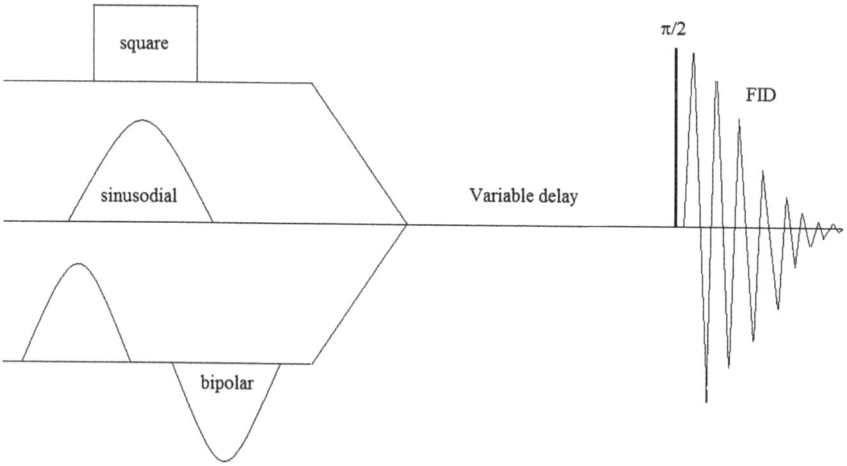

Fig. 3.14 A pulse sequence for evaluating the impact of eddy current transients on the FID from three different shapes of the gradient pulses; *square*, *sinusoidal*, and *bipolar* pair of sinusoidal gradients

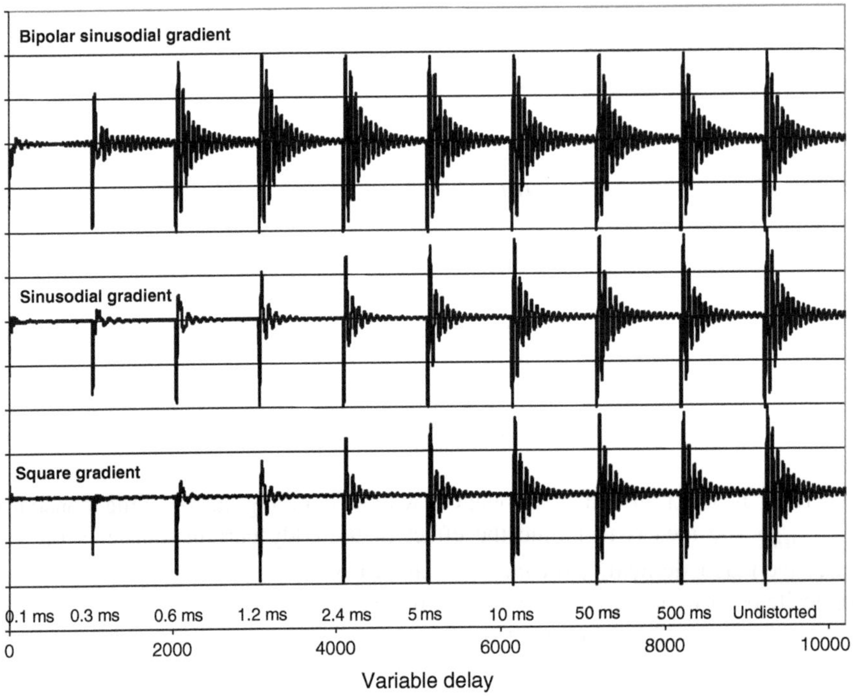

Fig. 3.15 Acquisition of FID's following the three types of gradient pulses shown in Fig. 3.14

off the current through the gradient coils, and not unexpectedly this generates the strongest eddy current transients. The time constant of the transients ranges from the order of μs to the order of seconds. The transients with a time constant of seconds do not affect the initial intensity of the FID significantly, but they increase the decay rate of the FID's, i.e. T_2^*, as compared to the undistorted FID. This may lead to phase distortions and the broadening of peaks in high resolution spectra [11].

For the sinusoidal shaped gradient pulse there is a small increase of the intensity of the FID's at the shortest delays, as compared to the square gradient pulse. This is most likely due to the controlled ramping of the current going through the gradient coils, which leads to a less and more continuous magnetic flux change when switching on/off the current. Still the transients with long time constant exist.

When applying a bipolar pair of gradient pulses, the eddy current transients with longer time constants cancel out. The FID's recorded from 1.2 ms and upwards seem to be unaffected by eddy current transients, but are reduced in intensity and T_2^* as the variable delay of 0.1 ms is approached. The transients with shorter time constants, the order of 1 ms or less, do not cancel out as their time constant is of the order of the gradient pulse length (0.5 ms). Thus, when adjusting the preemphasis parameters, changing the intensities and time constants for the controlled decaying current, one should focus on the region where the bipolar gradients do not cancel out the transients.

Without access to automatic adjustment of the preemphasis parameters [23], the procedure is to define a set of time constants going from approximately 10 ms and downward to the eddy current dead time, the shortest time where the eddy current transients do not influence the NMR signal in phase or amplitude. The adjustment procedure could be as follows:

- Start with one component and adjust its amplitude and time constant until the FID's with the longest waiting times have been optimized.
- Continue with a second component and adjust its amplitude and time constant and readjust the first component slightly until the FID's with the medium waiting times have ben optimized.
- If necessary introduce a third and fourth component to improve the FID's further.

Care should be taken not to overcompensate for the eddy current transients. That is to fit the FID's at the shortest delays well using several time constants without looking at the FID's at intermediate variable delays (≥10 ms). Then there is the danger of having FID's of less intensity and shorter T_2^* for this region. It is important to use the delays that cover the region of interest and to do the adjustment with as few time constants as possible.

In Fig. 3.16, a set of FID's are shown using preemphasis adjustment of the diffusion probe that generated the eddy current transients shown in Fig. 3.9. By varying three controllable decaying current components and their time constants it was possible to achieve an eddy current dead time of 200 μs.

3.2.3 Preparatory Gradients and Eddy Current Settling Time

Before the application of preemphasis adjustment was introduced to the PFG NMR experiments as well as the development of actively shielded gradient coils,

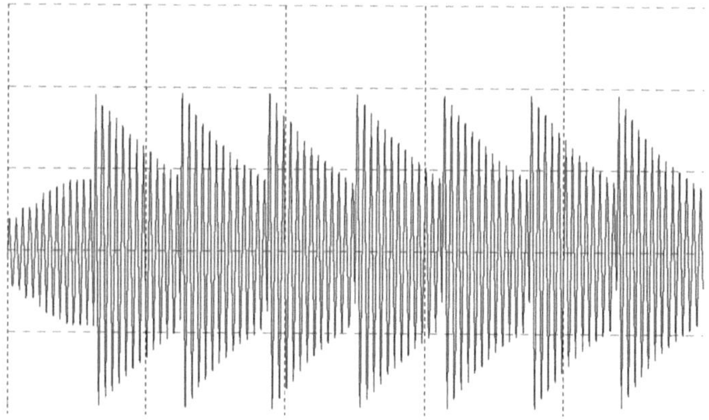

Fig. 3.16 Acquisition of FID's following a 100 G/cm gradient pulse. The variable delays after the gradient pulse is from *left to right*; 100, 150, 200, 300, 500, 700, 1,000 μs and 10 ms

other approaches had to be employed in order to produce an attenuating signal undistorted by eddy current transients. As already mentioned, the extra strength on the applied gradient in the prepare interval can correct for the mismatch in gradient area between motional encoding and decoding. However, this time consuming approach requires a unique correction factor for each gradient strength. An alternative is to use preparatory gradients in order to optimize the performance of the PFG NMR experiment. The preparatory gradient pulses are a set of pulses applied before the initiation of the first $\pi/2$ excitation pulse or before any set of applied gradients used for motional encoding/decoding. Figure 3.17 shows a PFG NMR sequence with preparatory gradients both prior to the first $\pi/2$ RF-pulse and during the z-storage interval. The preparatory gradient pulse length and dead times are the same as in the actual PFG NMR experiment, and a steady state is reached after a certain number of preparatory gradient pulses.

Having applied a sufficient number of preparatory gradient pulses (C1) the eddy current transients become saturated, such that the accumulated phases are equal in the read and prepare interval of the sequence ($\phi_2 = \phi_1$ in Fig. 3.10). However, when saturating the eddy current in this manner, the FID following the spin echo will be strongly affected, and in the Fourier transformed (FT) spectrum there will be phase errors and baseline distortions. This is solved by adding a z-storage delay to the PFG-sequence, starting at the time of the echo, and called the eddy current settling time [24]. However, it is difficult to wait for the settling of the eddy current when transients with time constants up to the order of a second exist (see Fig. 3.15). The solution then is to apply bipolar pairs of gradients in combination with an eddy current settling time of 10–50 ms depending on the duration of the gradient pulses. Such a sequence is shown in Fig. 3.18, the bipolar 13 interval sequence with preparatory bipolar gradient pulses repeated C1 times and an eddy current delay, Δ_{eddy}.

After introducing the bipolar version of the PFG-experiment, the use of preparatory gradients has become more beneficial in high resolution experiments [11]. As the gradient pulses come in pairs and with opposite polarity, the eddy current more or less cancel out. By applying preparatory gradients, one no longer saturates the eddy current transients, but achieves a steady state where the transients are much weaker. The results from the impact of this reduction in the eddy-current field, is that both phase and baseline errors in the FT spectra should be reduced.

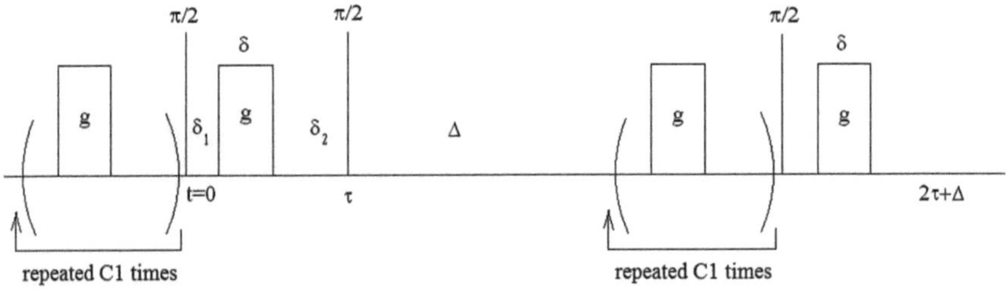

Fig. 3.17 The PFGSTE sequence with preparatory gradients prior to the first and last $\pi/2$ RF-pulse

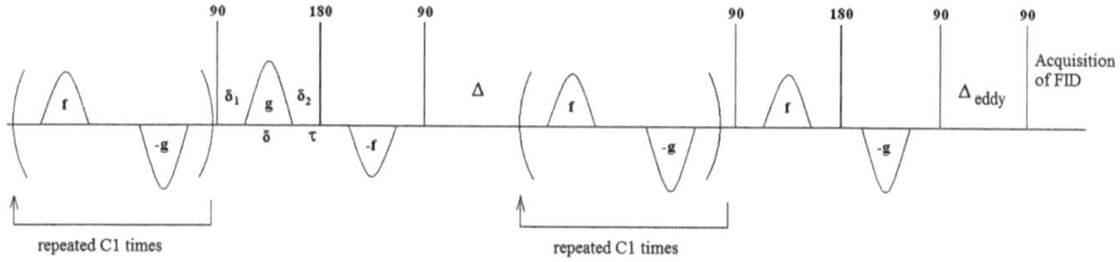

Fig. 3.18 The bipolar 13-interval PFGSTE sequence with preparatory bipolar gradient pulses repeated C1 times and an eddy current delay, Δ_{eddy}

Figure 3.19 shows two high-resolution FT spectra that were recorded with and without the use of preparatory gradients. With the preparatory gradients in use, it is evident that baseline and lineshape distortions are reduced due to a significant reduction of the eddy current transients. Depending on the experimental conditions, one may still have to apply the z-storage for eddy current settling time, but this delay is significantly reduced compared to the monopolar version of the PFG experiment. In addition, if a lock signal exists, the lock is not lost during the PFG experiment due to frequency shifts or T_2^* relaxation [11].

In work involving low resolution NMR equipment, where the spectral resolution is of limited interest, the use of preparatory gradients is beneficial for most purposes. Using the preparatory gradients in a constructive way, eddy current dead times down to the order of a few hundreds of μs can be achieved without actively shielded gradient coils and without the requirement for preemphasis adjustment. Then the implementation of the PFG sequences becomes less complicated and the maximum gradient strength increases as compared to having a set of actively

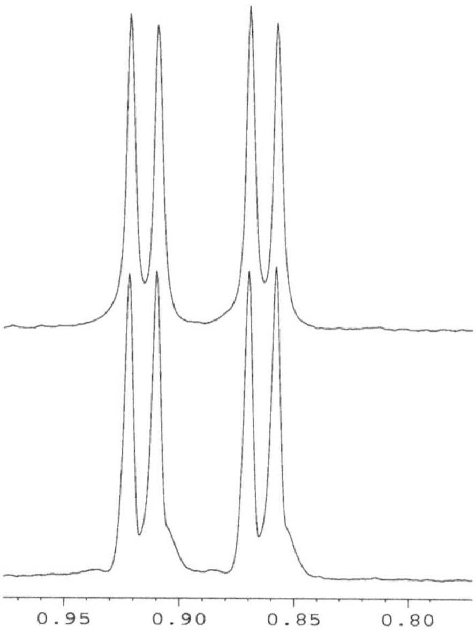

Fig. 3.19 Four peaks arising from two CH3 groups in the amino acid valine. The lower FFT spectrum was recorded with the ordinary double PGFSTE sequence and the upper spectrum using the double PFGSTE with preparatory gradient pulses. The experiment was recorded on a Bruker Avance DRX600 spectrometer and the spectra are displayed in units of ppm

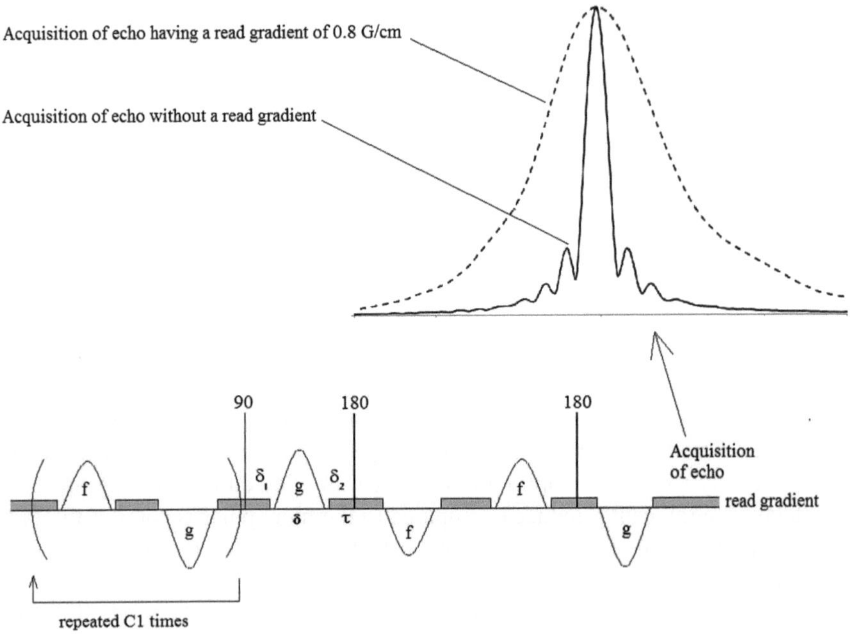

Fig. 3.20 The 11 interval bipolar PFGSE sequence with preparatory and read gradients. The two spin echoes at the *upper right* are acquired with (*dotted line*) and without (*solid line*) a read gradient

shielded gradient coils. The major drawback is that the width of the spin echo will narrow as the applied gradient strength is increased. When acquiring the echo signal, using a set of discrete measuring points, the probability of missing the maximum peak intensity will increase as the strength of the gradients is increased. However, by adding a read gradient throughout the experiment of opposite polarity to the last gradient pulse applied, as shown in Fig. 3.20 for the 11 interval bipolar PFGSE sequence with preparatory gradients, the width of the echo signal is increased and the recording of the echo peak point becomes more reliable [25].

The read gradient is equivalent to a primitive preemphasis adjustment with a constant current on all the time.

3.2.4 The Diffusion Dependent Impact of Residual Eddy Current Transients on PFG-NMR

The handling of eddy current transients is one of the most challenging tasks when setting up a PFG NMR diffusion experiment. The most subtle effect encountered is that they affect the diffusing molecules differently depending on the diffusion coefficient. If settings are optimized on a water sample, there is no guarantee that these settings will apply on a sample containing mineral oil, why? The molecular mobility is measured by a phase change due to movement from position (a) to (b) during the z-storage interval.

$$I(t) = I_0 \int\int G(\vec{r}, \vec{r}_0, t) e^{i[\gamma\delta\vec{g}(\vec{r}(t)-\vec{r}_0)+\varphi_{eddy}]} d\vec{r} d\vec{r}_0 \tag{3.5}$$

As was seen for the preemphasis adjustment, there could be a terms that are not dependent on the molecular movement during a z-storage interval, but on the experimental settings. If it is assumed that this effect has been minimised for a water sample it may be stated that

$$\gamma\delta\vec{g}(\vec{r}(t) - \vec{r}_0) + \varphi_{eddy} \approx \gamma\delta\vec{g}(\vec{r}(t) - \vec{r}_0) \tag{3.6}$$

The molecular displacement during the sequence controls whether the term representing the dephasing from eddy current transients can be neglected or not. When turning to an oil sample where the displacement is much less, (3.6) may not apply. Consequently, the net dephasing is reduced, but at the same time there exist components arising from the eddy current transients that remain unaltered by the change in diffusivity of the system investigated. Then the assumption made in (3.6) is no longer valid. In Fig. 3.21 the 11 interval bipolar gradient pulse is applied on two bulk samples, one with pure water ('•') and another with pure oil ('x'), and the experimental settings were identical for the two experiments. While the logarithm of the decay of the water signal fits nicely to a linear slope, there is an early departure from a linear decay of the logarithm of the oil signal. Furthermore, the difference in diffusion coefficient between the two systems is ~100, but the difference in initial slope of the oil attenuation and the water indicates a difference of a factor 10 only. Thus, there should hardly be any

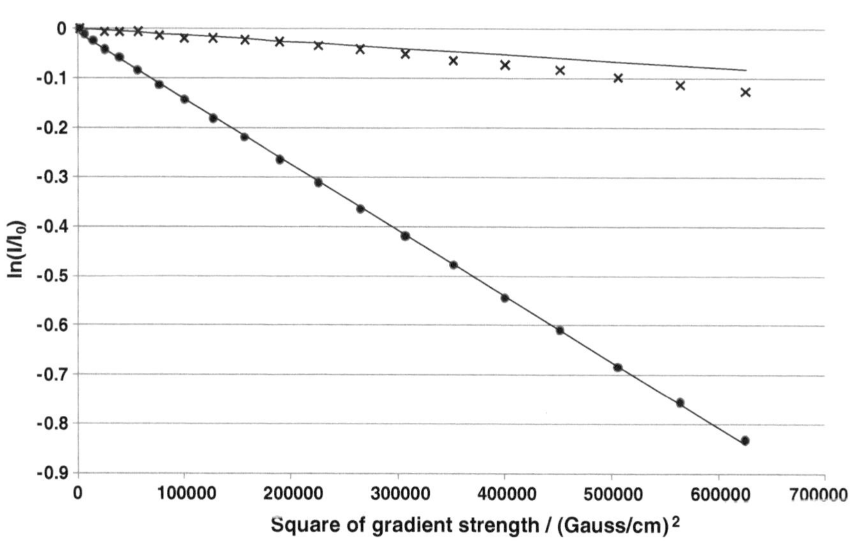

Fig. 3.21 The 11 interval bipolar PFGSE applied on two bulk samples; pure water ('•') and pure olive oil ('x')

decay of the oil signal, but due to ϕ_{eddy} an apparent and 10 times higher diffusion coefficient is observed for the oil signal.

The overall conclusion is that the impact of eddy current transients is something to carefully consider and take steps to minimise if the goal is to measure true molecular movement using the PFG NMR experiment.

3.2.5 Analysing the Frequency Components Affecting the PFG-NMR Experiment

When pulsing current through the gradient coils in a PFG NMR experiment the coil is also sensitive to any ripple current produced by the gradient amplifier. This ripple current is dependent on the mains stability where the spectrometer is located, the performance of the gradient amplifier and any other electronic instrumentation nearby and connected to the mains supply. Depending on the frequency at which the ripple current is oscillating, it may affect the PFG NMR as follows; if a ripple current produces effective gradients of opposite polarity in the read and prepare intervals, the mismatch between ϕ_1 and ϕ_2 will be at a maximum. If for example the τ-value in the ordinary PFGSE sequence is set to 10 ms, a ripple current having a frequency of 50 Hz (also the frequency of the mains) may have maximum negative impact on the phase stability of the actual PFGSE experiment.

The frequency components present in the PFG NMR experiment may be analysed by running a steady-state free precession sequence, and perform a Fourier transform of the acquired data [26]. In Fig. 3.22 the measured frequency components from a low

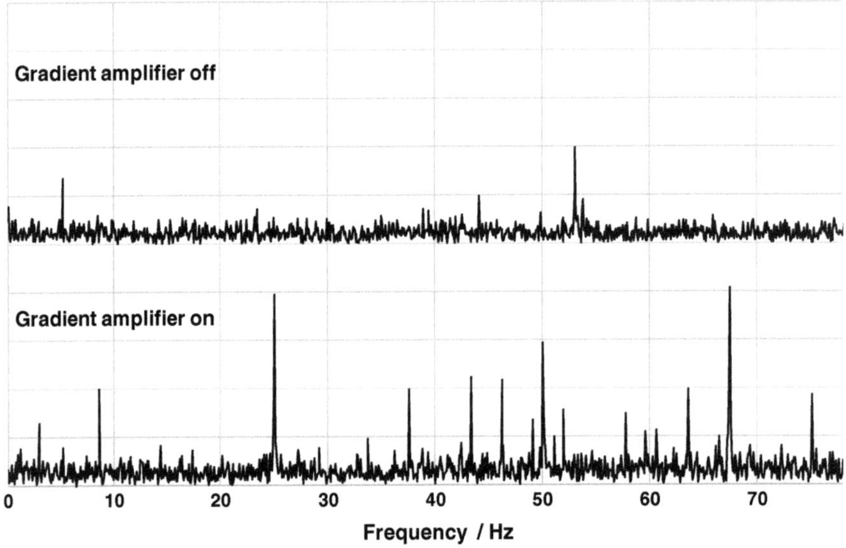

Fig. 3.22 Fourier transformed data acquired with the steady-state free precession sequence (*SSFP*) [26]

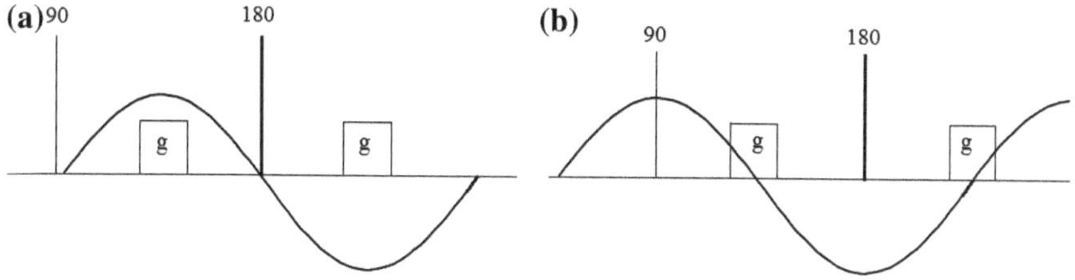

Fig. 3.23 The coupling of the PFGSE to a ripple current (*the solid sine curve*) either in a destructive way (**a**) where the echo is affected ($\phi_1 > \phi_2$) or in a constructive way (**b**) where the echo is unaffected ($\phi_1 = \phi_2$) by the ripple current. The mains amplitude is over emphasised relative to the gradients in the PFGSE

field spectrometer is shown when the current amplifier is switched on (lower figure) and off (upper figure). When having the current amplifier on it produces frequency components both in the low frequency (<50 Hz) and higher frequency region (>50 Hz), but when it is switched off there is hardly any noise peaks in the measured region.

In Fig. 3.23 a ripple current (the solid sine curve) is shown together with the ordinary PFGSE sequence. In (a) the period of the ripple equals 2τ, and the effective gradient induced by the ripple current through the gradient coils will dephase the nuclear spins in both the prepare and read intervals. Thus, it acts in a destructive way where time of the appearance of an apparent echo peak will be shifted away from the expected position. In (b) the effective gradient induced by the ripple current will cancel out both in the prepare and read interval, and the echo is thus unaffected by the ripple current.

Due to the potential errors ripple currents might produce, it is important to insulate the amplifier of NMR-spectrometer as much as possible from external electrical noise. However, ripple current generated internally within the amplifier itself is difficult to avoid. If necessary, the use of batteries instead of power amplifiers to generate the current through the gradient coils can be used to reduce the impact due to ripple currents.

By applying bipolar gradients in the prepare and read intervals of the 11 interval PFGSE sequence, ripple currents with a frequency significantly smaller than $1/(2\tau)$ will be cancelled out. When the above condition is satisfied the polarity of the ripple remains the same over the period of 2τ, and the gradient generated by the ripple may be treated as a constant background gradient. When optimizing low field instrumentation for fat content determination [27], there have been laboratory environments encountered where it was crucial to apply bipolar gradients instead of monopolar ones, in order to suppress the effect from ripple currents generated by the gradient amplifier as a result of contaminated mains power supplies.

3.3 Internal Magnetic Field Gradients

When subjecting a heterogeneous sample, such as a water saturated rock core plug, to an external magnetic field, the flux lines going through the sample will have a discontinuity in direction at the interfaces, according to the continuity of the Maxwell equations [28]. This will generate internal magnetic field gradients that varies depending on the material of the sample and its geometry or heterogeneity. The magnitude of the internal gradient in a system of a heterogeneous material as macroscopic grains surrounded and/or infiltrated by the probing substance is approximately given by the following [29, 30]

$$|g_{internal}| \approx \frac{\Delta H}{d_{grain}} \approx \frac{3\chi_0 H_0}{d_{grain}} \tag{3.7}$$

where H_0 is the external field, χ_0 is the magnetic susceptibility of the solid matrix, and d_{grain} is the typical dimension of the grains. Thus, the strength of the internal gradient will vary with the physical properties of the different components within the sample and with the degree of heterogeneity within the sample. In the following section, it will be assumed that the value of the internal magnetic field gradient as experience by the individual nuclear spins is randomly distributed around 0. The question then is how the existence of such an internal spatially varying magnetic gradient affects the diffusion measurements.

For the spins in position z experiencing an internal gradient of the same polarity as the applied gradient of duration δ, the phase encoding ϕ during the τ interval is written

$$\phi = \gamma (g_{applied}\delta + g_{\mathrm{internal}}\tau) z \tag{3.8}$$

Thus the nuclear spins that experience an internal gradient of the same polarity as the applied gradient will be more dephased than the nuclear spins experiencing an internal gradient of opposite polarity to the applied gradient. Consequently, at the time of acquisition of the echo signal in the PFG NMR experiment, the nuclear spins experiencing an internal gradient of opposite polarity to the applied one will contribute the most to the echo signal. Then the total gradient experienced by the ensemble of nuclear spins in the encoding/decoding will be less than the applied gradient strength. The slope of the echo attenuation as a function of the actual applied gradient strength will then be less than it would have been without the internal gradients present, and the extracted diffusion coefficient will be lower than the actual one [31].

As for the challenges arising due to eddy current transients, several approaches have been developed for avoiding the impact of internal magnetic field gradients on the measured diffusivity [7–9, 32, 33]. As the magnitude of the internal magnetic field is essentially proportional to the external magnetic field, and an obvious solution to the problem is therefore to work with spectrometers employing low magnetic fields, as 0.5 T or less. However, as found by Hürlimann et al. [34], there exist numerous heterogeneous systems that can produce strong internal magnetic field gradients even at 0.05 T, and experiments conducted at low fields suffer from poor signal to noise ratio.

Another approach is to reduce the influence from internal magnetic field gradients either by rotating the applied magnetic field gradient or physical rotation of the sample being investigated. A rotation of the applied magnetic field gradient implies the use of RF gradients [35], which sets restriction on the available strength of the applied gradient. A physical rotation of the sample at for example the magic angle [36] restricts the sample size that can be measured on while physical rotation in general may change the properties of a heterogeneous sample as compared to a stationary one.

When having access to strong applied magnetic field gradients, the effect from the coupling between applied and internal gradients can be significantly suppressed [33]. The challenge then is to match the encoding and decoding gradients in the presence of strong eddy current transients. Unfortunately, especially in rock cores, even gradient strengths of the order of 1,000 G/cm are not enough to allow for neglecting the effects of internal gradients, as such systems can contain ferromagnetic impurities which produce strong, localized gradients of over 1,000 G/cm, even when placed in a relatively low static field environment [34].

The most common solution to the problem is to apply a set of bipolar gradients separated by 180° RF-pulses in the PFG NMR [7–9], also shown in Chap. 1. Provided the internal magnetic field gradient experienced by the diffusing molecules does not vary during the PFG NMR sequence, the impact of an internal magnetic field gradient may be suppressed or reduced to an insignificant level. However, there are two important pitfalls to be aware of when applying the bipolar version of the PFG NMR experiment: When introducing more RF-pulses into the PFG NMR experiments, the number of unwanted coherence transfer pathways increases. Thus, there is a need for more sophisticated phase sequences [11], or the use of crusher gradients in the prepare and read intervals [3]. In Fig. 3.7 the effect from unwanted coherence transfer pathways is shown, and how the application of crusher gradients removes the contribution from the unwanted signal.

The other important pitfall to notice is the effect from diffusion in spatially varying internal magnetic field gradients. If the diffusing molecule experiences an internal gradient in the prepare interval that is different both in amplitude and polarity from the internal gradient experienced in the read interval, the attenuation from the bipolar PFG NMR experiments will again be affected by the internal magnetic field gradients. The consequence is that when a diffusing molecule experiences the heterogeneity of the sample, the measured apparent diffusion coefficient will be a function of the spatially varying internal magnetic field gradients [37].

3.3.1 Correcting for Spatially Varying Internal Magnetic Field Gradients

When the observation time is such that the diffusing molecules are experiencing internal magnetic field gradients in the prepare and read intervals that are practically uncorrelated, the measured apparent diffusion coefficient will be lower than the actual diffusion coefficient. This is the actual situation when the observation time is so long that the tortuosity limit is reached [38]. The root mean squared displacement then exceeds the typical dimension of the heterogeneity of the sample, and the diffusion coefficient approaches asymptotically a limiting value D_∞ which defines the tortuosity T:

$$\frac{1}{T} = \frac{D_\infty}{D_0} \tag{3.9}$$

However, due to the spatially varying internal magnetic field gradients, the measured tortuosity limit is wrong. Assuming that the absolute amplitude value of the internal gradient is constant, but the polarity changes between the prepare and read interval, the surviving cross terms in the attenuation from a bipolar13 interval sequence as shown in Chap. 1, assuming f = g, is written [37]

$$\ln\left(\frac{I}{I_0}\right)_{cross} = -4\gamma^2 D\delta\tau^2 g G_0 \tag{3.10}$$

Thus, there are cross terms between the applied and internal gradients present despite the application of bipolar gradient pulses, and in this cross term a τ-dependency is found. In order to correct for the spatially varying internal magnetic field gradients Seland et al. [37] proposed to run diffusion measurement at the longest observation time (tortuosity limit) but with varying the τ-value, and fit the measured apparent diffusion coefficients to a series expansion in τ as follows

$$\frac{D(t)}{D_0} = \frac{D(t)}{D_0}_{(\tau=0)} + c_1\tau + c_2\tau^2 + \Phi(\tau^3) \tag{3.11}$$

The extrapolation of the fitted data to $\tau = 0$ should provide the true diffusion coefficient at the tortuosity limit.

Figure 3.24 shows how the correct diffusion coefficient may be found using the series expansion up to τ^2. In this case, the influence of spatially varying gradients has been corrected for, and the fitted diffusion coefficient will reflect the true tortuosity of the system. The use of bipolar magnetic field gradients is an effective approach for suppression of the effect from internal magnetic field gradients on the measured apparent diffusion coefficient. However, at longer observation times one must correct for the existence of spatially varying internal magnetic field gradients. The approach can also be used at intermediate observation times, and it applies as long as the assumption of constant internal gradients in the prepare and read intervals apply. When this is not the case the solution is to increase the duration of the gradient pulse length, δ, such that

$$\begin{aligned}\phi &= \gamma \int_0^\tau (g_{applied} + g_{internal}(t))\, r(t)\, dt \\ &= \gamma \int_0^\delta g_{applied}\, r\, dt + \gamma\tau\langle r\, g_{internal}\rangle \approx \gamma\delta r g_{applied}\end{aligned} \tag{3.12}$$

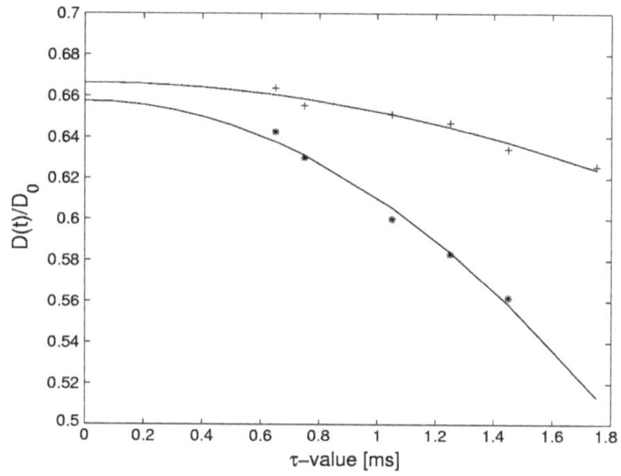

Fig. 3.24 Measurements of $D(t)/D_0$ of water immersed in compact mono-sized glass spheres using the 13-interval bipolar PFGSTE sequence with different τ values but keeping the observation time constant ($\Delta = 800$ ms). The τ dependence of $D(t)/D_0$ was recorded using two spectrometers at different field strengths, a 4.7 T Bruker DMX200 (+) and a 14.1 T Bruker DRX600 (*). At an observation time of 800 ms, the measured D(t) will correspond to $D\infty(\tau)$ as shown in (3.9)

where the following relation is assumed

$$\langle r\, g_{internal} \rangle = \int_0^{\delta} g_{internal}(t)\, r(t) dt \approx 0 \tag{3.13}$$

If (3.13) applies, the internal gradient fluctuates over distances much shorter than the root of the mean squared displacement for the diffusing molecule during the applied gradient pulse, and thus it does not affect the PFG NMR measurement. However, it is important to note that even if a zero contribution from (3.13) applies for a water molecule diffusing within in a heterogeneous sample; it does not necessarily apply for a crude oil. As the mobility of crude oil is lower than the water mobility, there can be a situation, for example in a rock core plug containing both water and crude oil, where the cross terms exist for the attenuation of the crude oil signal while they are absent in the attenuation of the water signal.

3.4 Non-brownian Motion

When measuring the true self-diffusion coefficient of liquids, the molecules are travelling distances of the order of μm during the experiment, through random or Brownian motion [39]. Any non-Brownian motion within the sample, for example convection set up by a temperature gradient, will increase the distance travelled significantly, and the attenuation of the PFG NMR experiment will appear non-linear [40]. The true self-diffusion coefficient is then not extractable from the initial slope of the attenuation. The non-Brownian motion within the system may originate from gradients in physical parameters over the sample, as for example the existence of temperature gradients when measuring at elevated temperatures by heating the sample with gas from below. Mechanical vibration can also induce an unwanted movement within the sample, as in connection with gradient switching that may cause vibrations of the gradient coils. This can result in transmission of vibrations from the gradient coil to the sample [41]. Figure 3.25 shows the recorded pulsed field gradient spin-echoes in a multi-PFGSE experiment from a sample that is subjected to vibration effects, and it is apparent that the odd echoes do not fit into the same decay as the even ones.

3.4.1 Compensating for Non-brownian Motion

The solution for avoiding convection within the sample could be to apply smaller sample dimensions. This can significantly reduce convection set up by temperature gradients, as the conditions for the onset of flow (i.e. changing the critical Rayleigh number Ra_C) would require higher temperature gradients than available. Another approach would be to apply the double PFG NMR sequences for convection compensation [13, 40], also given in Chap. 1. When assuming laminar flow the double PFG NMR sequences cancel out the term containing flow, and leaves the term that describes Brownian motion only in the equation for the attenuation. Figure 3.26 shows single and double PFGSE experiments performed

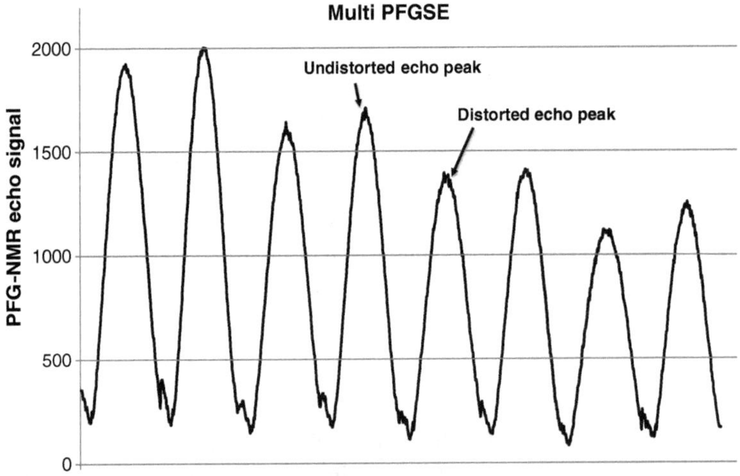

Fig. 3.25 A multi-PGFSE experiment performed under the influence of mechanical vibration. Along the x-axis only the acquisition of the spin-echoes is shown. Each echo was acquired over 0.13 ms and the duration between each spin-echo was 4.4 ms

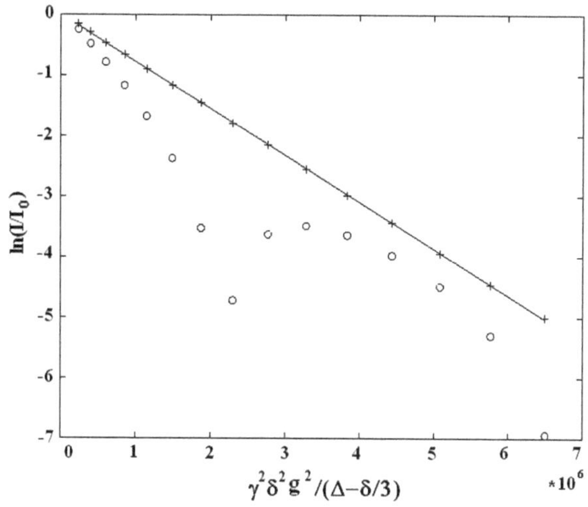

Fig. 3.26 A single pulsed field gradient experiment (E) and a double-pulsed field gradient experiment performed on acetontrile at 15 °C [13]

on acetonitrile at 15 °C. It is evident that the use of the double PFGSE sequence yields a mono-exponential decay that is unaffected by the laminar flow within the sample, whereas the single PFGSE yields an attenuation that is non-exponential and may even return a negative NMR signal [13].

It is shown that by combining the use of the double PFGSE experiment with preparatory gradient pulses, the performance of a high-resolution PFG experiment is improved, with respect to both convection and eddy current field transients. Then, the number of RF-pulses can be minimized and a less sophisticated phase sequence is required compared with the bipolar double PFGS(T)E sequences. When studying diffusion in liquids, where there is no need for bipolar sequences to suppress the effect from internal gradients, a double PFGSE experiment with preparatory gradients is then a simpler but just as effective experiment to employ.

The effect from mechanical vibration, either from the vibration of the gradient coils or any other vibration producing equipment in direct or indirect contact with the spectrometer, may also be cancelled out by the use of double PFG NMR sequences. That is, provided the vibrations generate laminar flow within the sample only. This is not necessarily correct, and efforts should be made to physically separate the gradient coils from the sample tube. For the 0.5 T low resolution spectrometer configuration developed by AMR Ltd. and Anvendt Teknologi (www.antek.no), the gradient coils are mounted on the pole pieces of the permanent magnet. In this configuration, the mechanical vibration of the gradient coils during current pulsing do not couple directly into the sample tube, thereby avoiding the vibrational effect.

3.5 The Second Cumulant Approximation and Finite Gradient Pulse Lengths

In PFG NMR work on any fluid confined within restricting geometries, the question arises to which degree the assumption of a Gaussian diffusion propagator is valid. The moving spins will sense restrictions during the PFG NMR experiment. The diffusion propagator can no longer be assumed Gaussian, as for diffusion in a homogeneous system

$$I = I_0 e^{-\gamma^2\delta^2 g^2 D(\Delta-\delta/3)} \tag{3.14}$$

Higher order moments will then have to be included in (3.14), and this will lead to difficulties extracting vital information such as the apparent diffusion coefficient in heterogeneous media. However, to be able to analyze the PFG NMR data from heterogeneous media an assumption usually referred to as the second cumulant approximation can be used. For the diffusion in restricting geometries, and assuming unrestricted diffusion during the gradient pulses, the attenuation of the PFG NMR experiment is written

$$I(t) = I_0 \int\int G(\vec{\mathrm{r}}, \vec{\mathrm{r}}_0, \mathrm{t}) e^{i\gamma\delta\vec{\mathrm{g}}(\vec{\mathrm{r}}(\mathrm{t})-\vec{r}_0)} d\vec{r} d\vec{r}_0 \tag{3.15}$$

where $G(\vec{\mathrm{r}}, \vec{\mathrm{r}}_0, \mathrm{t})$ is the diffusion propagator. The second cumulant approximation states that the initial decay of as a function of the gradient strength $\boldsymbol{g}$ is most dependent on the lowest order of $\boldsymbol{g}$, for which is $\boldsymbol{g}^2$ for a Gaussian propagator. As the contribution from higher orders of $\boldsymbol{g}$ is much smaller than the contribution from the second order, the initial decay should reflect a Gaussian distribution. In this regime, (3.14) can still be used to fit the experimental data and find a value for the apparent diffusion coefficient in heterogeneous media [42].

When the root of the mean squared displacement during the gradient pulses is comparable to the restricting geometries, one cannot disregard restricted diffusion during the gradient pulses as well. Still one may apply the second cumulant approximation to extract an apparent diffusion coefficient, but the restricted diffusion during the gradient pulses may alter the interpretation of the extracted apparent diffusion coefficients. This is especially important when the diffusion measurements are used to extract parameters as surface to volume in water saturated rock core plugs or in emulsions. When the typical cavity size is of the order of 1 μm, the water molecules are probing the same dimensions during a gradient pulse of 0.5 ms. Then it is important to have knowledge on the effect of restricted diffusion during the gradient pulses, and in the following section the concept of center of mass will be applied to discuss the effect of restricted diffusion during gradient pulses [43].

3.5.1 Center of Mass

Assuming a finite gradient pulse and molecules that diffuse during the pulse, the phase ϕ accumulated during a gradient pulse in the PFG NMR sequence is written [43]

$$\phi = \gamma \int_0^\delta \mathrm{gr(t)}dt \tag{3.16}$$

where ϕ is the accumulated phase, g is the applied magnetic field gradient experienced by the diffusing molecules and r(t) is the instantaneous position of the molecule at time t. This can be interpreted as if the molecule is labelled at rest during the gradient pulse at a mean position given by

$$\langle r \rangle = \frac{1}{\delta} \int_0^\delta \mathrm{r(t)}dt \tag{3.17}$$

The phase is then written

$$\phi = \gamma \delta g \langle r \rangle \tag{3.18}$$

where $\langle r \rangle$ is the average position or the center of mass for the molecule during the gradient pulse. The center of mass distribution function is then defined as the normalized density function for the center of mass including all the molecules

$$\langle z_{all} \rangle = \frac{1}{n\delta} \sum_{i=1}^{n} \int_0^\delta \mathrm{r}_i\mathrm{(t)}dt \tag{3.19}$$

Provided no correction is made for the restricted diffusion during the gradient pulses, the distance diffused by the molecules will be the distance between the two centres of mass for the read and prepare intervals

$$\left\langle z_{all}^{measured} \right\rangle = \left\langle z_{all}^{read} \right\rangle - \left\langle z_{all}^{prepare} \right\rangle \tag{3.20}$$

In a homogeneous system this will not affect the experiment as the centre of mass (CM) distribution will be identical to the real distribution of molecules within the sample. For a heterogeneous system there will be a discrepancy between the two distributions, and the effect of the restrictions will be visualized by random walk simulations in a one dimensional cavity: Assume a random walker in one dimension, with equal probability of jumping a distance li $= \pm 1$. The mean squared displacement is then written [32]

$$R^2 = \left\langle \left(\sum_{i=1}^{n} l_i \right)^2 \right\rangle \tag{3.21}$$

where n equals the number of moves. When the moves are uncorrelated this is simplified to

$$R^2 = n\left\langle l^2 \right\rangle \tag{3.22}$$

where $l = |l_i|$. The correlation time is defined as the time between the successive moves, $\tau_c = t/n$, where t equals the diffusion time. Employing the mean squared displacement in one dimension one finds

$$n\left\langle l^2 \right\rangle = 2Dt = 2Dn\tau_c \tag{3.23}$$

Direct comparison of actual diffusion in restricting geometries during a gradient pulse and a random walk simulation can then be done through the mean squared displacement. In Fig. 3.27, the CM distribution function is shown for a one dimensional system where the root of the mean squared displacement is 0.1. Then only a small fraction of the molecules, those close to the walls, will have experienced the restricting geometry during the random walk. The center of mass for these molecules will be biased towards the center of the one-dimensional box, and the CM distribution function will increase close to the walls. This is the actual situation when the root of the mean squared displacement is significantly smaller than the cavity size in which the molecules are diffusing. Then the second cumulant approximation can be applied and restricted diffusion at short observation times may be conducted [44]. However, due to the bias of the CM distribution

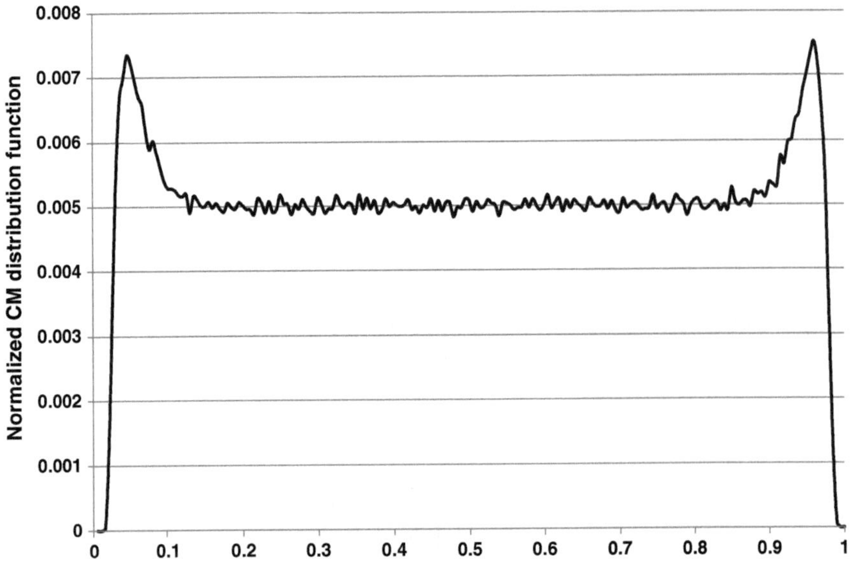

Fig. 3.27 Normalized CM distribution function for random walkers with a root mean squared displacement equal to 0.1 in a box of unit length

towards the center of the cavity the fitted surface to the volume ratio using the short observation time expansion by Mitra et al. [44] will result in higher values than the actual one. Fordham et al. [45] corrected for restricted diffusion during the gradient pulses by applying the short observation time expansion. This introduced the concept of corrected effective diffusion times, which provided more accurate results for the diffusion measurements at short observation times [46]. In Fig. 3.28, the difference between using standard diffusion times and corrected ones (that takes into account restricted diffusion during the gradient pulses [45]) is shown for water diffusing amongst compact mono-sized spheres. From the slopes in Fig. 3.28, it can be concluded that a reduction of the measured diffusion is reached at a later time using the standard diffusion times, which is equivalent to measuring a smaller surface to volume ratio (S/V). This can be interpreted as if the molecules are probing larger compact spheres than actually present. Using the procedure from Sørland [46] the size of the compact mono-sized spheres (98.8 μm) is found from the two different slopes to be 105.5 μm from standard diffusion times and 100.6 μm from corrected diffusion times. Conclusively the corrected diffusion times that takes into account restricted diffusion during the gradient pulses provide the result closest to the actual sphere diameter.

The left part of Fig. 3.29 shows the CM distribution function when the root of the mean squared displacement is 0.5. Such a CM distribution function during the motional encoding/decoding in the PFG NMR diffusion experiment complicates the interpretation significantly. If possible, using a duration of the gradient pulses

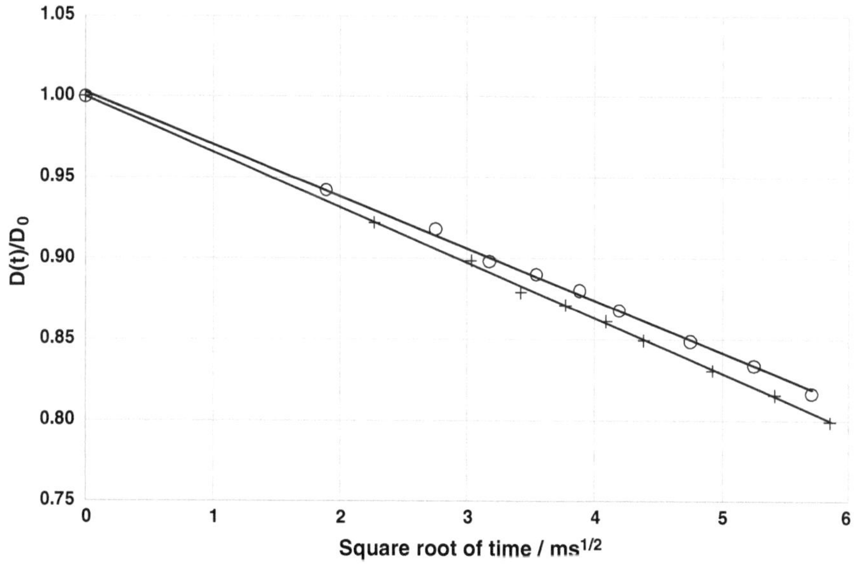

Fig. 3.28 Normalized diffusion coefficient as a function of diffusion times. The 'o' symbols represent the data set using standard diffusion times, while the '+' symbols represent the data set using corrected diffusion times

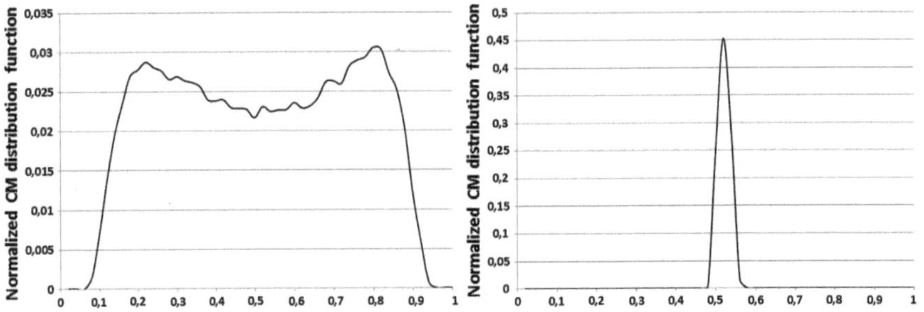

Fig. 3.29 Normalized CM distribution function for random walkers with a root mean squared displacement equal to 0.5 (*figure to the left*) and 7.0 (*figure to the right*) in a box of unit length

that results in such CM distributions should be avoided. An example of such is the diffusion measurements of water confined in closed cavities of size 1 μm at room temperature. Using a gradient pulse of 0.5 ms the root of the mean squared displacement is of the same size as the cavity. Thus both the use of the short observation time expansion by Mitra et al. and the asymptotic limit approximation ($\sqrt{D_0\delta} \gg 1\,\mu\text{m}$) do not apply.

As it is difficult to reduce the duration of the gradient pulse below 0.1 ms without having access to infinitely strong gradient amplifiers, the solution to the problem shown in the left part of Fig. 3.29 is to increase the duration of the gradient pulse length such that the asymptotic limit approximation is fulfilled. The right part of Fig. 3.29 shows the CM distribution when the root of the mean squared displacement is 7 times the size of the one-dimensional cavity. Here the mean position for all molecules is at or very close to the center of the cavity. The result from a PFG NMR diffusion measurement for this setting is an asymptotic attenuation as follows [47]

$$\ln\left(\frac{I}{I_0}\right) = -\frac{16\gamma^2 g^2 \delta R^4}{175 D_0} \tag{3.24}$$

As the CM approaches a δ-function at the center of the cavity, it approaches the situation for molecules confined in porous material where the cavities are of the order of Ångstrøms, as in zeolites [32]. For ethane confined in a H-ZSM5 crystallite [48] the root of the mean squared displacement after a gradient pulse of duration 1 ms is approximately 10,000 times the size of the cavities within the zeolite. The crystallites may then be treated a homogeneous system, that is provided the size of the crystallites is significantly larger than the root of the mean squared displacement of the ethane during the PFG NMR experiment. This is not always that case, but then one may treat the zeolites as a homogeneous material and use the short observation time expansion by Mitra et al. to fit a true intra crystalline diffusion and the average surface to volume ratio of the crystallites [48].

Appendix: Generating a Linear Magnetic Field Gradient Along an Axis

To be able to build a set of coils that generates a magnetic field gradient, one must evaluate the magnetic field, using the following two approaches

- Starting with the law of Biot-Savart [28], and make a Taylor expansion of the magnetic field
-

$$\vec{H} = \frac{I}{4\pi} \int \frac{d\vec{s} \times \vec{r}}{r^3} \tag{3.25}$$

where ds is a small vector element of the conducting rod, and I is the current passing through it
- Starting with the magnetic vector potential

$$\vec{H} = \frac{1}{\mu_0} \nabla \times \vec{A} \tag{3.26}$$

The first approach will be used to evaluate the magnetic field from a quadropolar coil (applied in low field NMR systems comprising permanent magnets), while the second approach will be used to evaluate the field from a set of circular coils (anti-Hemlholz coils used in high field NMR comprising superconducting magnets)

Magnetic Field Gradients Generated by a Quadropolar Coil

For an infinitely long conducting rod, the magnetic field in the x-direction is written

$$H_x = \frac{I}{4\pi} \left[\int_{-\infty}^{\infty} \frac{(b-y)dz - zdy}{\left[z^2 + (a-x)^2 + (b-y)^2\right]^{3/2}} \right] \tag{3.27}$$

The conductor is located along the z-axis, thus dy $= 0$. By substituting $c = [(a-x)^2 + (b-y)^2]^{1/2}$ and $x = c \tan\theta$, and recognizing that equation is an even function of z, the magnetic field is found to be [49]

$$H_x = \frac{I}{4\pi} Re \left[\frac{1}{(a+ib) - (x+iz)} \right] \tag{3.28}$$

Here i denotes the imaginary number, and Re means that the real part of the expression yields the magnetic field. When positioning 4 infinitely long conducting rods as shown in Fig. 3.30, a quadropolar coil is produced. Substitution of

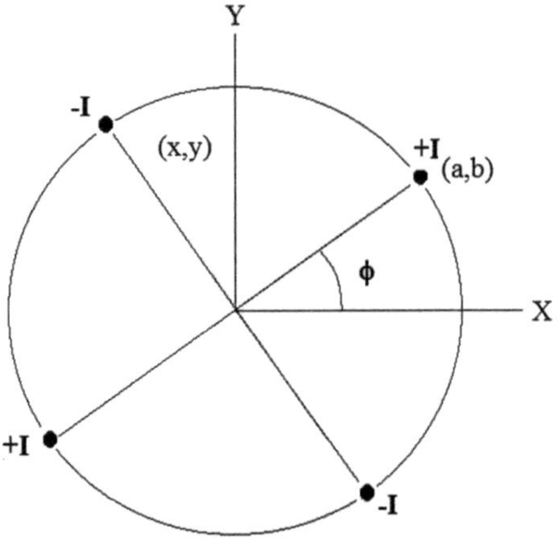

Fig. 3.30 A quadropolar coil as seen from above

ς = x + iy and re(iϕ) = a + ib leads to the following relation for one conducting rod

$$H_x = \frac{I}{2\pi} Re\left[\frac{1}{re^{(i\phi)} - \zeta}\right] \tag{3.29}$$

Then applying a Taylor expansion ((ς/r) < 1)

$$H_x = \frac{I}{2\pi r} Re \sum_{n=0}^{\infty} \left(\frac{\zeta}{r}\right)^n e^{(-i(n+1)\phi)} \tag{3.30}$$

For the 4 conductors with current polarity as shown in Fig. 3.28, the resulting magnetic field is written

$$H_x = \frac{I}{2\pi r} Re \sum_{n=0}^{\infty} \left(\frac{\zeta}{r}\right)^n \left[e^{-i(n+1)(\phi)} - e^{-i(n+1)(\phi+\pi/2)} + e^{-i(n+1)(\phi+\pi)} - e^{-i(n+1)(\phi+3\pi/2)}\right] \tag{3.31}$$

When performing the summation, the nonzero contributions are for n = 1, 5, 9 …

$$H_x = \frac{I}{2\pi r} Re\left[4 \cdot \left(\left(\frac{\zeta}{r}\right) e^{-2i\phi} + \left(\frac{\zeta}{r}\right)^5 e^{-6i\phi} + \left(\frac{\zeta}{r}\right)^9 e^{-10i\phi} + \cdots\right)\right] \tag{3.32}$$

In the area where ((ς/r) ≪ 1), the dominating contribution will be the linear one. Substitution of ς with x + iy and $e^{-2i\phi}$ with cos2ϕ − i sin2ϕ, and neglecting the imaginary terms, the magnetic field is written

$$H_x = \frac{2I}{\pi r^2}(x \cos 2\phi + y \sin 2\phi) \tag{3.33}$$

By choosing the right ϕ, the desired linear magnetic field gradient can be set to be either in the x or y-direction [49].

Magnetic Field Gradients Generated by an Anti-helmholz Coil

When calculating the magnetic field from a circular current loop, it is convenient to use spherical coordinate system. In the following the spherical coordinates will be used to evaluated the magnetic field from an anti-Helmholz coil, starting with the vector potential $B = \nabla \times A$. For a spherical system the vector potential reduces to one component, A_ϕ, which is parallel with the direction of the current [50]. The area in which we are interested in, close to the z-axis and the junction of the axes, is without any current sources, so the differential equation for A_ϕ is written (Fig. 3.31)

$$\nabla \times \nabla \times A_\phi = \frac{1}{r}\left(\frac{\partial^2}{\partial r^2}(rA_\phi) + \frac{\partial}{r\partial\theta}\left[\frac{\partial \sin\theta A_\phi}{\sin\theta\partial\theta}\right]\right) = 0 \tag{3.34}$$

The solution to this equation involves separation of the variables

$$A_\phi(r,\theta) = R(r)\Theta(\theta) \tag{3.35}$$

The solution of the radial part is written

$$R(r) = \sum_n \left[C_{1n}r^n + C_{2n}r^{-(n+1)}\right] \tag{3.36}$$

The differential equation for the angular part is written

$$\frac{\partial}{\partial\theta}\left[\frac{\partial}{\sin\theta\partial\theta}(\sin\theta\Theta)\right] = -n(n+1)\Theta \tag{3.37}$$

Performing the partial derivation and making the substitution $u = \cos\theta$ the equation is on the form of associated Legendre functions of the order n and degree 1 [51].

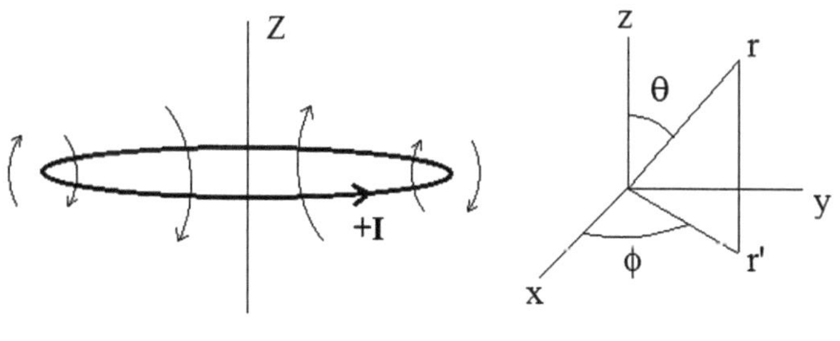

Fig. 3.31 Lines of force from a current loop

$$\frac{\partial}{\partial u}\left[(1-u^2)\frac{\partial\Theta}{\partial u}\right]+\left[n(n+1)-(1-u^2)^{-1}\right]\Theta=0 \tag{3.38}$$

Thus the general solution to the vector potential A_ϕ is written

$$A_\phi=\sum_n\left(C_{1n}r^n+C_{2n}r^{-(n+1)}\right)P^1_{n\cos\theta} \tag{3.39}$$

The magnetic field can now be found

$$H_r=\frac{1}{\mu_0}\frac{\partial\sin\theta A_\phi}{r\sin\theta\partial\theta}\sum_n n(n+1)(C_{1n}r^{n-1}+C_{2n}r^{-(n+2)})P_{n\cos\theta} \tag{3.40}$$

$$H_\theta=-\frac{1}{\mu_0}\frac{\partial(rA_\phi)}{r\partial r}\sum_n\left((n+1)C_{1n}r^{n-1}-nC_{2n}r^{-(n+2)}\right)P^1_{n\cos\theta} \tag{3.41}$$

These equations are generally valid for a symmetric magnetic field, i.e. A_ϕ parallel to the direction of the current. To find the magnetic field from a current loop it is convenient to start with the expressions in (3.40) and (3.41) and the constraints at r = c (see Fig. 3.32).

If the total current in the loop is I, then $K_\phi = I/(c\ d\theta)$. The distance from the centre to the loop is c, a is the radius of the current loop, and dθ is an infinitesimal change in θ (the current loop is assumed to be infinitely thin). The first condition ($H_{r1} = H_{r2}$) produces a relation between C_{1n} and C_{2n}

$$C_{1n}c^{n-1}=-C_{2n}c^{-(n+2)} \tag{3.42}$$

For the second condition, $H_{\theta1} - H_{\theta2} = \mu K_\phi$, the K_ϕ is expanded in an orthogonal set of associated Legendre functions $P^1_{n\cos\theta}$

$$K_\phi=\sum_n D_nP^1_{n\cos\theta} \tag{3.43}$$

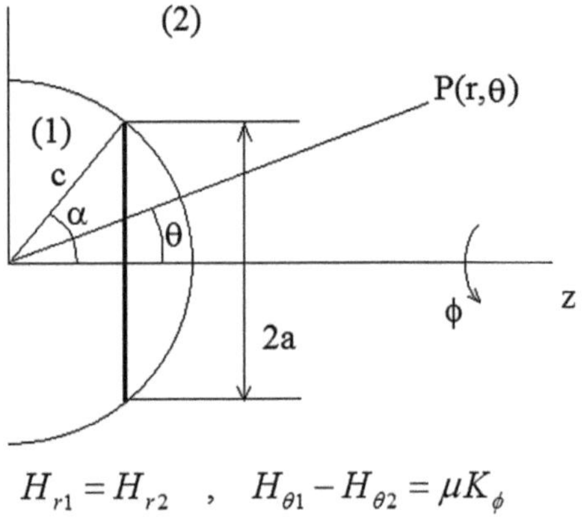

Fig. 3.32 Boundary conditions

The coefficients D_n are found using

$$N_n(1)D_n = \int_{\theta=0}^{\pi} K_\phi P^1_{n\cos\theta} \sin\theta d\theta \approx \frac{IP^1_{n\cos\alpha}\sin\alpha}{c} \tag{3.44}$$

where

$$N_n(1) = \int_{\cos\theta=-1}^{\cos\theta=+1} P^1_{n\cos\theta} P^1_{n\cos\theta} d\cos\theta = \frac{2(n+1)!}{(2n+1)(n-1)!} \tag{3.45}$$

C_{1n} and C_{2n} are then found using the second condition, $H_{\theta 1} - H_{\theta 2} = \mu K_\phi$, and the magnetic field is finally written

$$\begin{aligned}
H_r &= \tfrac{I}{2c}\sin\alpha \sum_{n=1}^{\infty} \left(\tfrac{r}{c}\right)^{n-1} P^1_{n\cos\alpha} P_{n\cos\theta}, & r < c \\
H_r &= -\tfrac{I}{2c}\sin\alpha \sum_{n=1}^{\infty} \left(\tfrac{c}{r}\right)^{n+2} P^1_{n\cos\alpha} P_{n\cos\theta}, & r > c \\
H_\theta &= -\tfrac{I}{2c}\sin\alpha \sum_{n=1}^{\infty} \left(\tfrac{r}{c}\right)^{n-1} \tfrac{1}{n} P^1_{n\cos\alpha} P^1_{n\cos\theta}, & r < c \\
H_\theta &= -\tfrac{I}{2c}\sin\alpha \sum_{n=1}^{\infty} \left(\tfrac{c}{r}\right)^{n+2} \tfrac{1}{n-1} P^1_{n\cos\alpha} P^1_{n\cos\theta}, & r > c
\end{aligned} \tag{3.46}$$

These equations are the starting point when aiming at generating a linear magnetic field in the centre of a coil configuration. The easiest way to generate such a field is to use two current loops separated by a certain distance and having current of opposite polarity. For c < r the magnetic field from the coil configuration shown in Fig. 3.33 is written

$$\begin{aligned}
H_r &= +\tfrac{I}{2c}\sin\alpha \sum_{n=1}^{\infty} \left(\tfrac{r}{c}\right)^{n-1} P^1_{n\cos\alpha} P_{n\cos\theta} \left(1 - (-1)^{(n+1)}\right), & r < c \\
H_\theta &= -\tfrac{I}{2c}\sin\alpha \sum_{n=1}^{\infty} \left(\tfrac{r}{c}\right)^{n-1} \tfrac{1}{n} P^1_{n\cos\alpha} P^1_{n\cos\theta} \left(1 - (-1)^{n+1)}\right), & r < c
\end{aligned} \tag{3.47}$$

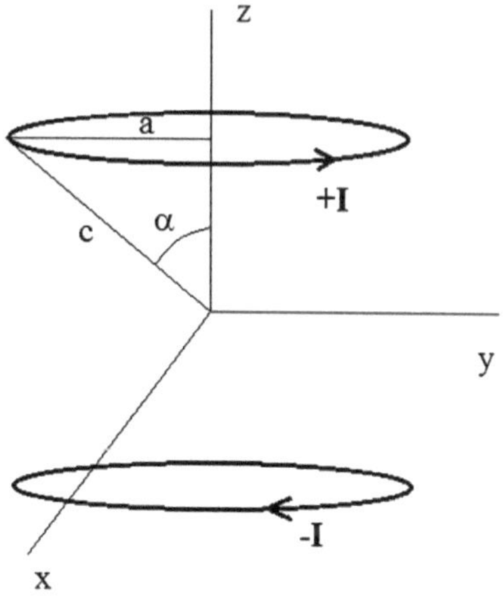

Fig. 3.33 The anti-Helmholz coil

Due to the symmetry of the current loops, every second term in the summation will cancel as

$$P^{1}_{n\cos(\pi-\alpha)} = (-1)^{(n+1)} P^{1}_{n\cos\alpha} \tag{3.48}$$

The summation may then start from n = 2, and the first part of the magnetic field will then be proportional to r, while the second and first nonlinear part will go as r^3. This first nonlinear term can be eliminated by choosing $\cos\alpha = (3/7)^{1/2}$. This is equivalent to letting the two current loops be separated by a distance $((3)^{1/2} \cdot a)$. Then the expression for the magnetic field along the z-direction can be written

$$H_z = +\frac{4I}{7a}\sum_{n=1}^{\infty}\left(\frac{r}{a}\right)^{2n-1}\frac{4}{7}^{\frac{2n-1}{2}} P^{1}_{2n\sqrt{\frac{3}{7}}}\left[P_{2n\cos\theta}\cos\theta - \frac{1}{2n}P^{1}_{2n\cos\theta}\sin\theta\right] \tag{3.49}$$

The first order correction to H_z along the z-axis ($\theta = 0$) will go as $(z/a)^5$.

References

1. A.D. Bain, Coherence levels and coherence pathways in NMR. A simple way to design phase cycling procedures. J. Magn. Reson. (1969) **56**(3), 418–427 (1984)
2. J.M. Fauth, et al., Elimination of unwanted echoes and reduction of dead time in three-pulse electron spin-echo spectroscopy. J. Magn. Reson. (1969) **66**(1), 74–85 (1986)
3. G.H. Sørland, B. Hafskjold, O. Herstad, A stimulated-echo method for diffusion measurements in heterogeneous media using pulsed field gradients. J. Magn. Reson. **124**(1), 172–176 (1997)
4. E.L. Hahn, Nuclear induction due to free larmor precession. Phys. Rev. **77**(2), 297–298 (1950)
5. E.L. Hahn, Spin echoes. Phys. Rev. **80**(4), 580–594 (1950)
6. N.E. Jacobsen, *NMR Spectroscopy Explained: Simplified Theory, Applications and Examples for Organic Chemistry and Structural Biology* (Wiley, New York, 2007)
7. R.M. Cotts, et al. Pulsed field gradient stimulated echo methods for improved NMR diffusion measurements in heterogeneous systems. J. Magn. Reson. (1969) **83**(2), 252–266 (1989)
8. R. Blinc, J. Pirš, I. Zupančič, Measurement of self-diffusion in liquid crystals by a multiple-pulse NMR method. Phys. Rev. Lett. **30**(12), 546–549 (1973)
9. R.F. Karlicek Jr, I.J. Lowe, A modified pulsed gradient technique for measuring diffusion in the presence of large background gradients. J. Magn. Reson. (1969) **37**(1), 75–91 (1980)
10. G.H. Sørland, D. Aksnes, L. Gjerdåker, A pulsed field gradient spin-echo method for diffusion measurements in the presence of internal gradients. J. Magn. Reson. **137**(2), 397–401 (1999)
11. D.H. Wu, A.D. Chen, C.S. Johnson, An improved diffusion-ordered spectroscopy experiment incorporating bipolar-gradient pulses. J. Magn. Reson. Ser. A **115**(2), 260–264 (1995)
12. A. Jerschow, N. Müller, Suppression of convection artifacts in stimulated-echo diffusion experiments. Double-stimulated-echo experiments. J. Magn. Reson. **125**(2), 372–375 (1997)
13. G.H. Sorland et al., Improved convection compensating pulsed field gradient spin-echo and stimulated-echo methods. J. Magn. Reson. **142**(2), 323–325 (2000)
14. P.D. Majors, et al., Eddy current compensation by direct field detection and digital gradient modification. J. Magn. Reson. (1969) **87**(3), 548–553 (1990)
15. R.S. Popovic, *Hall Effect Devices, Second Edition* (Taylor & Francis, London, 2010)

16. P. Mansfield, B. Chapman, Active magnetic screening of coils for static and time-dependent magnetic field generation in NMR imaging. J. Phys. E: Sci. Instrum. **19**(7), 540 (1986)
17. R. Turner, R.M. Bowley, Passive screening of switched magnetic field gradients. J. Phys. E: Sci. Instrum. **19**(10), 876 (1986)
18. R. Turner, A target field approach to optimal coil design. J. Phys. D Appl. Phys. **19**(8), L147 (1986)
19. R. Turner, Minimum inductance coils. J. Phys. E: Sci. Instrum. **21**(10), 948 (1988)
20. R. Turner, Gradient coil design: A review of methods. Magn. Reson. Imaging **11**(7), 903–920 (1993)
21. S.J. Gibbs, K.F. Morris, C.S. Johnson Jr, Design and implementation of a shielded gradient coil for PFG NMR diffusion and flow studies. J. Magn. Reson. (1969), **94**(1), 165–169 (1991)
22. J.J. Van Vaals, A.H. Bergman, Optimization of eddy-current compensation. J. Magn. Reson. (1969) **90**(1), 52–70 (1990)
23. V.J. Schmithorst, B.J. Dardzinski, Automatic gradient preemphasis adjustment: a 15-minute journey to improved diffusion-weighted echo-planar imaging. Magn. Reson. Med. **47**(1), 208–212 (2002)
24. S.J. Gibbs, C.S. Johnson Jr, A PFG NMR experiment for accurate diffusion and flow studies in the presence of eddy currents. J. Magn. Reson. (1969) **93**(2), 395–402 (1991)
25. P.T. Callaghan, PGSE-MASSEY, a sequence for overcoming phase instability in very-high-gradient spin-echo NMR. J. Magn. Reson. (1969) **88**(3), 493–500 (1990)
26. K. Scheffler, Fast frequency mapping with balanced SSFP: theory and application to proton-resonance frequency shift thermometry. Magn. Reson. Med. **51**(6), 1205–1211 (2004)
27. G.H. Sørland et al., Determination of total fat and moisture content in meat using low field NMR. Meat Sci. **66**(3), 543–550 (2004)
28. A. Pramanik, *ELECTROMAGNETISM: Theory and Applications* (PHI Learning, 2008)
29. L.E. Drain, The broadening of magnetic resonance lines due to field inhomogeneities in powdered samples. Proc. Phys. Soc. **80**(6), 1380 (1962)
30. J.S. Murday, R.M. Cotts, Self-diffusion coefficient of liquid lithium. J. Chem. Phys. **48**(11), 4938–4945 (1968)
31. J. Zhong, R.P. Kennan, J.C. Gore, *Effects of susceptibility variations on NMR measurements of diffusion.* J. Magn. Reson. (1969) **95**(2), 267–280 (1991)
32. J.R. Kärger, M. Douglas, *Diffusion in Zeolites and other Microporous solids* (Wiley-Intersciense, New York, 1992)
33. J. Kärger, F. Stallmach, NMR diffusion studies of molecules in nanoporous materials, in *Magnetic Resonance in Colloid and Interface Science*, eds. by J. Fraissard, O. Lapina (Springer, Netherlands, 2002), pp. 57–70
34. M.D. Hurlimann et al., Restricted diffusion in sedimentary rocks. Determination of surface-area-to-volume ratio and surface relaxivity. J. Magn. Reson. Ser. A **111**(2), 169–178 (1994)
35. M. Valtier, F. Humbert, D. Canet, Maps of self-diffusion coefficients by radiofrequency field gradient NMR microscopy. J. Magn. Reson. **141**(1), 7–17 (1999)
36. W.E. Maas, F.H. Laukien, D.G. Cory, Gradient, high resolution, magic angle sample spinning NMR. J. Am. Chem. Soc. **118**(51), 13085–13086 (1996)
37. J.G. Seland et al., Diffusion measurements at long observation times in the presence of spatially variable internal magnetic field gradients. J. Magn. Reson. **146**(1), 14–19 (2000)
38. L.L. Latour et al., Pore-size distributions and tortuosity in heterogeneous porous media. J. Magn. Reson. Ser. A **112**(1), 83–91 (1995)
39. P. Mörters, et al., *Brownian Motion* (Cambridge University Press, Cambridge, 2010)
40. A. Jerschow, N. Müller, Convection compensation in gradient enhanced nuclear magnetic resonance spectroscopy. J. Magn. Reson. **132**(1), 13–18 (1998)
41. N.K. Bar et al., Pitfalls in PFG NMR self-diffusion measurements with powder samples. J. Magn. Reson. Ser. A **113**(2), 278–280 (1995)

42. J. Stepišnik, Validity limits of Gaussian approximation in cumulant expansion for diffusion attenuation of spin echo. Physica B **270**(1–2), 110–117 (1999)
43. P.P. Mitra, B.I. Halperin, Effects of finite gradient-pulse widths in pulsed-field-gradient diffusion measurements. J. Magn. Reson. Ser. A **113**(1), 94–101 (1995)
44. P.P. Mitra, P.N. Sen, L.M. Schwartz, Short-time behavior of the diffusion coefficient as a geometrical probe of porous media. Phys. Rev. B **47**(14), 8565–8574 (1993)
45. E.J. Fordham, P.P. Mitra, L.L. Latour, Effective diffusion times in multiple-pulse PFG diffusion measurements in porous media. J. Magn. Reson. Ser. A **121**(2), 187–192 (1996)
46. G.H. Sørland, Short-time PFGSTE diffusion measurements. J. Magn. Reson. **126**(1), 146–148 (1997)
47. L.Z. Wang, A. Caprihan, E. Fukushima, The narrow-pulse criterion for pulsed-gradient spin-echo diffusion measurements. J. Magn. Reson. Ser. A **117**(2), 209–219 (1995)
48. L. Gjerdåker, G.H. Sørland, D.W. Aksnes, Application of the short diffusion time model to diffusion measurements by NMR in microporous crystallites. Microporous Mesoporous Mater. **32**(3), 305–310 (1999)
49. I. Zupancic, J. Pirs, Coils producing a magnetic field gradient for diffusion measurements with NMR. J. Phys. E: Sci. Instrum. **9**(1), 79 (1976)
50. E. Weber, *Electromagnetic Fields. Theory and Application* (Chapman & Hall, London, New York, 1950)
51. B. Spain, M.G. Smith, *Functions of Mathematical Physics* (Van Nostrand Reinhold Co, 1970)

Chapter 4
PFG NMR Spectrometer

Abstract This chapter provides an overview of the hardware required in order to conduct PFG NMR applications. Sections include a discussion on permanent magnets and their construction, magnetic field gradients, gradient amplifiers, RF amplifiers, probes and pre-amplifiers and NMR control systems including spectrometers, supervisor boards and host/control PCs. PFG NMR experiments have particularly demanding requirements in terms of gradient amplifier performance and practical factors that may impact this are discussed. A detailed discussion of spectrometer technology provides information on the individual spectrometer subsystems, as well as a description of common supervisor functions. The chapter concludes with a modular schematic which describes the layout of a modern PFG NMR system.

Since the initial discovery of the NMR phenomenon, hardware development for NMR and MRI applications has undergone significant progression. Early NMR experiments were often conducted with electromagnet based systems at relatively low polarising magnetic field strengths. In the modern laboratory research environment, NMR systems based around high field superconducting magnets are commonly encountered. These systems allow spectroscopic analysis due to their high frequency stability and ultra high homogeneity. The high magnetic field strengths available also lead to good peak separation in chemical shift experiments and in addition increase the signal to noise ratio significantly, which makes a wider variety of applications feasible that cannot be performed at lower magnetic fields. The cost, physical size and ongoing maintenance of these systems however makes them impractical for the routine analysis of samples in industrial quality control environments. Sample sizes in high magnetic field superconducting systems also tend to be limited to samples of 10 mm and downwards, which can be problematic if the sample is heterogeneous, although systems with larger sizes are available for some MRI applications with horizontal configuration magnets of the type used for small animal and whole body human imaging.

NMR has also become very popular as a characterisation method in industrial environments that neither tolerate the cost nor require the high performance

G.H. Sørland, *Dynamic Pulsed-Field-Gradient NMR*,
Springer Series in Chemical Physics 110, DOI 10.1007/978-3-662-44500-6_4

specifications found in the research environment. As a consequence NMR systems used in industrial product quality control/product characterisation environments have been devised that are usually based around permanent magnets for ease of use and cost, as installation, stray magnetic field and ongoing maintenance issues are much lower. These systems are known by a number of names, including benchtop NMR spectrometers, time domain NMR spectrometers, low field NMR spectrometers or low resolution NMR spectrometers. Most of these names derive from the fact that these systems have lower field stability, homogeneity and magnetic field strength in comparison with the more traditional superconducting magnets and do not typically allow spectroscopic experiments to be performed.

As benchtop NMR systems have been developed (Fig. 4.1), many of the features found on higher field superconducting magnets have been translated to the low magnetic field environment. Examples include the use of high performance magnetic field gradients which allow PFG NMR experiments to be performed, multi nuclear probes for some of the more sensitive nuclei (sodium and fluorine) and variable temperature apparatus. MRI applications have been extended to the benchtop systems by adding 3 axis magnetic field gradients as well as shaped RF capability.

Fundamentally, although the physical size, cost and capability of NMR systems can vary dramatically, the hardware required to perform the NMR experiment is largely similar and although the discussion in this chapter will focus on the

Fig. 4.1 A typical benchtop NMR system, based around a 20 MHz permanent magnet. This system has an 18 mm variable temperature probe and a gradient amplifier capable of generating magnetic field gradients up to 300 G/cm. Sophisticated control software allows the user to program pulse sequences, set parameters interactively while observing real time NMR signals and automate the acquisition of NMR data

hardware considerations of benchtop systems used for PFG-NMR experiments, much of the discussion is applicable to higher field systems.

Regarding the application of different NMR system configurations to PFG NMR experiments, low field or benchtop systems have a high degree of utility given their relatively low cost in this area, as they are often equipped with very strong magnetic field gradients which makes them ideal for conducting many of the experiments described in this book. PFG NMR experiments can also be coupled with relaxation contrast in two dimensional experiments to aid in sample characterisation.

The use of higher field systems based on superconducting magnets allows enhanced signal to noise ratio and also the characterisation/separation of diffusion information based on spectroscopic features, such as chemical shift. Examples of these more advanced techniques may be found in Chap. 6.

4.1 Permanent Magnets

Permanent magnets are used in many NMR applications in industrial and quality control environments. These magnets are characterised by relatively low cost, small physical dimensions and mass, low operating costs, relatively poor homogeneity and relatively low magnetic field strength (generally less than 1.5T or 60 MHz 1H frequency for most practical sample sizes). Permanent magnets are usually built from a steel yoke or frame design. Figure 4.2 shows the main elements of a typical permanent magnet design (there are several alternatives, all with relative strengths and drawbacks). The frame provides the return path for the magnetic flux, holds the pole pieces apart, and is normally made from well characterised low carbon steel.

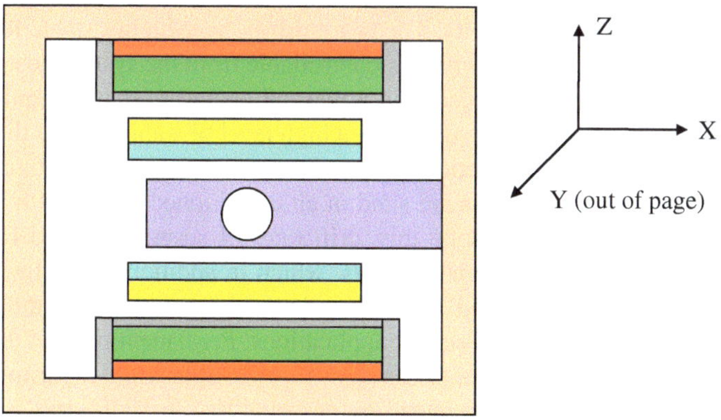

Fig. 4.2 A typical benchtop magnet geometry showing the frame (*pink*), backiron (*orange*), magnetic material (*green*), encapsulation (*grey*) and pole piece (*yellow*). The location of gradients and active shims is shown in *blue*. The cylindrical sample and probe chassis (*mauve*) is shown in the *centre*. The direction of magnetic field is usually defined as the z axis

To generate the polarising field, sections of magnetic material are usually glued in place onto a steel former to allow mounting to the frame (referred to as a backiron). The former provides mechanical support to the material, which can often be quite brittle. The sub assembly of the backiron and magnetic material (referred to as a lug) is usually encapsulated in order to prevent the individual magnetic blocks delaminating and to protect the brittle magnetic material from damage during assembly. A wide variety of different magnetic materials exist which vary in remanence (a measure of the amount magnetisation they retain following removal of the polarising magnetic field, which dictates the strength of the magnetic field produced), cost and temperature stability. The most common materials used in NMR magnets are the rare earth based materials Samarium Cobalt (SmCo, sometimes referred to as "samco"), Neodymium Iron Boron (NdFeB also referred to as "neo"), and AlNiCo (Aluminium Nickel Cobalt). In general NdFeB has a higher remanence and higher magnetic field variation as a function of temperature than SmCo. AlNiCo has a significantly lower remanence than SmCo and NdFeB, but has the advantage of a much more stable magnetic field as a function of temperature.

Pole pieces are placed in proximity to the lug in order to direct and homogenise the magnetic field. The pole piece orientation is often adjustable to allow correction for manufacturing tolerances and allows the magnet geometry to be accurately aligned as specified in the original magnetic model. This is necessary in order to produce the very homogeneous magnetic fields required for NMR applications. The pole piece geometry is often shaped to further improve the uniformity of the magnetic field, and the pole pieces themselves are usually made from magnetic material of high permeability (soft iron) in order to ensure that the maximum field strength is achieved. The entire design is usually analysed in a finite element analysis magnetic modelling program, which ensures that the appropriate magnetic field strength and uniformity (homogeneity) is attained from the assembly.

Often active shim coils (electrical coils that are either wound in specific geometries or laid out on printed circuit boards) are used as a final shimming method to improve the field homogeneity. These are driven by a highly stable low current power supply (the shim PSU) which is controllable from the system host PC.

The magnet itself is often placed in a thermal enclosure that is regulated via a power supply (magnet thermal supply) to a high level of stability (less than 0.1 °C) in order to maintain the magnetic field at a constant value. Manufacturers usually request that permanent magnets are sited in air conditioned rooms with well regulated temperature control. Despite this, drift rates of several hundred Hz per day are not uncommon in permanent magnets, which in addition to the low homogeneity of the magnetic field and low magnetic field strength limits the ability of the magnets to perform spectroscopic applications. Permanent magnet drift during long timescale NMR experiments can also lead to significant experimental artefacts. These can be largely eliminated by ensuring that the field drift is corrected between measurements, or by employing magnetic field locking techniques similar to those found in superconducting magnets.

Typically commercially available magnets vary in magnetic field strength from 1.5T (60 MHz) down to 0.05T (2 MHz). Magnets with lower field strengths allow larger sample sizes, with sample sizes typically up to 10 mm (tube diameter) available at the higher fields and larger than 100 mm from the lower field instruments. The choice of field strength for a particular application depends on the required signal to noise ratio, sample heterogeneity (which is important to consider in selecting an appropriate sample size) and additional experimental factors, for example in samples which contain large amounts of ferromagnetic impurities such as sandstone rock cores there are some benefits in working at lower rather than higher magnetic field strengths.

As described earlier, the design of NMR instrumentation for PFG NMR applications has a critical impact on the application performance. This extends to the magnet in many aspects. Of primary concern are the eddy currents that are induced in the magnet assembly following rapid switching of the gradient coils. Steps taken to minimise these include the use of actively shielded gradients, reduced proximity of the gradient set to the magnet assembly and special design of the magnet components such as pole pieces to reduce the potential for eddy current effects. For long time scale experiments magnet stability can be an issue when permanent magnets are used, the stability is improved by placing the magnet within a thermal enclosure within a temperature controlled laboratory environment. Mounting of the gradient coils is particularly important for PFG NMR applications, as these generate vibrations when pulsed. The vibrations can couple into the probe and sample, affecting the quality of the measurements and so gradients mounted on the pole pieces of the magnet and not directly on the probe chassis itself are preferred for PFG NMR experiments. Finally, the entire benchtop system should be sited on a bench that is free from vibration and preferably in a ground floor location, as often floor vibration in the upper floors of modern buildings can have a significant impact on the quality of the PFG NMR results. For further information regarding permanent magnet design, including a detailed discussion of the mathematical principles involved, readers should refer to the book by Moskowitz [1].

4.2 Magnetic Field Gradients

Dynamic or pulsed magnetic field gradients (from this point onwards referred to as gradients) are a fundamental requirement of any NMR system needed to perform PFG NMR applications. Gradient set design in many systems is fundamentally a compromise between several competing factors, including maximum magnetic field strength, rise time to maximum magnetic field strength, magnetic field screening (tendency to produce eddy currents in surrounding materials such as the magnet assembly and the NMR probe chassis) and linearity. The capabilities of the gradients and accompanying gradient amplifiers are often a limiting factor in performing PFG NMR applications.

A gradient set is usually comprised of several coils wound in specific geometries (see Sect. 3.6). These may be formed from wire wound on a former or on printed circuit boards. Wire formed gradient sets tend to have higher current

Fig. 4.3 A finite element model showing the major components of a permanent magnet (*green*) and field evaluation (*pink*) during the design process. Magnet design parameters are varied and the design re-evaluated in order to produce a central region of the required magnetic field strength and homogeneity (model generated with Cobham Opera Electromagnetic Design Software)

carrying capacity and thus can achieve higher gradient strengths without significant heating, although winding coils accurately in the complicated geometries required in order to produce a linear magnetic field variation can be challenging. Printed circuit board designs allow more accurate placement of the conductor elements, allow distributed conductor elements, are easier to fabricate and cheaper to produce. The X and Y (transverse) gradients (see Fig. 4.2) are symmetric by rotation and are often slightly weaker in strength than the Z (longitudinal) gradient, which is a more efficient field generating coil. Magnets of the design illustrated in Figs. 4.2 and 4.3 will often be fitted with a Y gradient if only a single gradient coil is present, due to the added utility of this gradient in conducting one dimensional profiles along the sample length.

The gradient set is often comprised of a sandwich of gradient coils and water cooling pipes which may be used to extract heat generated during the operation of the gradient set. This mechanism of heat dissipation allows the gradient set to run at higher duty cycles than would otherwise be possible, an important consideration for PFG NMR experiments where high duty cycle large magnitude current pulses are often used. Finally, thermal temperature sensors are included to allow the control software to monitor the temperature of the gradient set in real time during the PFG NMR experiment. These sensors can be used to shut down the gradient set if too much heat is generated during operation. The entire set is often encapsulated in epoxy to provide mechanical stability and insulation. Gradient sets in permanent magnets are generally easier to fabricate due to their planar geometry, and can be fabricated out of single PCB's mounted to the magnet pole faces, or planar assemblies made from a sandwich of primary coils, secondary coils and cooling elements.

Gradients can be considered as two general classes, actively shielded and unshielded. Actively shielded gradients are designed with magnetic screening, the purpose of which reduces eddy currents induced in the surrounding conductive structures that lie outside the gradient set (such as pole pieces) of the magnet assembly. The eddy currents act to limit the gradient rise time, and unless dealt with can have a devastating impact on the performance of the PFG NMR system. The use of actively shielded gradients significantly improves the gradient rise time and its ability to faithfully follow the gradient input waveform from the gradient amplifier. The penalty for this is design complexity, reduced maximum gradient strength and reduced space between the pole pieces of the magnet available for the sample, further limiting the maximum available sample size. Some actively shielded gradient sets have additional modes which allow the secondary (mostly shielding) coils to be removed from the gradient circuit to produce an additional operating mode with greater gradient strengths for experiments when absolute maximum achievable gradient strength is more important than gradient rise time. Although in this mode gradient strength is increased, gradient linearity often suffers in these configurations as in general the primary (mostly gradient generating) and secondary (mostly shielding) coils are designed holistically so that both coils contribute to the shielding and gradient performance.

Gradients used in PFG NMR experiments must be linear over a defined volume consistent with the sample size. This specification is directly competitive with gradient strength, as the most linear gradients are produced by coils that are distant from the sample. Linearity can be improved by accurate placing of the conductor elements, and using shaped conductive regions formed on printed circuit boards rather than discrete wires that only approximate to the ideal current distribution required to generate the gradient fields.

Gradient strength is an important parameter in most PFG NMR experiments, particularly if slow diffusion coefficient measurements are required. Gradient strengths are normally quoted in Teslas per metre per amp (T/m/A) which removes the dependency on the performance of the gradient amplifier used. Stronger gradients in general have a greater inductance and slower rise times. Strong gradients are also optimised in terms of active volume (the closer matched the gradient set to the sample volume, the stronger the gradients it can be designed to produce over that sample volume). This requirement competes directly with linearity.

Gradients must be capable of operating at a duty cycle consistent with the PFG NMR experiment being performed. Current pulses from the gradient amplifier cause resistive heating in the gradient coils, this can either be removed by conduction, or by circulation cooling. Circulation cooling can be performed by closed circuit water cooling which is very efficient at removing excess heat from the gradient set, especially when coupled with a chiller unit. Actively shielded gradient coils can utilise the space between the primary (gradient generation) and secondary (shielding) coils to accommodate circulation cooling tubes, which is an efficient use of space inside the magnet bore/pole gap.

Due to the high gradient strengths and duty cycles used in PFG NMR experiments, gradients should be hardware protected against overheating by thermal sensors. RTD sensors (such as PT100's) can be placed on the most resistive elements of the gradient set. Measurement circuits independent from the spectrometer should be used to measure the temperature of the gradients continuously and shutdown both the gradient amplifiers and spectrometer if a particular temperature is exceeded. Information from these sensors is often available to the system operator in real time, as they are a useful tool to investigate the ability of the system to perform aggressive PFG NMR sequences.

Thermal protection via resistance temperature detector (RTD's) or thermocouples works well for long timescale build up of heat in the gradient set, as there is normally some thermal inertia in the temperature detection system (the thermal sensor takes time to register increases in temperature in the gradient set resistive elements). Unfortunately, this method cannot protect well against short timescale high current pulses which may cause rapid temperature rises to occur before the thermal protection interlocks can react. This scenario is best dealt with using fuses in the gradient output lines that are set to fail if certain criteria are exceeded, for example if a gradient pulse of specified current and duration is used. Careful consideration of fuse type (breaking capability) needs to be made for this selection. In addition, if constant current amplifiers are used the amplifier needs to be capable of tolerating a load change from relatively low to infinite resistance when the fuse breaks. As well as hardware protection mechanisms, software limitations can also be placed on the duration of gradient pulses allowed in the pulse sequence. Systems which use gradient cooling systems will also have interlocks on the cooling circuit which will stop the spectrometer from continuing with the acquisition in the event of a cooling circuit failure. This is normally achieved with an in line flow sensor in the cooling circuit.

A detailed review of the methods of gradient coil design can be found in the review by Turner [2], which discusses many of the compromises that must be made when designing a gradient set.

4.3 Magnetic Field Gradient Amplifiers

Magnetic field gradient amplifiers are used to produce the high current pulses required to perform PFG NMR experiments. Specification of the gradient amplifier is critical in obtaining optimised gradient waveforms. Important parameters for amplifiers include maximum current output (which ultimately governs the strength of gradient pulses that can be produced), maximum voltage output (which decreases the gradient rise times associated with the gradient set inductance), amplifier bandwidth (or slew rate, which may limit the gradient rise times if insufficient), the maximum available gradient pulse duration and the maximum duty cycle available.

Amplifiers are constant current and usually driven via a low voltage input from the spectrometer hardware. The input can either be single ended or common mode, the common mode input rejects noise present on both signal input lines and leads to improved performance. This can be especially important in PFG NMR experiments where noise present on the amplifier signal inputs can propagate to the amplifier output, resulting in echo phase instability. Some spectrometer systems utilise remote DACs from the main spectrometer hardware and close to the amplifier signal inputs in order to reduce the possibility of external noise affecting the gradient drive signal.

Amplifiers are usually designed to produce maximum power into a load (in this case the gradient coil) of well specified impedance. In order to match the amplifier to the load impedance, it is often necessary to compensate the gradient amplifier with internal configuration options available on the amplifier itself. This can be especially important in achieving optimised rise times, as the amplifiers used in PFG NMR applications are often designed for general purpose applications (such as audio loudspeaker applications often loudspeaker drive) rather than as specific dedicated amplifiers for NMR applications.

Amplifiers can be either linear or switched mode. Switched mode amplifiers have become increasingly popular in recent years, as improved switched mode technology for power supplies has developed. However even with today's high switching rates switched mode amplifiers usually have compromised rise times/bandwidths compared with the more traditional linear designs, which can affect their capability to perform PFG NMR experiments. Switched mode amplifiers generally are more efficient than their equivalent linear type, but tend to generate more high frequency noise both on the gradient outputs and via emission from the amplifier control systems that may result in poorer signal to noise ratio and noise spikes in the PFG NMR experiment (Fig. 4.4).

Gradient power lines are often filtered to prevent RF noise conducted down the gradient power lines coupling into the RF probe. This filtering should be selected to eliminate RF noise but prevent modulation of the gradient waveform by the filtering circuit in order to maintain the gradient rise times. In addition, the filtering used should be capable of handling the high currents and voltages produced by the gradient amplifier.

Fig. 4.4 A 7224 AE Techron NMR/MRI Gradient Amplifier, a linear amplifier capable of performing PFG NMR experiments on low field permanent magnet based systems (courtesy of AE Techron Inc)

The output noise level of the amplifier is a critical parameter for NMR experiments, as even low levels of output noise can lead to phase instabilities in the echoes produced in NMR experiments. Mains ripple at the fundamental frequency (often 50 or 60 Hz) is almost always present on the outputs of amplifiers to some level even with no input signal present, and this can be particularly problematic and difficult to filter.

Amplifier mains supply requirements vary depending on the particular design, but high power amplifiers often require 3 phase mains inputs. Mains purity is of important consideration as many amplifiers do not always contain the necessary internal filtering in order to remove the most persistent mains harmonics and spikes present on local mains supplies. Industrial environments in particular may have very noisy mains supplies due to the operation of heavy equipment, which may lead to the amplifier producing spurious output signals which can lead to phase unstable behaviour in the PFG NMR echoes. Careful verification of the local supply integrity and on site filtering of the mains supply is often required in order to produce the best results.

As explained in Chap. 3, echo phase stability is a crucial element in the acquisition of high quality PFG NMR data, due to the phase sequence programs used to remove unwanted coherences and the addition of data in order to improve signal to noise ratio. The gradient amplifier is normally the dominant source of echo phase instability, usually due to the imperfections in the mains supply described earlier.

PFG NMR experiments are usually run asynchronously to the external mains supply, which leads to the pulse sequence starting at a different point each time in the mains cycle. This results in the sample experiencing a slightly different cumulative magnetic field effect due to the mains cycle signal propagating through the amplifier leading to phase fluctuation in the PFG NMR echo data. By synchronising the execution of the pulse sequence to the local mains period (usually 50 or 60 Hz) the effect of mains cycle on the echo phase can be made consistent. The synchronisation process is usually performed by sampling the amplifier mains input signal and deriving a low voltage trigger signal that is used to trigger the NMR spectrometer to commence the experiment (Figs. 4.5 and 4.6).

The large power capability of gradient amplifiers means that gradient sets can often be run to thermal destruction if inappropriate duty cycles are used without temperature safety mechanisms in place. In addition, the gradient amplifier hardware itself can be damaged by excessive duty cycles, although most amplifiers have internal thermal sensing to prevent this. Safety mechanisms, including software calculation and protection, devices which measure the temperature at the most resistive point in the gradient coil circuit and interlock to stop the gradient amplifier if an excessive temperature rise is observed are often used.

Gradients are protected against high energy pulses often using fuses in the main gradient drive lines. These fuses must be chosen of an appropriate type to ensure the gradient coils are protected adequately. In addition, the constant current amplifier must be capable of handling the change in load impedance when the fuses in the gradient circuit break.

A discussion of many of the principles of high power audio amplifier design may be found in the book by Self [3].

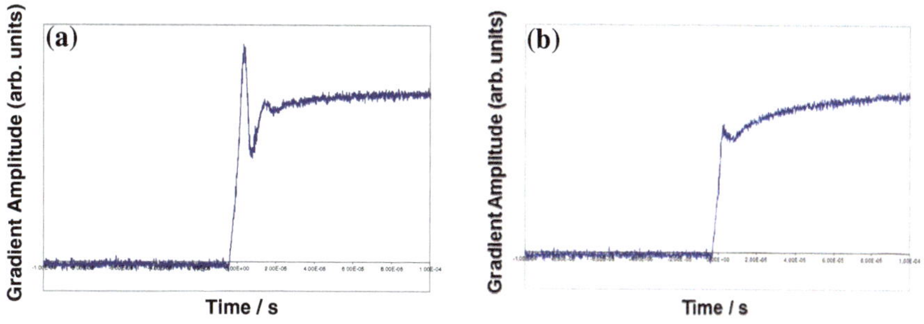

Fig. 4.5 An example of adjusting amplifier compensation on a gradient amplifier and a gradient set. In Fig. 4.5a the amplifier compensation achieves a fast rise time with considerable overshoot. After a change in compensation Fig. 4.5b is produced, demonstrating a reduction in overshoot but with an increase in rise time. Gradient waveforms which match the input amplifier signal more precisely can be achieved via the use of shaped gradients

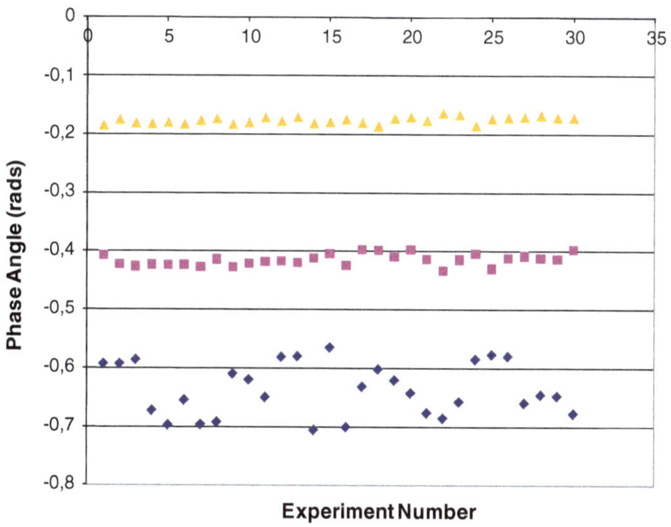

Fig. 4.6 The effect of gradient amplifier output noise on echo phase stability. In this experiment, successive spin echoes are acquired with the gradient amplifier switched on but with no current pulses generated. The effect on echo stability can be seen at echo times of 20 ms (*blue*) and 40 ms (*pink*) with the gradient amplifier on and 20 ms (*yellow*) with the gradient amplifier off for a 50 Hz mains cycle time. It is clear the echo stability at 20 ms is dramatically improved. Similar improvements can be made by gating the amplifier to the mains cycle

4.4 RF Amplifiers

The RF amplifier is used to generate the RF pulses necessary for the NMR excitation. Important amplifier parameters include frequency range/bandwidth (typically RF amplifiers working within the range of frequencies and power levels required for permanent magnets are specialist devices and are not readily available), power, duty cycle and output noise level.

Sufficient output power is required to perform the NMR excitation in a reasonable time. This is not normally a critical parameter for most PFG NMR experiments, but for some NMR experiments (such as the CPMG sequence used to measure T_2 relaxation processes) higher power levels lead to shorter possible echo times, which can allow better characterisation of short T_2 signals. Typical RF amplifier powers used in PFG NMR systems range from as low as a few watts to over 1 kW.

In CPMG experiments where both solid and liquid type signals are present, large numbers of echoes performed at short echo times are necessary in order to accurately characterise the relaxation properties of the entire system. These types of experiment are particularly demanding in terms of RF duty cycle and it is important that the RF amplifier is capable of meeting these demands.

Any general purpose RF amplifier used for permanent magnet applications needs to cover a range of frequencies, normally in the 100 kHz to 60 MHz range. It is especially important to consider wideband operation if multinuclear experiments are anticipated.

RF amplifiers are usually controlled by two input signals, a waveform and a gate signal generated by the NMR spectrometer. The gate signal normally blanks the output of the amplifier when the RF output is not required. RF amplifiers which do not use blanking often require passive elements such as crossed diodes or active switches in the amplifier output in order to prevent noise from the amplifier corrupting the very low level NMR signal during the acquisition period.

The input signal from the NMR spectrometer to the amplifier is typically in the 0 dBm range. If amplifiers are to be used in conjunction with gradients for slice selective experiments or in MRI applications the amplifier must faithfully reproduce the phase and amplitude of the input waveform. In PFG NMR experiments the use of phase sequences to cancel unwanted coherence pathways means that the amplifier must be capable of accurately reproducing changes in the RF phase present in the input waveform.

Modern RF amplifiers often provide a control interface to the spectrometer hardware which allows monitoring of the amplifier performance. Typical monitoring parameters include mismatch, which signals when the RF probe impedance does not correctly match the output impedance of the amplifier (usually 50 ohms, this parameter can be used to signal a failure in the RF probe circuit), a measure of the reflected power level which can be used to tune the probe, an amplifier temperature and duty cycle indicator (whether the amplifier has exceeded its maximum duty cycle) and a power good signal. These types of signals are useful for PFG NMR applications in industrial environments, as they allow remote diagnostic and automated system checking to ensure that the RF component of the NMR system is working correctly. For a detailed discussion of RF Amplifier design principles, readers can refer to the book by Kazimierczuk [4] which in particular discusses the Class AB RF power amplifiers commonly used in NMR applications (Fig. 4.7).

Fig. 4.7 A TOMCO BT-AlphaS Series RF Amplifier suitable for PFG NMR applications. This amplifier operates from 100 kHz to 30 MHz and has a nominal output power of 500 W. Important status parameters such as load mismatch and amplifier temperature status are available via a digital interface (photo courtesy of TOMCO Technologies)

4.5 RF Probes and First Stage Pre Amplifiers

All NMR systems require RF coils or probes in order to perform both the stimulation (RF pulse) and detection of the NMR signal. The probe configuration is critical, as the NMR experiment is inherently insensitive, so optimising the probe for a particular experiment/sample can lead to significant increases in the quality of the results or in the time required to perform experiments of a necessary quality. This is particularly applicable at low frequencies.

There are many possible probe configurations and geometries for NMR experiments. Some configurations use a single coil for both RF excitation and detection. Others use a separated configuration with separate coils for transmit and receive. A single probe configuration generally has the advantage of being easier to implement and is less expensive. However in NMR systems, the excitation and detection characteristics of the probe required for optimal system operation are to some degree orthogonal.

Figure 4.8 shows a simple RF probe configuration used in many NMR spectrometers. The transmit signal from the RF amplifier is shown on the left. It is usual to route the transmitter output through a blanking device to prevent noise propagating from the transmitter to the receiver during the NMR signal acquisition. This can be as simple as a pair of crossed diodes, which let the high power

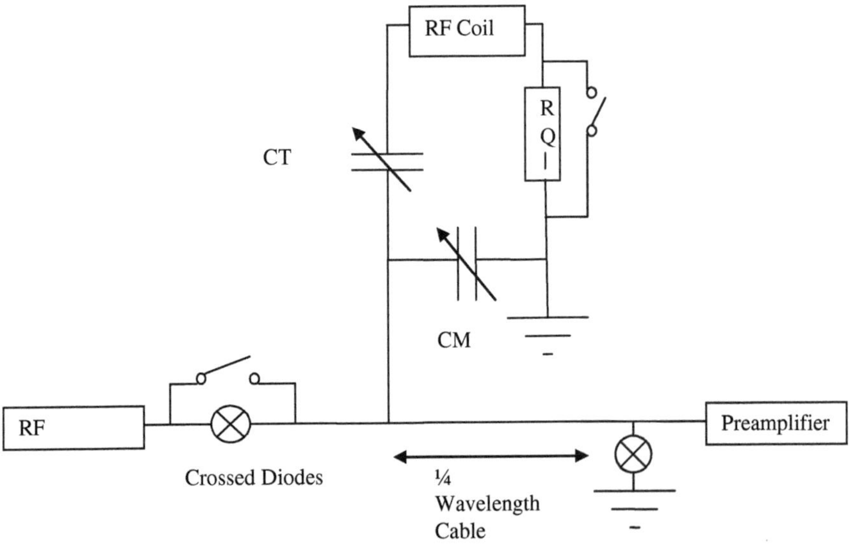

Fig. 4.8 Simple RF probe configuration used in NMR spectrometers. The crossed diodes in the transmit line and Q spoil resistor (RQ) may be switched during the NMR experiment in order to improve the probe performance

transmit pulse through but stop any low level noise. However, use of diodes introduces non linearity into the RF transmit path, which can affect the ability of the system to perform accurate shaped RF pulses, such as those used for slice selection. Instead of crossed diodes an active switch based on a PIN diode can be used, which is cable of being switched rapidly in order to blank the transmitter while being able to handle the high output power levels of the RF amplifier.

The probe circuit in this example is a simple tank circuit, tuned to match the input impedance of the preamplifier and the output impedance of the RF transmitter. The circuit is tuned and matched with the two capacitors (CT and CM). The path to the preamplifier is protected via a quarter wavelength element, which presents a high impedance to the transmit pulse. This can either be made from a length of co-axial cable or from appropriate inductors/capacitors. The preamplifier input is usually protected by a pair of crossed diodes to ground, which limit any high voltage signals from the transmit pulse reaching and damaging the sensitive first stage of the pre-amplifier.

The probe quality factor or Q is an important parameter in characterising probe performance. Probes with low Q have a relatively good impedance match (usually 50 ohms) to the RF amplifier/preamplifer over a wide range of frequencies, have short dead times (the period after the application of the RF pulse when the probe "rings" and does not contain any useful NMR signal) and are relatively insensitive to detuning by the sample (when a conductive sample is placed inside the probe it affects the probes' electrical characteristics). Probes with high Q match impedance over a narrow range of frequencies, have a long dead times and typically are very

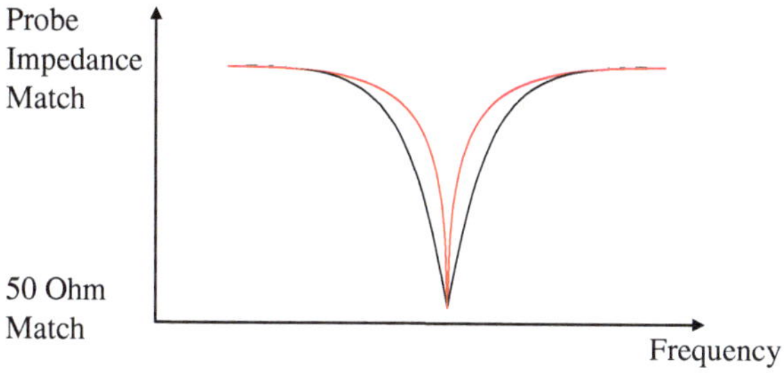

Fig. 4.9 Changing the probe Q for transmit and receive periods. During the transmit period with RQ in circuit the Q is low (*black*), leading to reduced probe ringing. Switching out the Q spoil resistor RQ increases the Q (*red*), leading to an increased SNR

sensitive to loading by the sample. High Q probes do however have one significant advantage, in the respect that their lower bandwidth means the signal to noise ratio of high Q probes is higher than a low Q design. Due to these properties, a probe with a low Q is generally preferred during the RF pulse transmit period to reduce ringing, whereas a probe with a high Q is generally preferred during the NMR signal acquisition period to increase the signal to noise ratio. Simple RF probe designs have a constant Q, which is often limited by a series resistor, RQ (Fig. 4.8). The choice of Q is a compromise determined by the application requirements.

More modern designs allow active switching of the resistor RQ during the RF pulse transmit and receive period. The effect of this can be seen in Fig. 4.9. Normally significantly more complex schemes than the simple switch illustrated in Fig. 4.8 are required to achieve the switching process, due to the speed of switching required and the RF powers involved. In addition, active elements may be used in place of the quarter wavelength cable in order to improve the probe ringing characteristics.

Along with the probe, the first preamplifier has a critical impact on the signal to noise ratio of the NMR experiment. Optimum design in order to produce the best results is essential.

The first stage of any NMR pre-amplification system needs to be able to amplify the original voltages induced in the RF coil which are at the microvolt level while being able to tolerate high voltage RF pulses that may leak through from the transmitter.

Typically in NMR systems amplification of the NMR signal is performed in several different stages. It can be seen from the Friis equation that the noise figure of the first stage makes the dominant contribution to the noise figure of the entire signal amplification chain.

Several different types of device can be used as a first stage, but typically microwave monolithic integrated circuits (MMICs) and GAsFETS are used. Typically MMICs with 1 dB noise figures are available that offer 30 dB or more of gain over the typical NMR frequency ranges. MMICs are generally more robust than GAsFET devices, which tend to have lower noise figures. For more information on NMR probe design, the book by Misplelter et al. [5] is recommended.

4.6 The NMR Spectrometer, NMR Supervisor System and Host PC

Modern NMR spectrometers make extensive use of customisable integrated circuits such as field programmable gate arrays (FPGA's), which effectively allow the electronic engineer to create a software defined application specific microprocessor for NMR experiments rather than use generic devices. Single chip NMR spectrometer solutions have become possible as the size and speed of the FPGA integrated circuits available has increased dramatically. The advances in digital technology have led to another significant development, the ability to implement an entire NMR receiver digitally without the need for down mixing using analogue components. This has led to significant improvements in data handling and filtering.

In addition to the spectrometer which performs the NMR experiment, a second control system is often used in order to perform the less time critical monitoring and control of the NMR system. Less time critical tasks include monitoring the status interfaces of the various amplifiers, thermal monitoring of the various system components and shutting down the spectrometer if temperature limits are exceeded, monitoring and controlling the temperature of the sample and performing system configuration switching. We refer to this control system as the NMR supervisor, and the main supervisor tasks are normally performed by a simple microprocessor.

The user normally interfaces to the NMR spectrometer and the NMR supervisor via a host PC, which operates a user friendly environment to allow NMR experiments/pulse programs to be designed, interactive setup of the NMR experiment to be performed and finally once the system is set correctly the experimental data to be acquired. The host PC can also be used to process the acquired data.

4.6.1 The Host PC and Pulse Programming

The host PC is the point at which the user interacts with the spectrometer system and performs several tasks. Firstly, a pulse program defines the events that take place in an NMR experiment, including the RF pulses and phase programs, data acquisition and gradient waveforms. The pulse program instructs the spectrometer what actions to perform during an event, and how long that particular event should take in time.

A typical generic pulse program for an FID experiment might be as follows:
Start Pulse Program;
Initialise System;
Set RF pulse Amplitude Frequency and Phase;
Enable/Gate RF transmitter;
Perform RF Pulses.
Wait for Probe Ring down and Set Receiver Phase;
Acquire Data;

Wait for Repetition Time;
Increment Phase Lists to perform Phase Cycle and Return to Perform Next Scan;
Return Data to Spectrometer;
End Pulse Program;

Implementation of the above sequence varies depending on the spectrometer manufacturer, but an example of generic syntax might be:

```
StartProgram;
Event(10000*us, SystemInitialisation);
ScanLoop:
Event(1*us,SetPhase(RF,Phase1)+SetRFAmplitude(Amp1)+SetRFFrequency
  (Freq1));
Event(20*us,TxEnable);
Event(P90*us,RF+TxEnable);
Event(Deadtime*us,SetPhase(REC,Phase2);
Event(AcquireTime*us,AcquireDa.ta);
Event(RD*us,0);
Event(1*us,IncPhase(Phase1)+IncPhase(Phase2));
Event(1*us, Loop(ScanLoop));
Event(ReturnDataTime*us,ReturnData);
EndProgram;
```

In the above example, each event has a time (first argument) plus actions to be performed during that time associated with it. The time is either specified either explicitly, or by a variable (for example RD) that may be altered interactively via the user interface. Multiple events occurring during the same time period are specified using the addition operator. The pulse program is usually written and compiled within its own IDE (integrated development environment). The binary file produced can then be downloaded to the spectrometer for execution. The host PC runs a control program that allows the pulse sequence to be loaded, parameters specified and data acquisition initialised. This program may run a simple command line interface, or may have more complex parameter specification options available to the end user. In order to set up the experiment correctly and efficiently parameters are usually specified interactively while observing the data on a real time display. Once the parameters are set up correctly the interactive parameter setup process can be stopped and an acquisition process can be started. After the acquisition process is complete data is returned to the host PC and may be processed in a variety of ways, which may include post acquisition digital filtering, Fourier transforms and curve fitting/data modelling. Finally, after the post processing is performed, the data can be saved to disk.

Often PFG NMR experiments will need to be performed under different experimental conditions (such as varying temperature) and the host PC should provide a high level language scripting environment to facilitate this process, allowing the user to sequence for example multiple experiments with different parameters or repeat experiments at regular time intervals to measure dynamic processes.

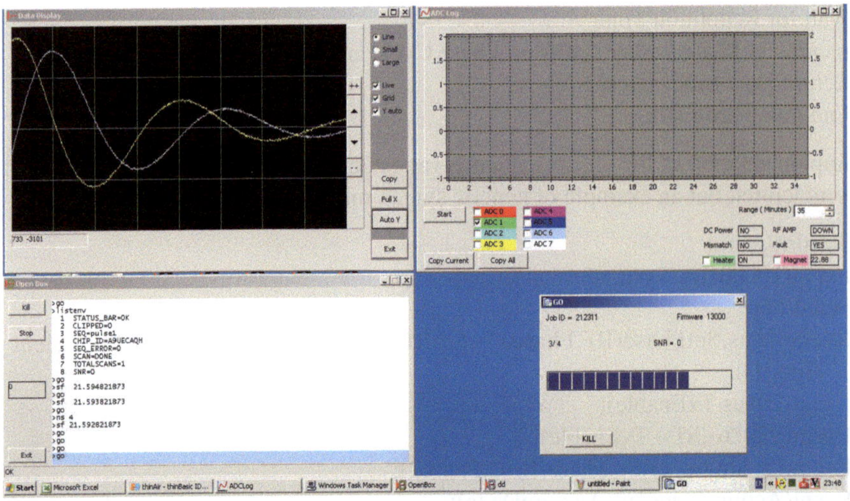

Fig. 4.10 A typical control interface to a benchtop PFG NMR spectrometer for research users. Pulse sequence parameters can be varied by typing on the command line (*bottom left*). The effect on the acquired data can be observed in real time (*top left*), allowing adjustment of the parameters depending on the data acquired. System status such as gradient set temperature (*top right*) can be monitored as the sequence progresses (*bottom right*)

The host PC usually links to the proprietary spectrometer electronics via a fast communications data link. The communications protocol may be Ethernet or increasingly modern systems use USB. The important point is to have a fast, reliable link that can transmit relatively large quantities of data that may be acquired in a short timescale during NMR experiments.

Due to the precise timing required in the NMR experiment on the order of microseconds, it is very important that communications between the host PC and the spectrometer electronics does not in any way affect the timing of the pulse sequence, as this may lead to spurious results. The spectrometer, once data acquisition has been initialised, must be capable of performing the exact timing of the pulse sequence without interruption, and must operate independently of the host PC state (Fig. 4.10).

4.6.2 Spectrometer System

At the heart of the modern NMR system lies the spectrometer, which sequences and controls the NMR experiment and outputs various control waveforms to the RF and gradient amplifiers described earlier. The spectrometer also acquires the NMR data, filters and averages it and stores it ready for download to a host PC, where the data can be further processed and visualised. Spectrometer electronics are usually proprietary because of two requirements, one because they need to perform tasks specific to the NMR experiment (such as generating shaped gradient waveforms) and the other is that the tasks normally need to be performed on a consistent and well defined timescale, which excludes the use of many modern operating systems which are often multi tasking rather than real time and would introduce unacceptable jitter into the precise timing required by all NMR experiments.

Key tasks include controlling the devices that generate RF waveforms (DACs), acquiring the NMR data (via ADCs), controlling analog filters, mixers, frequency generators and amplifiers, managing the data (either adding it to an internal memory store for data averaging or saving it separately) and maintaining the link to the Host PC. Traditionally these functions have been performed by microprocessors or digital signal processing (DSP) IC's, although increasingly in modern spectrometers many of these functions are implemented in soft logic on a single field programmable gate array (FPGA) integrated circuit. The spectrometer may run a commercial real time operating system, or run its own proprietary operating system specifically designed for NMR applications. In addition it is now common to perform significant amounts of data processing (digital mixing and filtering) within the digital electronics rather than via analog components.

The RF transmit sub system should be capable of generating low voltage RF waveforms for driving the high power RF transmitters described in Sect. 4.4. These waveforms are usually generated by a direct digital synthesizer (DDS), either internal to the microprocessor or on a dedicated integrated circuit. Data is then routed to a corresponding digital to analog converter in order to produce the RF output signal. The use of the DDS allows precise control of RF pulse phase, amplitude and frequency, which is essential for conducting NMR experiments. The transmit sub system should be capable of amplitude/phase modulation of the RF pulses for conducting MRI experiments and experiments that involve slice selection.

Often multiple RF output channels will be present, and auxiliary channels may be used for advanced techniques such as high power decoupling, magnetisation transfer, lock channels, probe tuning via directional couplers or for driving analog mix schemes. Parameters important for NMR include frequency, phase, and amplitude stability of the RF output, spectral purity/phase noise and the dynamic range available in frequency, phase and amplitude settings. For NMR experiments the ability to generate stable frequencies down to sub Hz level is particularly important, especially in applications involving spectroscopy.

The RF receive sub system normally consists of an analog to digital converter (ADC) and a signal processing chain. Important specifications for the ADC include the bandwidth and dynamic range (between 14 and 16 bits is typical). Following digitisation the RF signal is digitally mixed to base band with an internal frequency generated by another DDS. This produces the two quadrature data channels.

The data as a result of the digital mix process contains wanted and unwanted frequency components, as well as out of band noise components. As the data acquisition rates produced by NMR experiments are reasonably high, the filtering process is often designed as a cascaded filter made from different filter types. The first stage of the filter is usually generated from a digital filter called a comb integrate comb

(CIC) filter. This filter can handle high input data bandwidths but has a relatively broad frequency response. Once data is decimated to an appropriate level using this filter, the data rates become more compatible with filters that have more desirable response characteristics over the pass band, such as FIR filters. The application of these filters both improves the signal to noise ratio of the data and reduces the amount of data required for transfer over the Spectrometer-Host PC link.

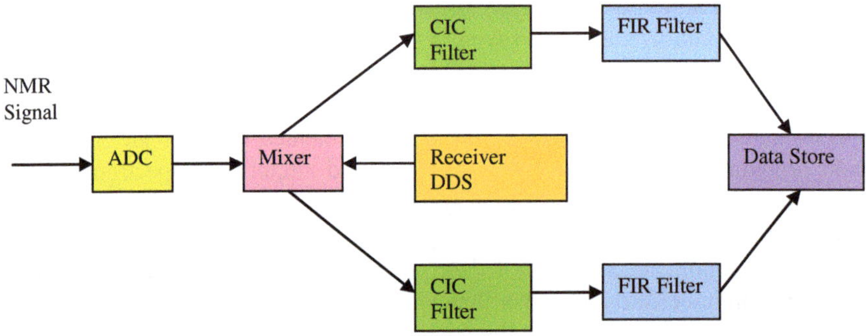

Fig. 4.11 An example RF receiver subsystem. The NMR signal is digitised with the ADC before being digitally down mixed using a frequency generated by a direct digital synthesizer. This generates two data channels. The broadband CIC filters remove the unwanted mix components and decimate the data bandwidth to a rate which can be handled by the FIR filters, which perform narrowband filtering of the data to optimise the signal to noise ratio. Finally the data is stored in the data store ready to be transmitted back to the host PC

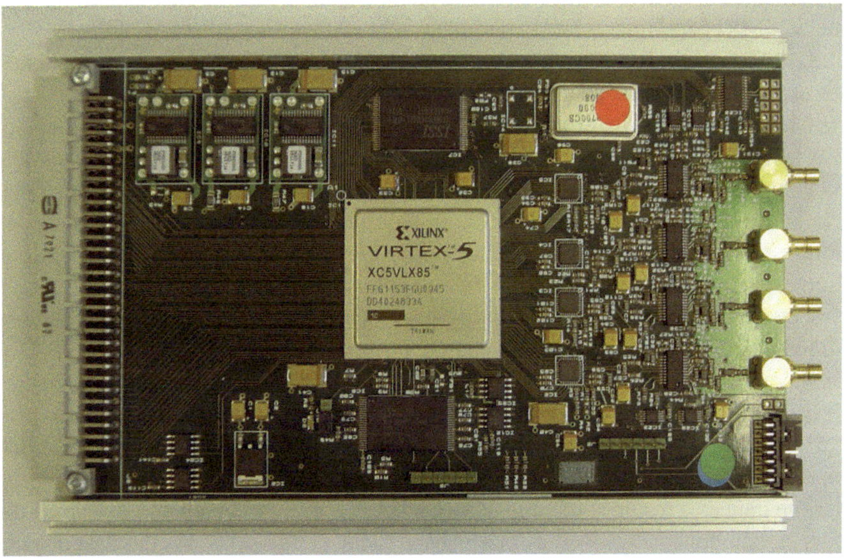

Fig. 4.12 Advanced Magnetic Resonance Ltd NMR Spectrometer Board. This board implements a 4 channel NMR spectrometer using an FPGA (*centre*). The entire spectrometer logic is software defined and implemented on the FPGA. The ADCs for data acquisition can be seen to the right of the board

Fig. 4.13 Advanced Magnetic Resonance Ltd NMR Supervisor Board. This board uses a microcontroller to implement a complete NMR Supervisor System. Controls included RTD temperature monitoring, high current relay switching, voltage rail monitoring, output voltages for shim power supplies, digital IO lines, SPI, I2C and RS232 interfaces. The supervisor board interfaces to the host PC via a USB communications link

The gradient waveform subsystem generates the low voltage level gradient waveforms which are used to drive the gradient amplifier. Important specifications include the dynamic range/resolution of the DAC's used, which can be particularly important for PFG NMR experiments and the ability to use shaped rather than simple gradient pulses for minimisation of eddy currents and preemphasis. In addition to gradient pulse shaping, gradient amplitudes can also be variable on a scan per scan basis in order to provide spoiler gradient capability (Figs. 4.11 and 4.12).

4.6.3 NMR Supervisor System

Due to the complexity and number of different interacting components in the modern PFG NMR system, a supervisor system is required. As supervisor functions do not require exact and reproducible timing, it is more appropriate to

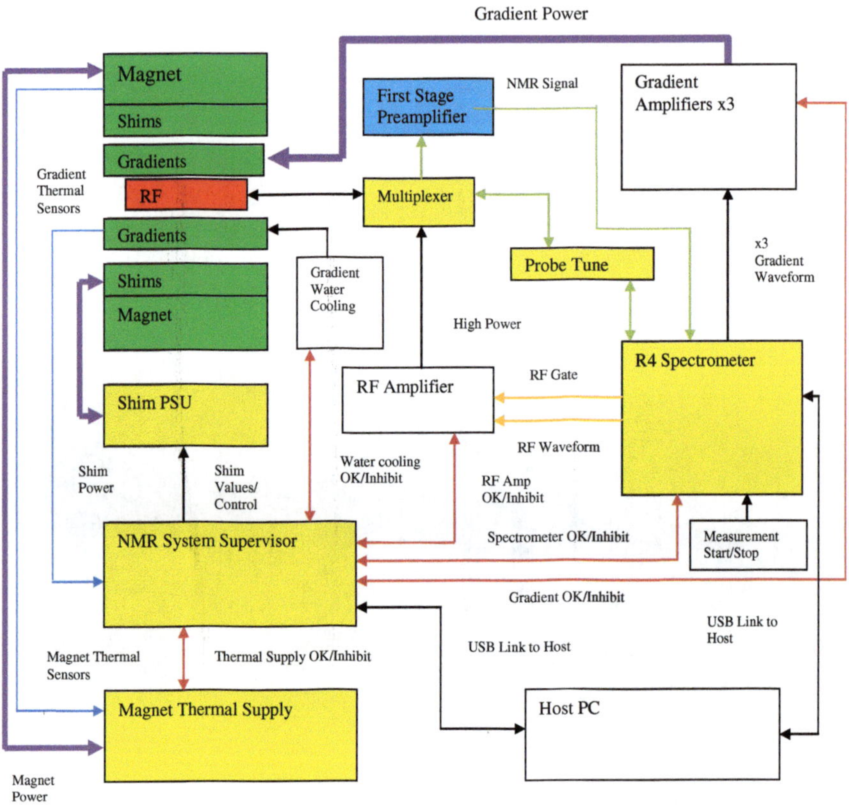

Fig. 4.14 Schematic showing the high level components of an NMR system described in this chapter

implement these functions outside of the spectrometer hardware. One key task of the supervisor system is to monitor the gradient thermal performance for safety shutdown. This is important, as high pulse sequence duty cycles can lead to excessive gradient heating and damage to the gradient set. Any water cooling unit used to cool the gradients will usually have interlocks for flow monitoring and control which are handled by the supervisor. Most RF amplifiers have interfaces to indicate power good, and also inhibit lines to switch the amplifier off in the event of a fault condition, these interfaces should be monitored by the supervisor. The supervisor is also used to power relays for RF coil tuning/diagnostics, as switching the spectrometer from an acquisition mode to a probe tuning mode often requires an rf relay and electronics. The supervisor should provide an independent means of shutting down the spectrometer. This is particularly applicable in the event of gradient thermal overload, where it is prudent to stop the pulse sequence as quickly as possible as well as simultaneously inhibiting the gradient amplifiers in the event of a fault condition. If the supervisor is to control the shim PSU (used in conjunction with the magnet shim coils in order to improve the magnet homogeneity) then several high resolution DAC channels are required in order to specify the necessary shim currents. The ability to monitor the system components constitutes a full diagnostic check of the NMR/MRI system, which is required in environments where non-expert users are present and can assist in validating the results in quality control environments (Fig. 4.13).

4.7 Summary

Figure 4.14 shows the complete modular schematic of a typical PFG NMR system with the components described in this chapter. The system may be highly integrated, as in the case of benchtop NMR systems, or be formed from separate components, such as a separate magnet unit and 19″ rack to hold the spectrometer, amplifiers and other components. As mentioned in the earlier sections, siting can be particularly critical in order to achieve good experimental results, and particular emphasis should be placed on siting the system in a place where there is good mains integrity (mains supply free of noise spikes and unwanted harmonics), a low RF noise environment and a low vibration environment. In particular, vibrations transmitted though building structures by vehicles or other large moving objects close to the magnet site can lead to devastating effects on the quality of the PFG NMR results, so siting the system in a low vibration area is critical in order to produce good results.

References

1. L.R. Moskowitz, *Permanent magnet design and application handbook.* (Cahners Books International, Boston,1976)
2. R. Turner, Gradient coil design: a review of methods. Magn. Reson. Imaging **11**(7), 903–920 (1993)
3. D. Self, *Audio Power Amplifier Design Handbook*, 5th edn. (Focal Press, Elsevier, Boston, 2009)
4. M.K. Kazimierczuk, *RF Power Amplifiers*. (Wiley, New York, 2008)
5. J. Mispelter, M. Lupu, A. Briguet, *NMR Probeheads for Biophysical and Biomedical Experiments: Theoretical Principles & Practical Guidelines* (Imperial College Press, London, 2006)

Chapter 5
Analysis of Dynamic NMR Data

Abstract An analysis of one and two-dimensional data sets using the so-called inverse Laplace transforms (ILT) and a discrete approach called Anahess is presented. Actually, the ILT is in mathematical terms an ill posed numerical inversion of Fredholm integrals with smooth kernels, meaning that several and very different solutions may fit the experimental data equally well. This inversion is referred to as an ILT in this book. The ILT and Anahess are fundamentally different methods for analysing the data, and the two methods are evaluated against each other on synthetic data sets. These data sets will represent the NMR data typically found from analysing real systems such as cheese or fluid saturated porous rock core plugs. Singular components and distribution are represented in the synthetic data sets. While the ILT algorithm fits the experimental data to a large set of fixed points in a grid, the Anahess algorithm minimizes the number of points, which are not fixed due to user defined grid as for the ILT algorithm. From the comparison, it will become obvious that both approaches have their benefits and drawbacks. Ultimately, user interpretation is required to decide which particular model fits the system better, either the Anahess algorithm, which fits a limited number of discrete components that are continuously variable, or the ILT algorithm, which assigns a larger set of components from a fixed grid.

The analysis of NMR data arising from measurements of dynamic parameters such as diffusion and relaxation in heterogeneous systems is a challenging task. As discussed in Chap. 3 there are several factors besides the dynamic parameters that influence the attenuation of the NMR signal while incrementing the applied magnetic field gradient. The most important ones are the restricting geometries experienced during the prepare and read intervals in the PFG NMR sequence and the possible existence of internal magnetic field gradients. Due to these influences, it is not straightforward to assign a deviation from the mono exponential to a multi exponential behaviour in the measured diffusivity, thus being indistinguishable from experimental artefacts [1–6].

In the following section a method will be provided to analyse one-dimensional data sets that minimises the afore mentioned artefacts. Afterwards, an analysis of one and two-dimensional data sets using the so-called inverse laplace transforms

G.H. Sørland, *Dynamic Pulsed-Field-Gradient NMR*,
Springer Series in Chemical Physics 110, DOI 10.1007/978-3-662-44500-6_5

(ILT) and a discrete approach called Anahess will be presented. Actually, the ILT is in mathematical terms an ill posed numerical inversion of Fredholm integrals with smooth kernels, meaning that several and very different solutions may fit the experimental data equally well [7]. This inversion will be referred to as an ILT in the remainder of this book. The ILT and Anahess are fundamentally different methods for analysing the data, and the two methods are evaluated against each other on synthetic data sets. These data sets will represent the NMR data typically found from analysing real systems such as cheese or fluid saturated porous rock core plugs. Singular components and distribution will therefore be represented in the synthetic data sets. The two approaches for analysing the PFG NMR data provide different results due to their initial assumptions. While the ILT algorithm fits the experimental data to a large set of fixed points in a grid, the Anahess algorithm minimizes the number of points, which are not fixed due to a user defined grid as for the ILT algorithm. From the comparison, it will become obvious that both approaches have their benefits and drawbacks. Ultimately, user interpretation is required to decide which particular model fits the system better, either the Anahess algorithm, which fits a limited number of discrete components that are continuously variable, or the ILT algorithm, which assigns a larger set of components from a fixed grid.

5.1 Linear Analysis of One Dimensional Data

Consider the logarithm of PFG NMR echo attenuation from a water sample immersed in a porous medium as shown in Fig. 5.1. The observation time is such that the molecules are experiencing the restrictions of the pore walls, and thus a fraction of the diffusion molecules will have their root of mean squared displacement reduced as compared to the value achieved from bulk diffusion. Under these conditions, a deviation from the usual mono exponential decay should occur as

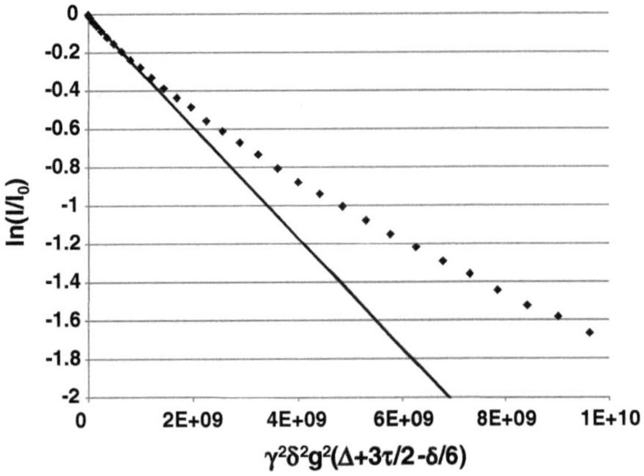

Fig. 5.1 The attenuation of ethane confined in a H-ZSM5 zeolite, using a Bruker DMX 9.4 T

the gradient is increased. However, as discussed in Chap. 3 the deviation from a mono exponential decay may also be due to coupling to internal magnetic field gradients, a coupling that is more difficult to avoid as the observation increases. A second potential effect arises from restricted diffusion during the motional encoding and decoding of the phases of the nuclear spins, as a mono exponential decay is valid only when there is Gaussian behaviour of the diffusion propagator. Due to the restricted geometry experienced by the diffusing molecules during the gradient pulses, non-Gaussian behaviour is expected, and one must assume higher order terms of the applied magnetic field gradient strength to affect the attenuation of the PFG NMR signal. The solid line in Fig. 5.1 represents a weighted linear fit to the data. The slope of this fit contains important information regarding the system under investigation, the apparent diffusion coefficient.

In the following discussion, it shall be assumed that the attenuation of the PFG NMR data is only dependent on the dynamic parameters Diffusion, T_1, or T_2-relaxation, and not due to internal magnetic field gradients or other experimental artefacts. When the mobile phase is experiencing restrictions, one must expect a nonlinear decay. The attenuation from the ordinary pulsed field gradient stimulated echo sequence as the gradient strength is increased for a fixed gradient duration can be described by a multi exponential decay

$$\mathrm{I} = \sum_{\mathrm{i}} \mathrm{I_i e}^{-\gamma^2 \mathrm{g}^2 \delta^2 (\Delta - \delta/3) \mathrm{D_i}} \tag{5.1}$$

i represents the fraction i with an apparent diffusivity D_i. The decay due to a finite relaxation is incorporated into I_i, as it only represents a constant factor when keeping the z-storage delay and gradient duration fixed. The initial attenuation of the echo signal can then be analysed via an expansion of the exponential coefficient as a function of g^2

$$\begin{aligned} I &= \sum_i I_i \mathrm{e}^{-\gamma^2 g^2 \delta^2 (\Delta - \delta/3) D_i} \\ &= I_0 \Big(1 - \frac{1}{I_0} \sum_i I_i \gamma^2 g^2 \delta^2 (\Delta - \delta/3) D_i + \Phi(g^4)\Big) \\ &= I_0 \big(1 - \gamma^2 g^2 \delta^2 (\Delta - \delta/3) \overline{D} + \Phi(g^4)\big) \end{aligned} \tag{5.2}$$

Thus by assuming the second cumulant approximation to be valid for small attenuations given the appropriate values of the applied magnetic field gradient strengths [8], i.e. the term containing g^4 to be insignificant, the attenuation is written

$$I \approx I_0 \mathrm{e}^{-\gamma^2 g^2 \delta^2 (\Delta - \delta/3) \bar{D}} \tag{5.3}$$

where

$$I_0 = \sum_i I_i \quad \text{and} \quad \overline{\mathrm{D}} = \frac{1}{\mathrm{I_0}} \sum_{\mathrm{i}} \mathrm{I_i D_i} \tag{5.4}$$

The initial decay is probably one of the most important parameters to extract from a single diffusion experiment in a porous system. It minimizes the effect from a non Gaussian behaviour during motional encoding and decoding and also yields the average diffusion coefficient that can be used to fit for example the expansion at short observation times [9]. Care should be taken when analysing one dimensional diffusion data, as sometimes it can be difficult to resolve diffusion driven non-exponential decay from deviations due to restricted diffusion occurring during the motional encoding/decoding interval. Thus, it is convenient to apply a weighted linear fit to the data, which can be performed by taking the logarithm of the data as follows

$$\ln(I) = -\gamma^2 g^2 \delta^2 (\Delta - \delta/3)\bar{D} + \ln(I_0) = k\bar{D}g^2 + \ln(I_0) \tag{5.5}$$

where $k = \gamma^2\delta^2(\Delta - \delta/3)$. The data is now of a format where a linear regression fit can be applied to the data. Denoting (5.5) on a matrix form ($y = [ln(I)]$, $V_{in} = [1, \mathrm{kg}^2]$ and $b = [ln(I_0), D]$)

$$y = \mathrm{V}_{in} b \tag{5.6}$$

The dimension of y is the number of gradient strengths applied, n, the dimension of V_{in} is then $(2 \times n)$ and b is (1×2). In order to weight the data a weighting function W is introduced. If the requirement is to weight the linear fit to the intensity of the measuring points, the expression $W = I^x$ can be used, where x is an exponent that may be chosen based on what information that needs to be extracted. If the exponent is assigned to be positive, the data is weighted towards the data points of the highest intensity, which is the region where the second cumulant approximation applies. Following matrix operations for solving out b [10], b is written

$$b = (\mathrm{V'V})^{-1}(\mathrm{V'})(\mathrm{Wy}) \tag{5.7}$$

where

$$\mathrm{V} = \mathrm{WV}_{in} \tag{5.8}$$

A Matlab program for the use of this procedure can be found in the appendix to this chapter. In this program, the option of weighting the data set is included and a data set for testing is provided.

When analysing binary systems using the PFG NMR experiment, the complexity increases. Again, the second cumulant approximation is related to the initial decay of the diffusion data. If there are two components of very different mobility (as moisture and fat in a cheese sample), the second cumulant approximation is valid in different regions of the applied gradient strength due to the significant difference in molecular mobility. Figure 5.2 shows two diffusion experiments; one from a cheese sample and one from a mixture of crude oil and brine. For the cheese, there is evidently a fast attenuating and slow attenuating component. A two component model yields an acceptable fit, but the residuals are not randomly distributed, which indicates a poor fit model. Besides the fact that there may be variation in molecular mobility, there is also the possibility that the attenuation of the water is affected by higher order $\boldsymbol{g}^2$ terms. Fortunately, as the diffusion

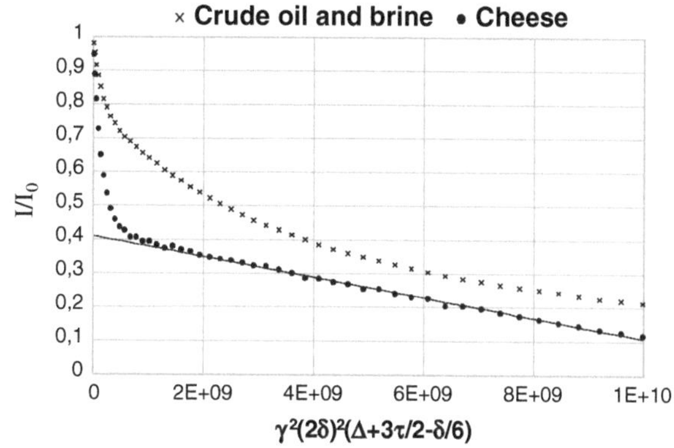

Fig. 5.2 The PFG-NMR echo attenuation from two samples; a cheese, and a mixture of crude oil and brine. The data were acquired on a low field magnet (0.47 T) using the spoiler recovery 13 interval PFGSTE sequence with bipolar gradient pairs

coefficient differs two orders of magnitude, the fit is still likely to reflect the diffusion coefficient of the two components quite accurately. The dotted line shows a weighted linear fit to the data, where the weighting is a function of the inverse of the measured intensities, i.e. ($1/I^x$). x is the exponent which can be increased by visual inspection until the water signal is being neglected in the fit. In Fig. 5.2, the situation is quite different for the crude oil and brine. This is a system where a fast decaying component from the brine still exists, but the oil phase does not return a clear mono exponential decay in the same way as fat present in the cheese. Instead, a broad distribution of diffusivities for the crude oil is expected, due to a variation in molecular composition of the oil, producing a span of diffusivities from 10^{-9} m^2/s down to 10^{-12} m^2/s, i.e. three orders of magnitude.

The reminder of this chapter will describe analysis methods that assume a multi-exponential decay of the NMR data. It should be stressed again that multi-exponential character in the data may arise as a consequence of experimental artefacts, as well as the dynamic behaviour of the system under investigation. Despite the presence of these artefacts (which can be minimised according to the procedures described in Chap. 3), the analytical methods used have significant value in producing a qualitative description of the system under investigation.

5.2 The 1- and 2-Dimensional Inverse Laplace Transforms

The fitting of a sum of exponentials to the measured response from a PFG NMR experiment, either in one or two dimensions, is a vital tool for interpretation of the outcome of the experiment. This fitting procedure we may call the Inverse Laplace transform (ILT). The goal of the ILT is to transform the data from a sum of exponentially decaying components to a set of parameters characterizing the exponential decay of each component. Depending on the experimental setup, the parameters have a physical interpretation such as spin-lattice relaxation time T_1, spin-spin relaxation time T_2 and/or diffusion coefficient D. Although such a transform is notoriously ill conditioned numerically [11], algorithms that attempt to perform this transform are in widespread use. In the following a brief review of the theory behind the numerical procedure for achieving processed data will be given.

5.2.1 The 1 Dimensional Inverse Laplace Transform (1D-ILT)

When dealing with experimental data, there is always some noise present in the experimental decay, which has to be accounted for in the data processing. In the following, a set of data that attenuate due to relaxation will be considered, and where experimental noise is present. The multi-exponential signal $S(t)$ can then be written as:

$$S(t_i) = \sum_k I(T_k) \exp\left(-\frac{t_i}{T_k}\right) + \varepsilon_i \qquad (5.9)$$

In this expression, T_k and $I(T_k)$ correspond respectively to the relaxation times and the relaxation time distribution of the system under study. ε represents the random noise of the measured signal. The general procedure to invert the data and determine the distribution of relaxation times $A(T_i)$ consists of applying the non-negative least square algorithm, as described in the Lawson and Hanson book [12]. This would consist of the minimization of:

$$\min\left\{\chi^2 = \sum_{j=1}^{n}\left(S(t_j) - \sum A(T_i)\exp(-t_j/T_i)\right)^2\right\} \qquad (5.10)$$

However, because the signal is noisy, the determination of the distribution by a simple non-negative least square fit leads to a multitude of solutions. Small fluctuations in the acquired data lead to many different solutions of (5.9), all of which provide good fits of the experimental data. The problem is thus defined as mathematically ill-posed [13, 14]. To avoid this problem, a so-called regularization function (or smoothing function) is added and weighted by the regularization parameter α [15–17]. As initially no information is known regarding the time constants or intensities, the regularisation could be specified by parsimony. That is, as the ill-posed problem has an infinite number of solutions, choose the simplest solution that reveals the least information that can be extracted from the fitted data, which still fit the experimental data equally well. Then it follows that the regularisation parameter should be chosen to be as small as possible and the number of peaks in the produced distribution will be minimized. The new least square function to minimize is displayed in the following equation.

$$\min\left\{\chi^2 = \sum_{j=1}^{n}\left(S(t_j) - \sum A(T_i)\exp(-t_j/T_i)\right)^2 + \alpha^2\sum_{i=1}^{m}(F(A))^2\right\} \qquad (5.11)$$

where F(A) varies depending on the choice of smoothing;

- *"Amplitude or Norm"*

$$\min\left\{\chi^2 = \sum_{j=1}^{n}\left(S(t_j) - \sum A(T_i)\exp(-t_j/T_i)\right)^2 + \alpha^2\sum_{i=1}^{m}(A_i)^2\right\}$$

- *"Slope"*

$$\min\left\{\chi^2 = \sum_{j=1}^{n}\left(S(t_j) - \sum A(T_i)\exp(-t_j/T_i)\right)^2 + \alpha^2\sum_{i=1}^{m}(A_{i+1} - A_i)^2\right\}$$

- *"Curvature"*

$$\min\left\{\chi^2 = \sum_{j=1}^{n}\left(S(t_j) - \sum A(T_i)\exp(-t_j/T_i)\right)^2 + \alpha^2\sum_{i=1}^{m}(2A_i - A_{i-1} - A_{i+1})\right\}$$

In Fig. 5.3, the results from a 1D-ILT is shown on a recording of a CPMG echo train data from a crude oil sample using curvature smoothing. A crude oil contains a large and continuous variety of hydrocarbon molecules of different chain lengths and molecular sizes, and thus the distribution of T_2 values tends to be continuous. If a too large a smoothing parameter is applied, the distribution will separate into three distinct peaks. However, then the level of detail is enhanced in the distribution such that what is seen is more likely to be an artefact than the existence of a well-fractionized crude oil. Consequently, the distribution with three distinct peaks is more likely to be a fitting artefact than due to the molecular properties of the sample. In Fig. 5.4 the residuals of the two fits are displayed. The shape appearance of the residuals provides a good estimate on whether the fit is good or not.

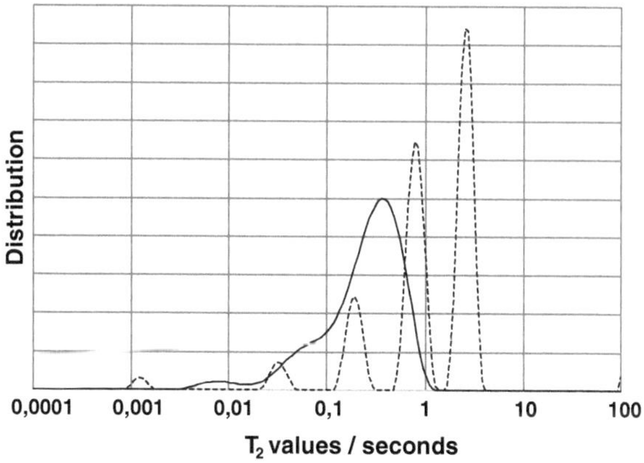

Fig. 5.3 The T_2-distribution from a crude oil sample using two different smoothing values, α_1 yielding the continuous distribution (*solid line*) and α_2 the numerous peaks (*dotted line*)

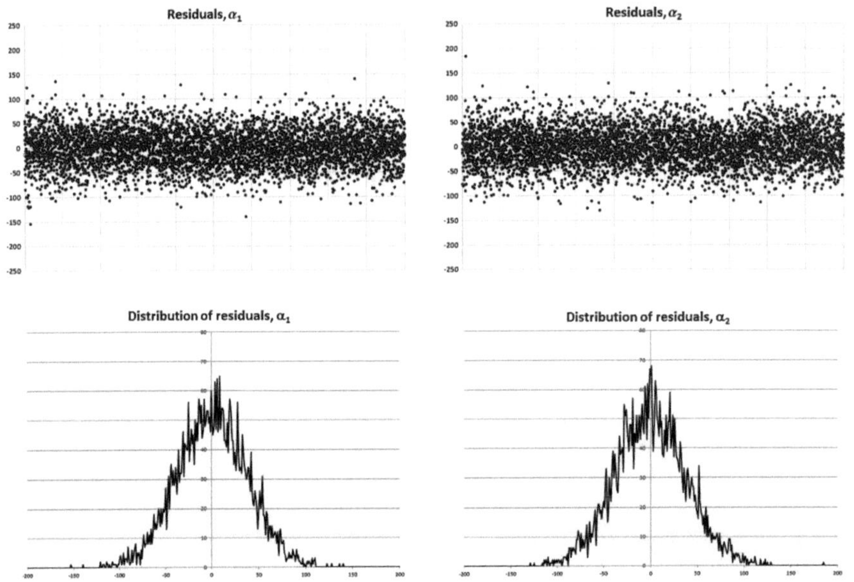

Fig. 5.4 The residuals from the fit to the CPMG-data and their distribution arising from two different values of α

When the residuals are randomly distributed about zero, the fit is good. Therefore, for both of the processed data with different smoothing, the fit seems to be fine, but only the one with the continuous distribution reflects the system investigated. As it is not likely that the crude oil consist of resolved components but a continuum of molecular sizes, it may therefore be concluded by parsimony that the best fit is the one with the lowest value on the regularization parameter in order to produce residuals distributed about zero, α_1.

If there is no basic knowledge on how the shape of the distribution should look like, the option is to calculate the chi square, χ^2, for a number of α values to provide an appropriate value where χ^2 reaches a threshold value (Fig. 5.5). Here it can be seen what values of α provide fitted distributions that are the least sensitive to noise in the experimentally determined data set. The calculated χ^2 as a function of α shown in Fig. 5.5 are from the CPMG data of for the crude oil as shown in Fig. 5.3. α_1 is 10^9 and α_2 is 10^{13} in the T_2-distributions, where α_2 produces the resolved peaks. It can be seen that α_2 is definitely in the region defined as ill-posed while α_1 is within the region where fit is the least sensitive to noise.

Another option is to apply the method proposed by [18], the so-called BRD-routine that uses amplitude smoothing to optimize the regularization parameter α. Another optimization routine has been proposed which, compared to commercial software using the BRD-routine, claims to provide more reliable distributions [19]. Using either of the methods for inverting the NMR data to distributions through an inverse Laplace transform, it is important to not to forget where the data originates from, knowledge by parsimony!

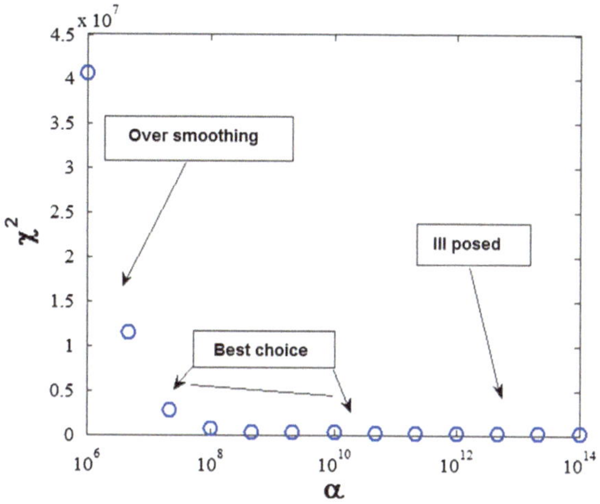

Fig. 5.5 Determination of a good value for α using curvature smoothing. In the region "Best choice of α" the fitted distribution is the least sensitive to noise. This figure also applies for determining α for a two-dimensional data set

5.2.2 The Two Dimensional Inverse Laplace Transform (2D-ILT)

Song et al. [20] have developed a Tikhonov regularisation routine, which is as in the 1D-ILT, based on the strategy given by Lawson and Hanson [12], in which a regularization function or smoothing function is added to the data before minimizing the squared sum of errors. A singular value decomposition is then used to reduce the size of the data. Instead of computing a specific set of number of components, *NCO*, with accompanying **a**, $\mathbf{T}_1$ and $\mathbf{T}_2$ vectors, the results are computed as a second order distribution and presented as a matrix of **A** values, with row and column indices corresponding to for example evenly spaced T_1 and T_2 values. Again the weighting of the regularization function is determined by the parameter α, chosen by the user by parsimony and/or calculating χ^2 as a function of α (see Fig. 5.5). In this model, any element $R_{i,j}$, $i = 1..NSA$, $j = 1..NSE$ in the response surface **R** is assumed to follow the function:

$$R_{i,j} = f(\mathbf{A}, \mathbf{T}_1, \mathbf{T}_2)_{i,j} + E_{i,j} \quad \text{with}$$
$$f(\mathbf{A}, \mathbf{T}_1, \mathbf{T}_2)_{i,j} = \sum_m^M \sum_n^N A_{m,n} e^{(-(1/T_{1,n})g_i - (1/T_{2,m})t_j)} \tag{5.12}$$

The regularized least squares function to be minimized now takes on the form:

$$SS_{res} = \sum_i^{NSA} \sum_j^{NSE} (f_{i,j} - R_{i,j})^2 + \frac{1}{\alpha^2}\left\{\sum_{m=2}^{M-1} \sum_n^N (2A_{m,n} - A_{m+1,n} - A_{m-1,n})^2 + \sum_m^M \sum_{n=2}^{N-1} (2A_{m,n} - A_{m,n+1} - A_{m,n-1})^2\right\} \tag{5.13}$$

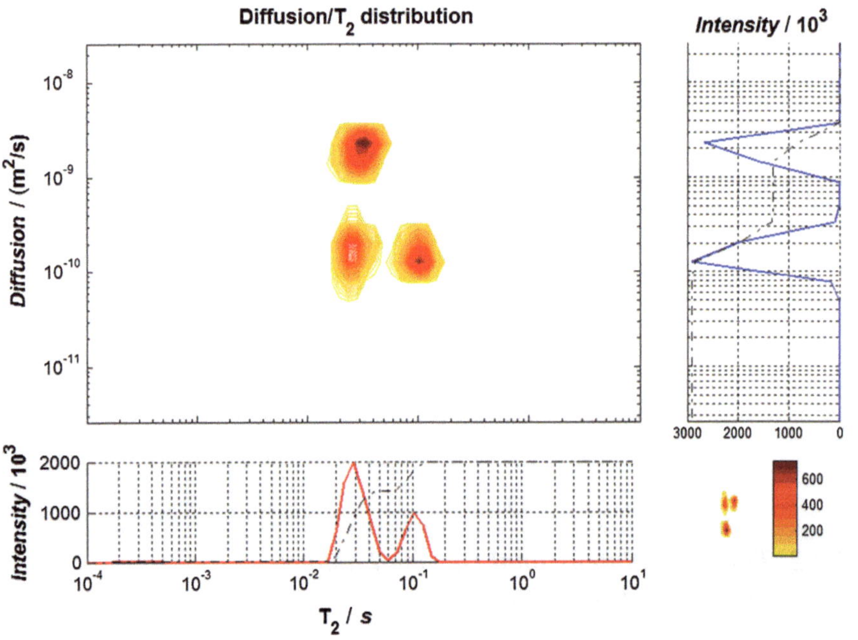

Fig. 5.6 The 2D-ILT distribution arising from crude oil and brine in a native state rock core plug

The first term in this equation is a standard least squares term, while the second term headed by $1/\alpha^2$ is a regularization term which adjusts the curvature of the solution in for example T_1-T_2 space. The first term inside the regularization term covers the second derivatives along the rows of **A**, while the second term inside the regularization term covers the second derivatives along the columns of **A**. Adjusting α will thus move the solution towards the optimum least squares solution (large α) or towards smoothing of the T_1-T_2 distribution (see Fig. 5.5). The actual implementation by Song et al. will be denoted the nonnegative least squares (NNLS) algorithm in the following algorithm comparison.

In Fig. 5.6, the results of applying the 2D-ILT transform using the NNLS algorithm on a dataset arising from a native state rock core plug containing crude oil and brine can be seen. To achieve this two dimensional plot it is important to set the regularization parameter according to Fig. 5.5, and to choose the region of interest, the grid, appropriately. Even though the 2D-ILT is said to return a distribution, this distribution consist of a set of discrete points, a mesh, on which the fitted values of the dynamic parameters T_1, T_2, and D must be found. If the true value of these parameters is somewhere in between the grid points, the 2D-ILT routine will compensate by distributing intensities that fall in non mesh points between adjacent points in the grid.

5.3 The Discrete Approach for 1- and 2-Dimensional Data Analysis, the Anahess

The Anahess approach for analysing 1- and 2-dimensional PFG NMR data is designed to find the smallest number of exponential components that fit the experimental data set. Assume an experiment which produces a second order matrix **R** having number of samples (*NSA*) rows and number of sensors (*NSE*) columns. Assume also that the measured response is the linear sum of *NCO* different components, each component decaying exponentially with both the row index and the column index. The row-wise exponential decay follows an axis g_i, $i = 1..NSA$, contained in a vector **g** of size *NSA*, while the column-wise exponential decay follows an axis t_j, $j = 1..NSE$, contained in a vector **t** of size *NSE*. Each component is assumed to exhibit a row-wise decay with a characteristic parameter $T_{1,p}$, $p = 1..NCO$, while the component's column-wise decay is assumed to have a characteristic parameter $T_{2,p}$, $p = 1..NCO$. The magnitude of each component is determined by a factor a_p, $p = 1..NCO$, while any permanent signal offset present is assigned to the variable a_0. The spacing between subsequent element in **g** and **t** may or may not be constant. Any element $R_{i,j}$, $i = 1..NSA, j = 1..NSE$ in **R** is then assumed to follow the function:

$$R_{i,j} = f(a_0, \mathbf{a}, \mathbf{T}_1, \mathbf{T}_2)_{i,j} + E_{i,j} \quad \text{with}$$
$$f(a_0, \mathbf{a}, \mathbf{T}_1, \mathbf{T}_2)_{i,j} = a_0 + \sum_{p}^{NCO} a_p e^{(-(1/T_{1,p})g_i - (1/T_{2,p})t_j)} \tag{5.14}$$

with a_p, $T_{1,p}$ and $T_{2,p}$ being the elements with the index $p = 1..NCO$ in the vectors **a**, $\mathbf{T}_1$ and $\mathbf{T}_2$, and subject to the following inequality constraints:

$$\begin{aligned} a_p &\geq 0, \quad p = 1..NCO \\ T_{1,p} &\geq 0, \ p = 1..NCO \\ T_{2,p} &\geq 0, \ p = 1..NCO \end{aligned} \tag{5.15}$$

a_0 is a baseline offset which may be positive or negative, and $E_{i,j}$ is the contribution from noise. The parameters a_p, $T_{1,p}$ and $T_{2,p}$ are thus characteristic properties of the component with index *p* out of all the *NCO* components. The objective is to perform a least squares fit of the function *f* to the data in the matrix **R**. This may be done by minimizing the function:

$$\begin{aligned} SS_{res} = F(a_0, \mathbf{a}, \mathbf{T}_1, \mathbf{T}_2) &= \sum_{i}^{NSA} \sum_{j}^{NSE} \{f(a_0, \mathbf{a}, \mathbf{T}_1, \mathbf{T}_2) - R_{i,j}\}^2 \\ &= \sum_{i}^{NSA} \sum_{j}^{NSE} \{a_0 + \sum_{p}^{NCO} a_p e^{(-(1/T_{1,p})g_i - (1/T_{2,p})t_j)} - R_{i,j}\}^2 \end{aligned} \tag{5.16}$$

Again, subject to the constraints given by the inequalities in (5.15). This problem is ill-posed, with very minute changes in the data leading to considerably different solutions, all of which may fit the data equally well.

5.3.1 An Inverse Hessian Algorithm

The minimization of the function given in (5.16) can be reformulated as an unconstrained minimization by introducing a new set of variables a_0, $\boldsymbol{\alpha}$, $\boldsymbol{\beta}$ and $\boldsymbol{\gamma}$ so that:

$$\begin{aligned} a_p &= \alpha_p^2, \quad p = 1..NCO \\ 1/T_{1,p} &= \beta_p^2, \; p = 1..NCO \\ 1/T_{2,p} &= \gamma_p^2, \; p = 1..NCO \end{aligned} \tag{5.17}$$

For all real values of a_0, $\boldsymbol{\alpha}$, $\boldsymbol{\beta}$ and $\boldsymbol{\gamma}$, minimizing $F(a_0, \boldsymbol{\alpha}, \boldsymbol{\beta}, \boldsymbol{\gamma})$ will yield a solution within the permitted region of $\mathbf{a}$, $\mathbf{T}_1$ and $\mathbf{T}_2$ which is defined by (5.15). If the variables a_0, $\boldsymbol{\alpha}$, $\boldsymbol{\beta}$ and $\boldsymbol{\gamma}$ are combined into one vector of variables $\mathbf{x}$ (of size $3NCO + 1$) as follows:

$$\mathbf{x} = \begin{bmatrix} a_0 \\ \boldsymbol{\alpha} \\ \boldsymbol{\beta} \\ \boldsymbol{\gamma} \end{bmatrix} \tag{5.18}$$

then an inverse Hessian (or Gauss-Newton) algorithm will be based on the equation:

$$\mathbf{x}_{k+1} = \mathbf{x}_k - \phi \mathbf{H}(\mathbf{x}_k)^{-1} \nabla(\mathbf{x}_k) \tag{5.19}$$

in which k is the index of the present iteration, $\mathbf{x}_k$ is the present position in the variable space, ϕ is the optimal scalar step length, $\mathbf{H}(\mathbf{x}_k)$ is the Hessian matrix of second derivatives of F at $\mathbf{x}_k$, and $\nabla(\mathbf{x}_k)$ is the gradient of F at $\mathbf{x}_k$. ϕ is found using a suitable univariate minimization algorithm. If the search direction of $-\mathbf{H}(\mathbf{x}_k)^{-1}\,\nabla(\mathbf{x}_k)$ is not a descending direction at the present location $\mathbf{x}_k$, a steepest descent search along $-\nabla(\mathbf{x}_k)$ is performed. See Press [7] for a detailed description of inverse Hessian algorithms. Analytical expressions of $\nabla(\mathbf{x}_k)$ and $\mathbf{H}(\mathbf{x}_k)$ are given in the next section.

5.3.2 Analytic Expressions of the Gradient and the Hessian Matrix

In order to simplify and clarify the expressions, the matrices $\mathbf{E}_p$, $p = 1..NCO$ and $\mathbf{M}$ are introduced. The matrix $\mathbf{M}$ as well as each of the $\mathbf{E}_p$ matrices has dimensions $NSA \times NSE$.

$$\begin{aligned} \mathrm{E_p} &= \{\mathrm{E_{i,j}}\}_\mathrm{p} = \{\mathrm{e}^{(-(1/\mathrm{T_{1,p}})\mathrm{g_i}-(1/\mathrm{T_{2,p}})\mathrm{t_j})}\}_\mathrm{p} = \{\mathrm{e}^{(-\beta_p^2 \mathrm{g_i}-\gamma_p^2 \mathrm{t_j})}\}_\mathrm{p} \\ \mathrm{M} &= \{\mathrm{M_{i,j}}\} = \{\mathrm{a_0} - \mathrm{R_{i,j}} + \sum_{\mathrm{p}}^{\mathrm{NCO}} \alpha_\mathrm{p}^2 \mathrm{e}^{(-\beta_p^2 \mathrm{g_i}-\gamma_p^2 \mathrm{t_j})}\} \\ &= 1_{\mathrm{NSA}\times 1} 1_{1\times\mathrm{NSE}} \mathrm{a_0} - \mathrm{R} + \sum_{\mathrm{p}}^{NCO} \alpha_\mathrm{p}^2 \mathrm{E_p} \end{aligned} \tag{5.20}$$

The function in (5.16) to be minimized can then be written:

$$F = \sum_{i}^{NSA} \sum_{j}^{NSE} \{a_0 - R_{i,j} + \sum_{p}^{NCO} \alpha_p^2 (E_p)_{i,j}\}^2 = \sum_{i}^{NSA} \sum_{j}^{NSE} M_{i,j}^2 \tag{5.21}$$

Let $\otimes$ denote element-by-element matrix or vector multiplication. Then, for all indices $p = 1..NCO$ and $r = 1..NCO$, define the following functions:

$$A_0 = 2 \cdot \sum_{i}^{NSA} \sum_{j}^{NSE} \mathbf{M} \qquad B_{0(p)} = 4 \cdot \alpha_p \cdot \sum_{i}^{NSA} \sum_{j}^{NSE} \mathbf{E}_p$$

$$A_{1(p)} = 4 \cdot tr(\mathbf{M}\mathbf{E}_p^{\mathrm{T}}) \qquad B_{1(p,r)} = 8 \cdot \alpha_p \alpha_r \cdot tr(\mathbf{E}_p^{\mathrm{T}} \mathbf{E}_r)$$

$$A_{2(p)} = -4 \cdot \alpha_p \cdot \mathbf{g}^{\mathrm{T}} diag(\mathbf{M}\mathbf{E}_p^{\mathrm{T}}) \qquad B_{2(p,r)} = -8 \cdot \alpha_p a_r \beta_r \cdot \mathbf{g}^{\mathrm{T}} diag(\mathbf{E}_p \mathbf{E}_r^{\mathrm{T}})$$

$$A_{3(p)} = -4 \cdot \alpha_p \cdot \mathbf{t}^{\mathrm{T}} diag(\mathbf{M}^{\mathrm{T}}\mathbf{E}_p) \qquad B_{3(p,r)} = -8 \cdot \alpha_p a_r \gamma_r \cdot \mathbf{t}^{\mathrm{T}} diag(\mathbf{E}_p^{\mathrm{T}} \mathbf{E}_r)$$

$$A_{4(p)} = -8 \cdot a_p \beta_p \beta_p \cdot (\mathbf{g} \otimes \mathbf{g})^{\mathrm{T}} diag(\mathbf{M}\mathbf{E}_p^{\mathrm{T}}) \qquad B_{4(p,r)} = 8 \cdot a_p a_r \beta_p \beta_r \cdot (\mathbf{g} \otimes \mathbf{g})^{\mathrm{T}} diag(\mathbf{E}_p \mathbf{E}_r^{\mathrm{T}})$$

$$A_{5(p)} = -8 \cdot a_p \gamma_p \gamma_p \cdot (\mathbf{t} \otimes \mathbf{t})^{\mathrm{T}} diag(\mathbf{M}^{\mathrm{T}}\mathbf{E}_p) \qquad B_{5(p,r)} = 8 \cdot a_p a_r \gamma_p \gamma_r \cdot (\mathbf{t} \otimes \mathbf{t})^{\mathrm{T}} diag(\mathbf{E}_p^{\mathrm{T}} \mathbf{E}_r)$$

$$A_{6(p)} = 8 \cdot a_p \beta_p \gamma_p \cdot \mathbf{g}^{\mathrm{T}} (\mathbf{M} \otimes \mathbf{E}_p)\mathbf{t} \qquad B_{6(p,r)} = 8 \cdot a_p a_r \beta_p \gamma_r \cdot \mathbf{g}^{\mathrm{T}} (\mathbf{E}_p \otimes \mathbf{E}_r)\mathbf{t}$$

$$C_{0(p)} = 2 \cdot NSA \cdot NSE$$

$$C_{1(p)} = -4 \cdot a_p \beta_p \cdot \sum_{j}^{NSE} \mathbf{g}^{\mathrm{T}} \mathbf{E}_p \qquad C_{2(p)} = -4 \cdot a_p \gamma_p \cdot \sum_{i}^{NSA} \mathbf{E}_p \mathbf{t} \tag{5.22}$$

With the above functions defined, the gradient ∇ and the Hessian matrix $\mathbf{H}$ may be calculated as follows:

$$\nabla = \begin{array}{|l|} \hline \dfrac{\partial F}{\partial a_0} = A_0 \\ \hline \dfrac{\partial F}{\partial \alpha_p} = \alpha_p A_{1(p)} \\ \hline \dfrac{\partial F}{\partial \beta_p} = \alpha_p \beta_p A_{2(p)} \\ \hline \dfrac{\partial F}{\partial \gamma_p} = \alpha_p \gamma_p A_{3(p)} \\ \hline \end{array} \tag{5.23}$$

$$\mathbf{H} = \begin{array}{|c|c|c|c|} \hline \dfrac{\partial^2 F}{\partial a_0 \partial a_0} = C_0 & \dfrac{\partial^2 F}{\partial a_0 \partial \alpha_p} = B_{0(p)} & \dfrac{\partial^2 F}{\partial a_0 \partial \beta_p} = C_{1(p)} & \dfrac{\partial^2 F}{\partial a_0 \partial \gamma_p} = C_{2(p)} \\ \hline \dfrac{\partial^2 F}{\partial \alpha_p \partial a_0} = \dfrac{\partial^2 F}{\partial a_0 \partial \alpha_p} & \begin{array}{c} \dfrac{\partial^2 F}{\partial \alpha_p \partial \alpha_p} = B_{1(p,p)} \\ + A_{1(p)} \\ \dfrac{\partial^2 F}{\partial \alpha_p \partial \alpha_{r,p \neq r}} = B_{1(p,r)} \end{array} & \begin{array}{c} \dfrac{\partial^2 F}{\partial \alpha_p \partial \beta_p} = B_{2(p,p)} \\ + 2\beta_p A_{2(p)} \\ \dfrac{\partial^2 F}{\partial \alpha_p \partial \beta_{r,p \neq r}} = B_{2(p,r)} \end{array} & \begin{array}{c} \dfrac{\partial^2 F}{\partial \alpha_p \partial \gamma_p} = B_{3(p,p)} \\ + 2\gamma_p A_{3(p)} \\ \dfrac{\partial^2 F}{\partial \alpha_p \partial \gamma_{r,p \neq r}} = B_{3(p,r)} \end{array} \\ \hline \dfrac{\partial^2 F}{\partial \beta_p \partial a_0} = \dfrac{\partial^2 F}{\partial a_0 \partial \beta_p} & \dfrac{\partial^2 F}{\partial \beta_p \partial \alpha_r} = \dfrac{\partial^2 F}{\partial a_r \partial \beta_p} & \begin{array}{c} \dfrac{\partial^2 F}{\partial \beta_p \partial \beta_p} = B_{4(p,p)} \\ + \alpha_p A_{2(p)} - A_{4(p)} \\ \dfrac{\partial^2 F}{\partial \beta_p \partial \beta_{r,p \neq r}} = B_{4(p,r)} \end{array} & \begin{array}{c} \dfrac{\partial^2 F}{\partial \beta_p \partial \gamma_p} = B_{6(p,p)} \\ + A_{6(p)} \\ \dfrac{\partial^2 F}{\partial \beta_p \partial \gamma_{p,p \neq r}} = B_{6(p,r)} \end{array} \\ \hline \dfrac{\partial^2 F}{\partial \gamma_p \partial a_0} = \dfrac{\partial^2 F}{\partial a_0 \partial \gamma_p} & \dfrac{\partial^2 F}{\partial \gamma_p \partial \alpha_r} = \dfrac{\partial^2 F}{\partial \alpha_r \partial \gamma_p} & \dfrac{\partial^2 F}{\partial \gamma_p \partial \beta_r} = \dfrac{\partial^2 F}{\partial \beta_r \partial \gamma_p} & \begin{array}{c} \dfrac{\partial^2 F}{\partial \gamma_p \partial \gamma_p} = B_{5(p,p)} \\ + \alpha_p A_{3(p)} - A_{5(p)} \\ \dfrac{\partial^2 F}{\partial \gamma_p \partial \gamma_{r,p \neq r}} = B_{5(p,r)} \end{array} \\ \hline \end{array} \tag{5.24}$$

Note that ∇ has size $3NCO + 1$ while $\mathbf{H}$ has size $(3NCO + 1) \times (3NCO + 1)$. If $\mathbf{H}$ is not positive definite, then a diagonal matrix $\mathbf{D}$ is calculated and added to $\mathbf{H}$ in order to make $\mathbf{H}$ positive definite. The calculation of $\mathbf{D}$ is performed using a square root free Cholesky factorization [21] developed by Borchers [22].

5.3.3 The Univariate Minimization

Each iteration in an inverse Hessian algorithm requires a univariate minimization along the search direction of $-\mathbf{H}(\mathbf{x}_k)^{-1}\ \nabla(\mathbf{x}_k)$. In the present Matlab implementation of Anahess, this is performed using the routine *fminbnd*, which is based on golden section search and parabolic interpolation [7]. In the present implementation, the optimal step length ϕ is always somewhere between a minimum of 10^{-20} and a maximum of 20. In addition, if the solution strays outside of the interval defined by the minima and maxima of $\mathbf{g}$ and $\mathbf{t}$, F is assigned a very high value.

5.3.4 Choice of the Initial Estimate for One Specific Choice of NCO

Kaufmann [14] has developed a non-iterative algorithm for the fitting of a sum of exponentials to first order data. The goal of the first order sum of exponentials fit is to fit a sum of exponentials to a vector $\mathbf{r}$ so that any element r_i, $i = 1..N$ in $\mathbf{r}$ follows the function:

$$
\begin{aligned}
r_i &= f(a_0, \mathbf{a}, \mathbf{b})_i + \mathrm{e}_i \quad \text{with} \\
f(a_0, \mathbf{a}, \mathbf{b})_i &= a_0 + \sum_{p}^{NCO} a_p \mathrm{e}^{b_p v_i}
\end{aligned}
\tag{5.25}
$$

in which v_i is the element having index $i = 1..N$ in the vector $\mathbf{v}$. Applying Kaufmann's algorithm on each row in R provides an estimate of ${\gamma_p}^2 = -b_p$, $p = 1..NCO$. Likewise, Kaufmann's algorithm applied on each column provides an estimate of ${\beta_p}^2 = -b_p$, $p = 1\ldots NCO$. Once the ${\beta_p}^2$ and ${\gamma_p}^2$ estimates are available, estimates of a_0 and a_p, $p = 1\ldots NCO$ may then be found using the method of linear least squares. If a 2D-ILT calculation with *NCO* components is attempted, and the results from a calculation on the same data set with *NCO*-1 components are available, then the *NCO*-1 calculation results are compared to the initial estimates from Kaufmann's algorithm, and taken into account when the *NCO* calculation is started.

5.3.5 Choice of Number of Components

A good criterion for choosing the correct number of components *NCO* depends on the descriptive power of the model and the parsimony of the model. The descriptive power of a model is the accuracy with which the full information content in

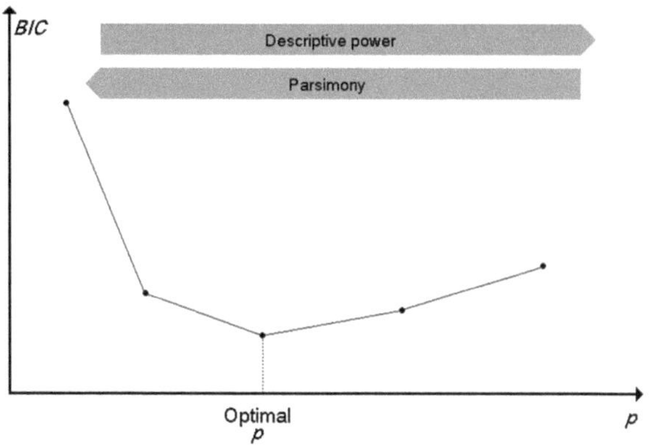

Fig. 5.7 The Bayesian information criterion and the trade-off between descriptive power and parsimony

the data is modelled. In this case, the descriptive power rests on the model's ability to minimize the sum of squared residuals, while the parsimony is the model's dependence on a low number of free parameters. When dealing with a demanding modelling problem, a model's descriptive power and the parsimony typically pull in opposing directions. Improving one property usually diminishes the other.

One criterion for choosing a model is the Bayesian information criterion (*BIC*), introduced by Schwartz [23]. Let n be the number of observed data points, let p be the number of free model parameters, and ss_{res} be the sum of squared residuals. Then if the residuals are normally distributed, *BIC* has the form:

$$BIC = n\ln(\frac{ss_{res}}{n}) + p\ln(n) \tag{5.26}$$

In this equation, a good model fit gives a low first term while few model parameters give a low second term. When comparing a set of models, the model with the minimal *BIC* value is selected. The overall concept is illustrated in Fig. 5.7.

An alternative but closely related information criterion is the Akaike information criterion (*AIC*), developed by Akaike [24]. Given the assumption that the residuals are normally distributed, *AIC* has the form:

$$AIC = n\ln(\frac{ss_{res}}{n}) + 2p \tag{5.27}$$

AIC is based on information theory, and the underlying Bayesian framework is very similar to the framework for *BIC*. *AIC* penalizes a high number of free model parameters p less severely than *BIC* does. In practice, *AIC* should be corrected for finite sample size (finite size of n). The corrected Akaike information criterion is [25]:

$$AICc = n\ln(\frac{ss_{res}}{n}) + 2p + \frac{2p(p+1)}{n-p-1} \tag{5.28}$$

AICc tends to penalize a large value of p less than *BIC*, but more than *AIC*. In the context of ANAHESS, n is $NSA \times NSE$, SS_{res} is given in (5.16), and p is equal to $3NCO + 1$, since a_0, $\mathbf{a}$, $\mathbf{T}_1$ and $\mathbf{T}_2$ are the free parameters in the model.

In Anahess, *BIC* and *AICc* are calculated for each successive value of *NCO*, and the model with the minimal *BIC* or the minimal *AICc* is chosen as the final model.

5.3.6 Summary of Anahess

The elements outlined above may now be combined into the following algorithm:

I. Acquire the necessary input data: **R**, **g** and **t**.
II. Choose a suitable maximum number of components $NCO_{\max}$.
III. For $NCO = 1$ to $NCO_{\max}$ do (component loop):

   (a) Find an initial estimate $\mathbf{x}_0$ using Kaufmann's algorithm along the rows and columns of **R**.
   (b) If $NCO > 1$ then compare $\mathbf{x}_0$ to the results from the previous calculation with *NCO*-1 components, and use the results to improve upon $\mathbf{x}_0$ if possible. While no convergence in F and $\mathbf{x}_k$ perform (inverse Hessian loop):

      (A) Calculate the various functions given by (5.22).
      (B) Calculate $\nabla(\mathbf{x}_k)$ according to (5.23).
      (C) Calculate $\mathbf{H}(\mathbf{x}_k)$ according to (5.24).
      (D) Check **H** for positive definiteness. Make **H** positive definite if necessary.
      (E) Perform a univariate optimization of ϕ in (5.19), and move from $\mathbf{x}_k$ to $\mathbf{x}_{k+1}$.
      (F) Check for convergence in F and $\mathbf{x}_k$.

   (c) End inverse Hessian loop.
   (d) Extract **a**, $\mathbf{T}_1$ and $\mathbf{T}_2$ according to (5.17) and (5.18), and store *NCO*, a_0, **a**, $\mathbf{T}_1$ and $\mathbf{T}_2$.
   (e) Calculate *BIC* and *AICc* for the given *NCO*.

IV. End component loop.

On completion, this algorithm produces a set of 2D-ILT solutions with *NCO* ranging from 1 to $NCO_{\max}$, with one solution for each *NCO* value. The solution with the minimal *BIC* or the minimum *AICc* is finally chosen as the optimal Anahess solution.

5.3.7 Anahess in the 1-Dimensional Case

In the 1-dimensional case, (5.14) takes on the simplified form (if the experiment is a T_2 relaxation experiment):

$$R_j = f(a_0, \mathbf{a}, \mathbf{T}_2)_j + E_j \quad \text{with}$$

$$f(a_0, \mathbf{a}, \mathbf{T}_2)_j = a_0 + \sum_p^{NCO} a_p e^{-(1/T_{2,p})t_j} \tag{5.29}$$

All previous descriptions of the Anahess algorithm still hold if we insert the following assumptions:

$$NSA = 1$$
$$\mathbf{T}_1 = \infty \quad \text{and thus } \boldsymbol{\beta} = \mathbf{0}$$

Equation 5.16 is then simplified to:

$$\begin{aligned} SS_{res} = F(a_0, \mathbf{a}, \mathbf{T}_2) &= \sum_{j}^{NSE} \{f(a_0, \mathbf{a}, \mathbf{T}_2) - R_j\}^2 \\ &= \sum_{j}^{NSE} \{a_0 + \sum_{p}^{NCO} a_p \mathrm{e}^{-(1/T_{2,p})t_j} - R_j\}^2 \end{aligned} \tag{5.30}$$

The vector of variables $\mathbf{x}$ now has the size $2NCO + 1$, and has the structure:

$$\mathbf{x} = \begin{array}{|c|} \hline a_0 \\ \hline \boldsymbol{\alpha} \\ \hline \boldsymbol{\gamma} \\ \hline \end{array} \tag{5.31}$$

The equations for $\mathbf{E}$ and $\mathbf{M}$ are now:

$$\begin{aligned} \mathbf{E}_p &= \{E_{i,j}\}_p = \{\mathrm{e}^{(-(1/T_{2,p})t_j)}\}_p = \{\mathrm{e}^{(-\gamma_p^2 t_j)}\}_p \\ \mathbf{M} &= \{M_{i,j}\} = \{a_0 - R_{i,j} + \sum_{p}^{NCO} \alpha_p^2 \mathrm{e}^{(-\gamma_p^2 t_j)}\} \\ &= \mathbf{1}_{NSA\times 1}\mathbf{1}_{1\times NSE}\, a_0 - \mathbf{R} + \sum_{p}^{NCO} \alpha_p^2\, \mathbf{E}_p \end{aligned} \tag{5.32}$$

Equation 5.21 then becomes:

$$F = \sum_{j}^{NSE} \{a_0 - R_j + \sum_{p}^{NCO} \alpha_p^2 (E_p)_j\}^2 = \sum_{j}^{NSE} M_j^2 \tag{5.33}$$

Given these modifications, the previous 5.22–5.24 for the calculation of the gradient and the Hessian matrix still hold.

5.4 Comparison Between 1D-ILT and Anahess

In order to compare the performance of 1D-ILT and Anahess algorithms, CPMG experiments were performed on crude oil, water in oil emulsion and tap water. The water is likely to result in a close to a mono exponential decay, as it is a well-defined single component system. For the emulsion, we would expect a

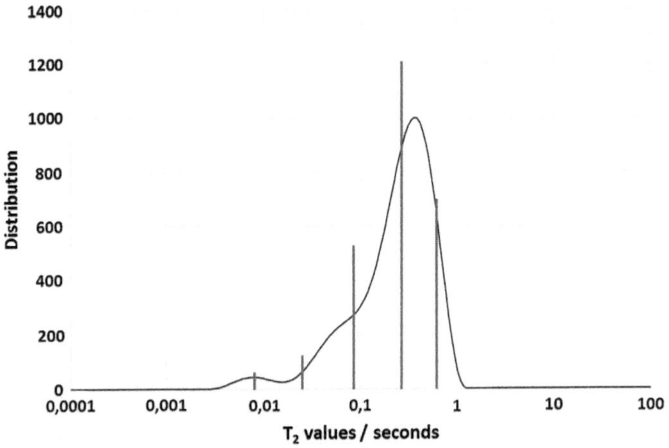

Fig. 5.8 Comparison of T_2 distribution of a sample of crude oil. The fitting methods are the 1D-ILT for the continuous distribution and Anahess giving a discrete set of T_2 values

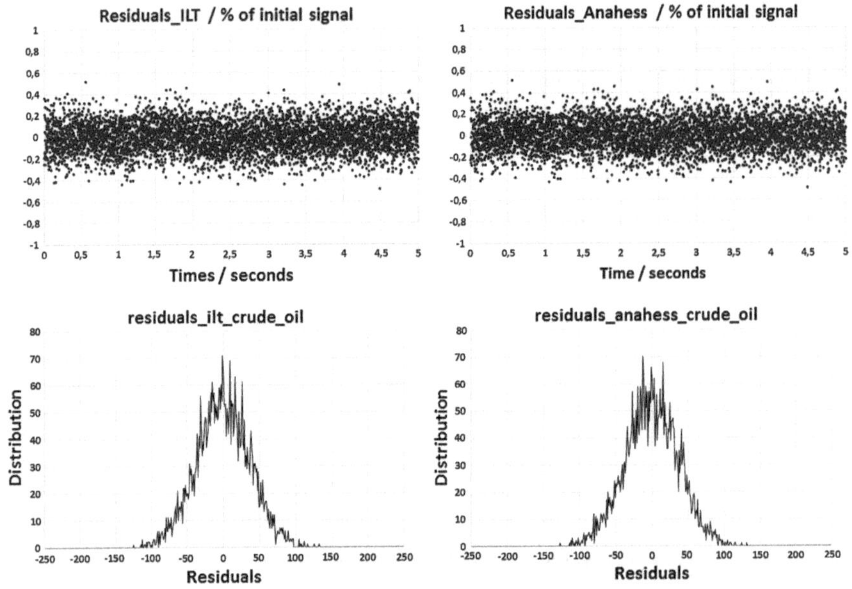

Fig. 5.9 The residuals and their distributions arising from two different fitting methods applied on the CPMG-data from crude oil

distribution from two phases, either completely separated or partly overlapping in T_2 values. The T_2 distribution of the brine in the emulsion is broadened and moved towards shorter T_2 values as compared to the bulk water. For the crude oil it is already argued that there ought to be a continuous distribution of T_2 values (see Figs. 5.3 and 5.5).

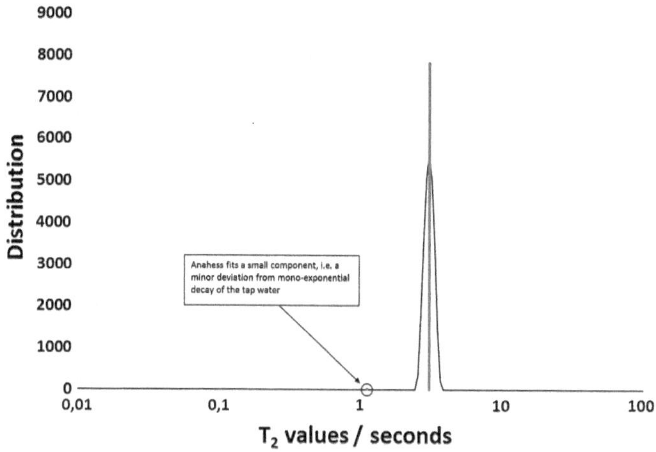

Fig. 5.10 Comparison of T_2 distribution of a sample of tap water. The fitting methods are the 1D-ILT for the continuous distribution and Anahess giving a discrete set of T_2 values

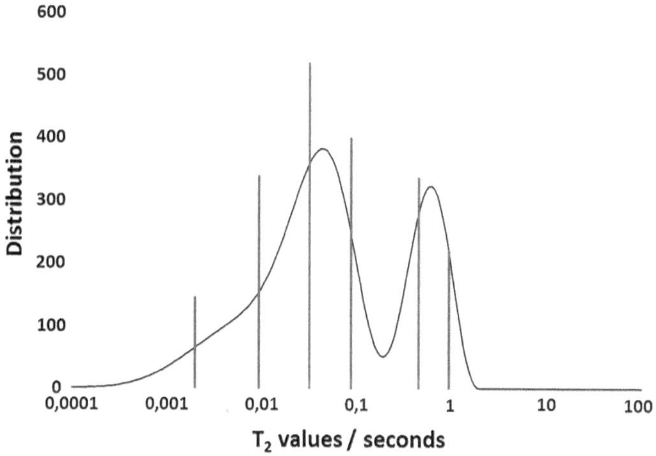

Fig. 5.11 Comparison of T_2 distribution of a sample of brine in crude oil emulsion. The fitting methods are the 1D-ILT for the continuous distribution and Anahess giving a discrete set of T_2 values

As seen in Figs. 5.8 and 5.9 both processing methods fit well to the experimental data. While the 1D-ILT returns 50 components of finite values and as a continuous distribution, Anahess fits the data to 5 discrete components only. For the crude oil we know that the Anahess result does not reflect the nature of the sample, a continuum of molecular sizes. Thus the 1D-ILT is the definitely the choice for being closest to representing the true T_2-values of the system, provided the regularization parameter is chosen correctly.

In Fig. 5.10 the results are shown for 1D-ILT and Anahess performed on CPMG data arising from tap water. As a mono exponential decay is expected, Anahess fits one component to 99.7 % of the signal, and one small (0.3 %) that results from a minor deviation from mono exponential behaviour. 1D-ILT spreads the fit across 9 components to fit the CPMG-data into a narrow distribution. Consequently, the Anahess algorithm proves to be the best representation of the true T_2-values to be expected from of the single component system, despite the fact that both models produce good fits.

Figure 5.11 shows the fitted T_2 distribution (1D-ILT) and a discrete set of T_2 values (Anahess) for a brine in oil emulsion. The brine droplets are distributed around 5 μm, and thus it is expected that the water phase relaxation data will be multi-exponential in T_2, as the T_2 values are proportional to the droplet size (Chap. 2). Again both methods applied fit the experimental data well, and once again, the 1D-ILT seems to provide the solution closest to the anticipated T_2 values. Another factor in favour of the 1D-ILT method for this data set is the ability to detect a local minimum around 0.2 s in the distribution. As the brine droplets are rather small, the T_2 values have decreased to shorter values such that that the distribution peaks from brine and crude oil partly overlaps (T_2 is proportional to the droplet radius [26]). Even though there is a larger gap between the discrete components separating the crude oil and brine signal, there is no obvious way to distinguish the crude oil and brine using the Anahess fit. It does not resolve the local minimum as 1D-ILT does.

Provided an appropriate choice of the regularization/smoothing parameter, the 1D-ILT yield the processed data that can be of best use for the largest variety of systems to be investigated by NMR. These are systems best described by a continuum of T_2 values with or without local minima or separated distributions, for example water in oils emulsions, crude oils, and fluid saturated porous rock core plugs. Exceptions are the systems that consist mainly of single discrete components, or where the initial decay is the one of greatest interest. The latter is most important when measuring diffusion in heterogeneous media, attempting to extract true diffusivity unaffected by artefacts.

5.5 Comparison Between 2D-ILT and Anahess

In this section a comparison of the robustness and reliability of the two different approaches for processing two-dimensional data, the NNLS and Anahess, is performed (Fig. 5.12). Based on the results from the one-dimensional data one may ask whether there is any no point in comparing the ILT routine in two dimensions against Anahess, as it would be expected that the distributions being continuous in one dimension are likely to be continuous in two dimensions as well. In this comparison, data sets where the distributions are known a priori will be used, i.e. synthetic data sets. This avoids complicating factors such as experimental/instrumental artefacts affecting the data and thus influencing the analysis. For all the generated synthetic data sets, randomly distributed noise has been applied where the half width of the noise is 2.75 in absolute values. The target values used to

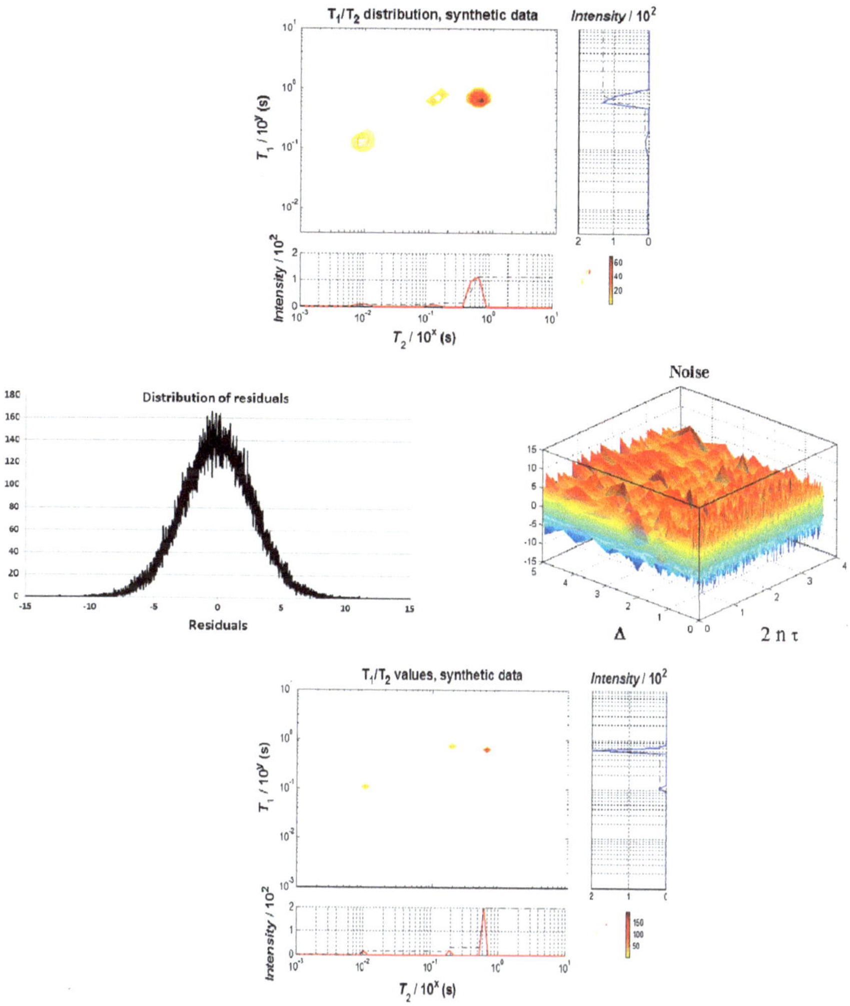

Fig. 5.12 Results from 2D-ILT (*upper distribution*) and Anahess (*lower distribution*) processing of synthetic data, and a representative visualization of the distribution of their residuals

generate the synthetic data are provided in the tables along with the fitted area/intensities of the peaks/components and their corresponding relaxation times, both from the NNLS and Anahess procedures.

There is one fundamental difference when processing the data either using NNSL or Anahess; the input required by the user of the routines. Assuming the correct kernel is used for processing the acquired data, the user has to specify the following parameters before running the NNLS-algorithm

- Define the mesh or grid in which the dynamic parameters will be fitted to. This involves setting the boundaries where the values of the dynamic parameters are expected to appear and number of values within this region. For example for a

T_1-T_2 experiment on a meat sample at 30 °C, one could set the region both in T_1 and T_2 to $[10^{-4}$–$10^1]$ s and one could specify 64 values in each direction. Then the grid points, the values for the dynamic parameters that the NNSL can fit to, are fixed.

- The regularization parameter α. After having specified the region of the grid correctly, an appropriate value for the smoothing needs to be specified. This can be done as described before for the 1-dimensional case (Fig. 5.5).

Even if these values are set in accordance with what one expects to find for the dynamic parameters, some adjustments of the grid may be required in order to achieve a fit from the NNLS that returns residuals that are randomly distributed. Thus, it is important to calculate the surface of the residuals in the two dimensions in order to judge whether the fitting parameters are appropriate. In the NNLS processing of the synthetic data, the grid and α has been varied such that the result is as close to the target values, both in intensities, T_1, and T_2's. For each synthetic data set fitted by NNLS the algorithm has been applied approximately 20 times; 10 times to establish the correct value for α and 10 times to optimize both the α value and the grid to provide randomly distributed residuals and values as close to the target values as possible within this number of runs of the algorithm. In reality the user could vary the α- and grid values to achieve even closer correspondence to the target values, but this is normally not practical, and identifies a key issue with applying the NNLS algorithm, namely that the result are dependent on the start/input conditions to the algorithm and to some degree are subjective based on the user input.

For the Anahess algorithm, the situation differs. Provided the appropriate kernel is being used in the algorithm, there is no more subjective input from the user required. After loading the data into the algorithm and starting it according to Sect. 5.3, the Anahess fits the data to a number of components that are in accordance with the stop criteria, the minimization of the BIC and/or AIC numbers. Then the output is the component intensities and their corresponding pairs of the dynamic parameters Diffusion, T_1, and/or T_2, for example pairs of (T_1, T_2) for each component. What is important to note is that the fitted dynamic parameters do not have to be placed on a grid of fixed points, but are allowed to take on any value that best fits to the data set.

In the following NNSL and Anahess will be used to fit synthetic data sets, which consist both of data of discrete points and of a distribution of points on a grid. The kernel of the data set is given as

$$I = \sum_{i,j} I_{i,j} \mathrm{e}^{-\frac{t_1}{T_1^i}} \mathrm{e}^{-\frac{t_2}{T_2^j}} \tag{5.34}$$

where the 32 Δ values and the 3,000 $2n\tau$-values spans the observation time in both dimensions until there is only noise left in the data at the longest observation times.

The NNLS algorithm used in this section was bought for research purposes by the University of Science and Technology (NTNU, Norway) from Professor

Paul Callaghan's group at Massey University (New Zealand) in 2003. It runs on Matlab and is based on the work done by Song et al. [20]. The routine for running Anahess is an in house routine (www.antek.no) that runs as a stand-alone application.

5.5.1 Three Discrete Components

In Table 5.2 the results are displayed from the fit of the synthetic data consisting of three target components of varying intensities. There are eight synthetic data sets where the target values of the (T_1, T_2) are the same, but where the intensities of individual components have values of either 20 or 200. This provides eight combinations for intensities for the three components, for which one processed result is displayed in Fig. 5.13 (Anahess) and one in Fig. 5.14 (NNLS). This combination had a target intensity of 20 for the two components with shortest T_2 and 200 for one with longest T_2.

Regarding the results using the Anahess algorithm the fits are considered as good, as there is little deviation from the target values for all fits (see Tables 5.2 5.3). When running the algorithm, it starts with one component, fits the data to one component and calculates the stop criteria (BIC and AIC numbers). Then it proceeds to two components and continues until the BIC and/or AIC have reached

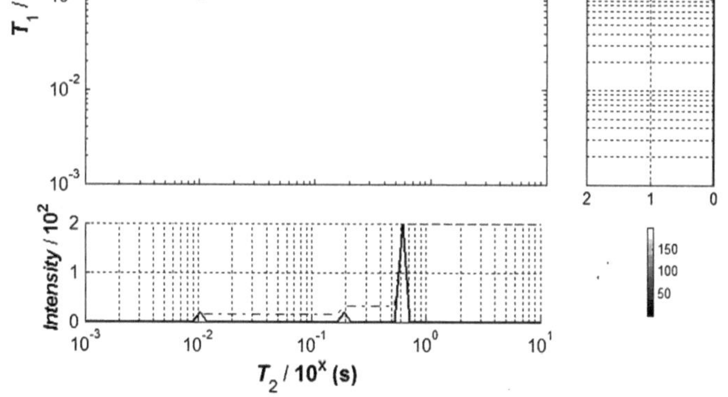

Fig. 5.13 The Anahess result for Run #2 in Table 5.2, represented as singular peaks in a log-plot

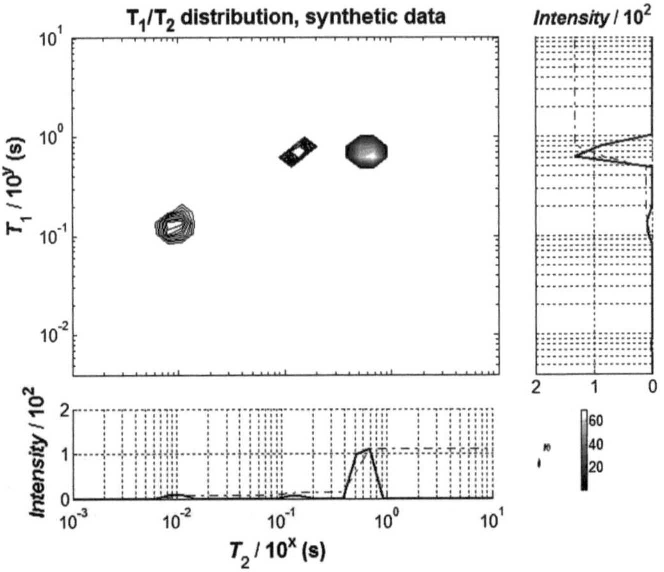

Fig. 5.14 The NNLS results for Run #2 in Table 5.2

their minima. For all the eight data sets, the stop criteria returned three components as the best fit of the data, which is consistent with the number of components in the target data. In Table 5.1, the BIC and AIC numbers are shown together with the sum of the squared residuals up to 4 components, where the routine halts as the minimum is reached. While the sum of the squared residuals continue to decrease with a relatively small amount, the BIC and AIC-numbers again reach the minimum at a number of components equal to 3.

For the Anahess processing it can be seen from Table 5.3 that the largest errors are found for the low intensity components, which is not unexpected due to the added noise in the synthetic data. It can also be seen that the relative deviation in T_1-values is generally larger than the deviation found for the T_2-values, which is a result of the less number of measurement points for T_1(32) as compared to T_2(3,000). The Anahess algorithm appears to have no difficulties resolving the two components that are relatively close in pairs of (T_1, T_2), (689.8, 193.1) and (689.8, 621.8) values. Even though the T_1's are identical and the T_2's only 3.2 times different, both the values for the intensities and the T_1's and T_2's are determined quite

Table 5.1 BIC and AIC numbers as a function of components fitted

Number of components fitted	BIC-number	AIC-number	Sum of squared residuals
1	202,555.2	202,517.3	791,413.1
2	194,938.2	194,871.9	730,784.1
3	193,733.8	193,639.1	721,414.5
4	193,768.0	193,644.8	721,412.4

close to the target values. Altogether, Anahess returns results that are in agreement with the target values, with the only subjective processing input required by the user being the use of the correct kernel.

The results from the NNLS processing were achieved by varying the number of grid points and the region of interest such that the distribution of residuals was Gaussian (see Fig. 5.12) and its values randomly distributed around 0. When running the NNLS algorithm focus was on producing results where the peak areas and average T_1's and T_2's were as close to the target values as possible. Some trial and error adjustment was necessary to achieve this, and it led to approximately 10 extra or more runs of the NNLS algorithm. The distributions representing the discrete components were constructed from 5 to 8 grid points, depending on the final settings for the user defined parameters in the NNLS routine. As can be seen from Tables 5.2 and 5.3, NNLS provides good fits for most of the data. However, there are some systematic discrepancies. The peak area for the low resolution signals (5.20) are less accurately fitted, especially for the component with the shortest (T_1, T_2) (Run # 1, 2, 3 and 4) and when there is a high resolution signal (200) present

Table 5.2 Fitted I_0, T_1 and T_2 using Anahess and NNLS, and the corresponding target values

Run #	Anahess I_0	T_1 (ms)	T_2 (ms)	NNLS I_0	T_1 (ms)	T_2 (ms)	Target I_0	T_1 (ms)	T_2 (ms)
1	20.4	726.5	197.3	18.8	730	192.8	20	689.8	193.1
	19.6	669.5	625.5	20.2	687.4	613.7	20	689.8	621.8
	19.4	121.6	10.4	16.8	127	11	20	116	10.4
2	20.4	727.6	197.3	11.5	705.4	135.1	20	689.8	193.1
	199.6	687.7	622.2	209.1	693.8	607.1	200	689.8	621.8
	19.4	121.5	10.4	17.9	128.8	9.4	20	116	10.4
3	200.4	693.7	193.5	194	695	191.2	200	689.8	193.1
	199.6	687.6	622.2	207.5	693	608.2	200	689.8	621.8
	19.4	121.3	10.4	17.7	124.7	8.5	20	116	10.4
4	200.4	693.6	193.5	202.6	697	196.2	200	689.8	193.1
	19.6	669.1	625.5	16	663.3	671.8	20	689.8	621.8
	19.4	121.4	10.4	18	126.5	9.4	20	116	10.4
5	199.4	116.5	10.4	200.2	118	10.4	200	116	10.4
	19.6	669.5	625.5	19.3	667.8	629	20	689.8	621.8
	20.4	726.6	197.3	20.7	733.5	202.3	20	689.8	193.1
6	199.4	116.5	10.4	191.2	120	10.2	200	116	10.4
	199.6	687.7	622.2	209	693.9	607.2	200	689.8	621.8
	20.4	727.7	197.3	11.5	702.9	139.8	20	689.8	193.1
7	199.1	116.5	10.4	199.1	117.8	10.3	200	116	10.4
	19.6	669.1	625.5	16.3	657.8	670	20	689.8	621.8
	200.4	693.6	193.5	204.3	697	196.9	200	689.8	193.1
8	199.4	116.5	10.4	198.7	117.7	10	200	116	10.4
	199.6	687.6	622.2	207.1	695.1	608.6	200	689.8	621.8
	200.4	693.7	193.5	194.1	695.1	191.4	200	689.8	193.1

Table 5.3 Deviations of fitted I_0, T_1 and T_2 using Anahess and NNLS

Run #	Anahess ΔI_0 (%)	ΔT_1 (%)	ΔT_2 (%)	NNLS ΔI_0 (%)	ΔT_1 (%)	ΔT_2 (%)	Target I_0	T_1 (ms)	T_2 (ms)
1	2.00	5.29	2.18	−6.0	5.8	−0.2	20	689.8	193.1
	−2.00	−2.97	0.60	1.0	−0.4	−1.3	20	689.8	621.8
	−3.00	4.83	0.00	−16.0	9.5	5.8	20	116	10.4
2	2.00	5.48	2.18	−42.5	2.3	−30.0	20	689.8	193.1
	−0.20	−0.30	0.06	4.6	0.6	−2.4	200	689.8	621.8
	−3.00	4.74	0.00	−10.5	11.0	−9.6	20	116	10.4
3	0.20	0.57	0.21	−3.0	0.8	−1.0	200	689.8	193.1
	−0.20	−0.32	0.06	3.8	0.5	−2.2	200	689.8	621.8
	−3.00	4.57	0.00	−11.5	7.5	−18.3	20	116	10.4
4	0.20	0.55	0.21	1.3	1.0	1.6	200	689.8	193.1
	−2.00	−3.00	0.60	−20.0	−3.8	8.0	20	689.8	621.8
	−3.00	4.66	0.00	−10.0	9.1	−9.6	20	116	10.4
5	−0.30	0.43	0.00	0.1	1.7	0.0	200	116	10.4
	−2.00	−2.94	0.60	−3.5	−3.2	1.2	20	689.8	621.8
	2.00	5.33	2.18	3.5	6.3	4.8	20	689.8	193.1
6	−0.30	0.43	0.00	−4.4	3.4	−1.9	200	116	10.4
	−0.20	−0.30	0.06	4.5	0.6	−2.3	200	689.8	621.8
	2.00	5.49	2.18	−42.5	1.9	−27.6	20	689.8	193.1
7	−0.45	0.43	0.00	−0.5	1.6	−1.0	200	116	10.4
	−2.00	−3.00	0.60	−18.5	−4.6	7.8	20	689.8	621.8
	0.20	0.55	0.21	2.2	1.0	2.0	200	689.8	193.1
8	−0.30	0.43	0.00	−0.7	1.5	−3.8	200	116	10.4
	−0.20	−0.32	0.06	3.6	0.8	−2.1	200	689.8	621.8
	0.20	0.57	0.21	−3.0	0.8	−0.9	200	689.8	193.1

together with the low resolution signal in the pairs of (T_1, T_2) closest to each other (Run # 2, 4, 6 and 7). Regarding the average T_1- and T_2-values for the peaks, they are in good agreement with the target values, but again with the same discrepancies as for the peak areas. Thus, it may be concluded that the NNSL has difficulties with fitting component with shortest (T_1, T_2) when its target intensity is low, i.e. an absolute value of 20. In addition, the NNLS has difficulties resolving the low resolution component from a high resolution component when they are close in pairs of (T_1, T_2). When optimizing the settings for the NNLS, a set of grid points was used where the target values for the pairs of (T_1, T_2) were present. This allowed the NNLS algorithm the optimum chance of arriving at the best solution. However, it turned out that the best fit was achieved when the target values were not on the grid points, and for all data presented in Table 5.2, none of the grid points match the target values.

When comparing the fitted results from Anahess and NNLS on a synthetic data set containing three discrete components, robustness and reliability is clearly in the favour of Anahess. Anahess reliably produces a three component fit and there the dynamic parameters established are found to be close to the target values. NNLS also finds most of the components correctly (although they are represented as distributions), but with some discrepancies where certain combinations of the target values makes it difficult to find good fits for them. A priori knowledge is necessary to achieve the results found in Tables 5.2 and 5.3 for the NNLS. The challenge with NNLS is that there exist so many solutions, and not all solutions produce randomly distributed residuals, even though the user defined input parameters are in the region where the solution is the least sensitive to noise. Thus, a proper use of NNLS requires a user that is experienced in handling two-dimensional data and has some advance knowledge of the data output expected.

The synthetic data were constructed to check the robustness of the two processing algorithms and how they behave when components are close to each other or when there are one or more components of low resolution. The results were not known on beforehand and the components intensities or pairs of (T_1, T_2) were not altered to favour either of the methods. The NNLS algorithm was favoured to some degree by applying the knowledge of what to look for, i.e. the target values, in order to achieve the best fit within 10 runs. This additional information was not necessary using Anahess. The next step is to compare a set of synthetic data where the NNLS algorithm would be expected to perform much better, namely a distribution of intensities rather than a set of discrete exponential components.

5.5.2 Distribution with One Peak

The synthetic dataset providing a distribution with one peak consisted of a peak constructed from 13 discrete points where the intensities were assigned symmetrical around the maximum peak intensity, as shown in Fig. 5.17. The user ran the Anahess routine without having any influence on the output while several user defined runs were made of NNSL to find the fit closest to the target values. The results for the fitted peak areas and average (T_1, T_2) are given in Table 5.4 while the distributions are shown in Figs. 5.15 and 5.16.

Table 5.4 Fitted I_0, T_1 and T_2 using Anahess and NNLS, and the corresponding target values, including deviations from the target values

	Anahess I_0	T_1 (ms)	T_2 (ms)	NNLS I_0	T_1 (ms)	T_2 (ms)	Target I_0	T_1 (ms)	T_2 (ms)
1 Distr.	129.5	217.7	145.6	128.9	216.5	145.3	130	216	145.1
	Anahess ΔI_0 (%)	ΔT_1 (%)	ΔT_2 (%)	NNLS ΔI_0 (%)	ΔT_1 (%)	ΔT_2 (%)	The key I_0	T_1 (ms)	T_2 (ms)
1 Distr.	−0.4	0.8	0.3	−0.8	0.2	0.1	130	216	145.1

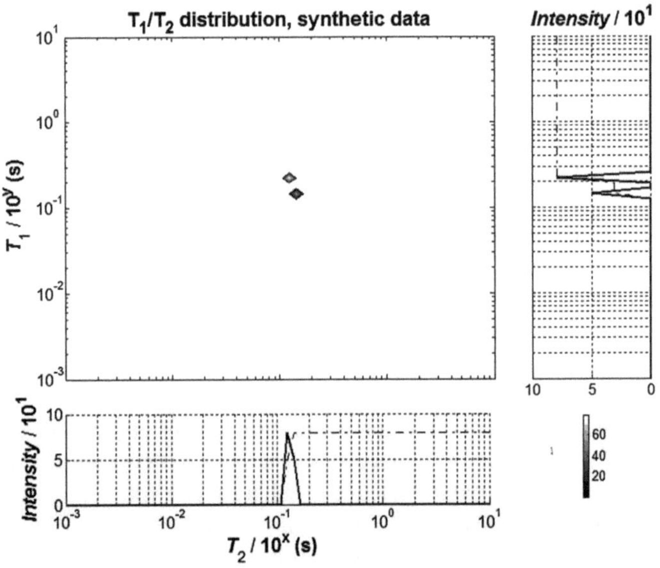

Fig. 5.15 The Anahess result for a distribution with one peak, represented as singular peaks in a log-plot

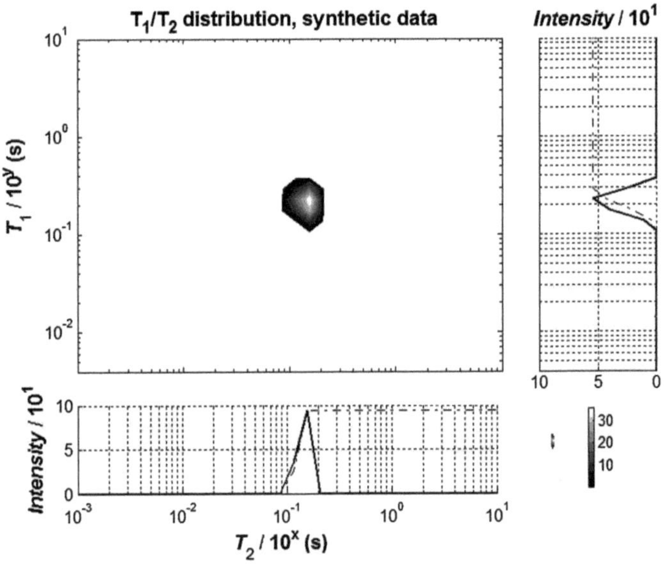

Fig. 5.16 The NNLS results for a distribution with one peak

The Anahess routine terminates the fit at a number of two components located in close proximity to each other (see Fig. 5.15). With two components, the stop criteria are fulfilled, as residuals are randomly distributed around zero. The fitted «peak» area and its average (T_1, T_2) are in accordance with the target values.

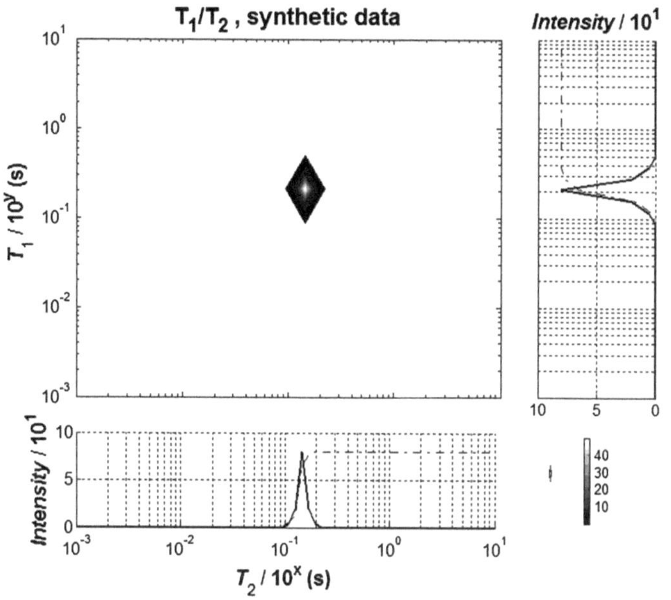

Fig. 5.17 The target distribution with one peak

However, the number of components that Anahess produces as the best fit is far from the 13 original grid points.

Regarding the NNLS result, the fitted «peak» area and its average (T_1, T_2) are also in good agreement with the target values. The best fit for the distribution within 10 trials lead to 7 grid points of varying intensity. The fitted peak intensity deviates by ~30 % and it does not reproduce the symmetry or the half width of the target distribution.

When comparing the Anahess and NNLS results it can be concluded that both methods fits the peak area and average (T_1, T_2) well. Both methods fail in fitting the correct distribution, but NNSL is much closer to fitting a distribution that resembles the target distribution.

5.5.3 Distribution with Two Peaks

The synthetic dataset providing a distribution with two peaks consisted of two identical peaks constructed from 13 discrete points each. The intensities were set symmetrical around the maximum peak intensity, as shown in Fig. 5.20. Again, the user ran the Anahess routine without having any influence on the output while several runs were made of NNSL to find the fit closest to the target values. The results for the fitted peak areas and average (T_1, T_2) are given in Table 5.5 while the distributions are shown in Figs. 5.18 and 5.19.

Table 5.5 Fitted I_0, T_1 and T_2 using Anahess and NNLS, and the corresponding target values, including deviations from the target values

Run	Anahess I_0	T_1 (ms)	T_2 (ms)	NNLS I_0	T_1 (ms)	T_2 (ms)	Target I_0	T_1 (ms)	T_2 (ms)
2	129.4	217.8	145.6	128.7	216.4	145.3	130	216	145.1
Distr.	128.6	64.7	4.4	122.7	68.1	4.2	130	65.8	4.3
Run	Anahess ΔI_0 (%)	ΔT_1 (%)	ΔT_2 (%)	NNLS ΔI_0 (%)	ΔT_1 (%)	ΔT_2 (%)	The key I_0	T_1 (ms)	T_2 (ms)
2	0.5	−0.8	−0.3	1.0	−0.2	−0.1	130	216	145.1
Distr.	1.1	1.7	−2.3	5.6	−3.5	2.3	130	65.8	4.3

The fit using the Anahess algorithm now terminates after only three components, but still the fitted «peak» area and its average (T_1, T_2) are in accordance with the target values. Thus, the distribution of two peaks of 26 intensities all together, can be fitted to only three discrete components where the noise is Gaussian, randomly distributed around 0 and the half width is the same as the half width of the random noise that was used when constructing the synthetic data sets.

Regarding the NNLS result, the fitted «peak» area and its average (T_1, T_2) are also in accordance with the target values. However, for the peak with the shortest (T_1, T_2), the discrepancy was larger than found with Anahess. The best fit for the

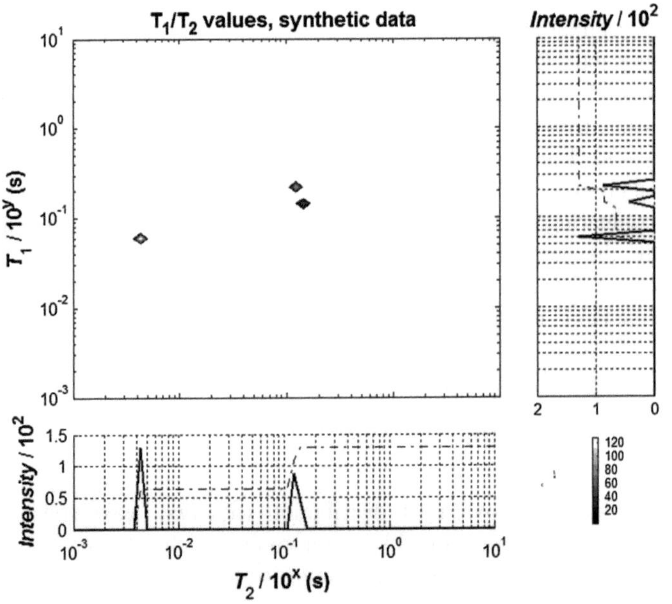

Fig. 5.18 The Anahess result for a distribution with two peaks, represented as singular peaks in a log-plot

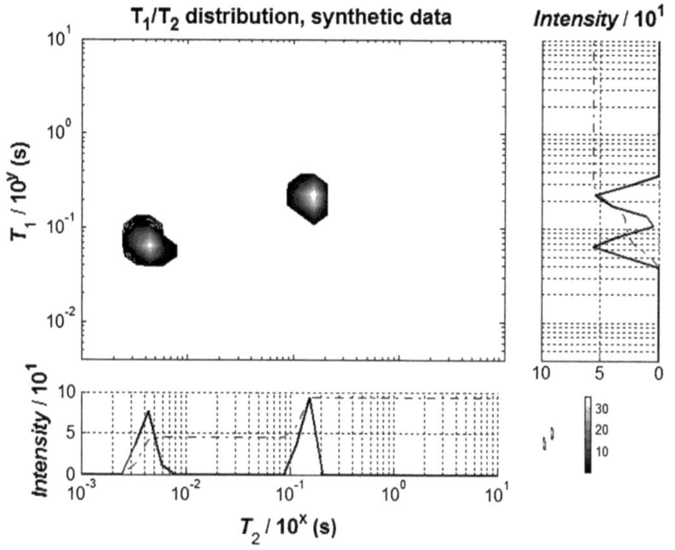

Fig. 5.19 The NNLS results for a distribution with two peaks

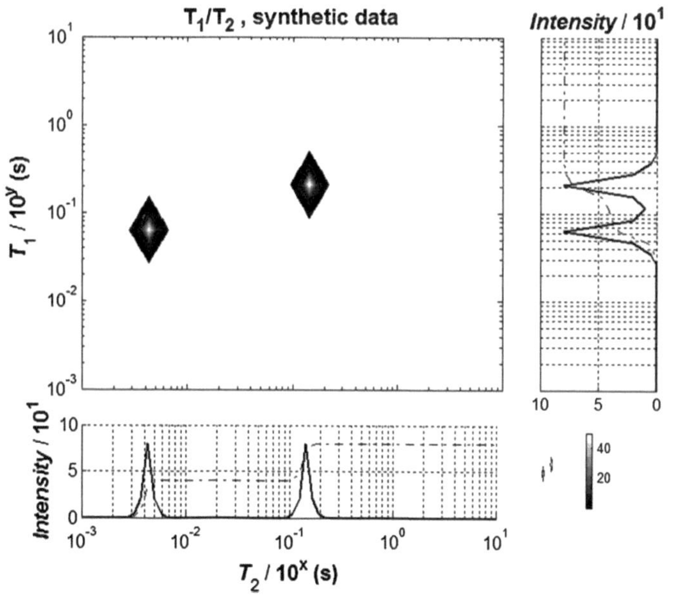

Fig. 5.20 The target distribution with two identical peaks

distribution within 30 trials lead to 7 grid points of varying intensity for the peak with the longest (T_1, T_2) and 13 non zero grid points representing the peak with shortest (T_1, T_2). The number of trials had to be increased as it was more complicated to provide well fitted peak areas for both peaks simultaneously. The fitted peak intensity deviates more than 30 % for both peaks and they do not reproduce the symmetry or the half width of the target distribution.

When comparing the Anahess and NNLS results it can be concluded that both methods fits the peak area and average (T_1, T_2) values reasonably well, but not the shape of the distributions. Again, NNLS appears to perform the closest fit to the target distribution. The results could probably be improved by performing more runs and varying the user input parameters.

5.6 Low Resolution Data

The final comparison against synthetic data was performed on sets of discrete points where the initial intensity of each of the components is 5, whilst the half width of the noise in the Gaussian distribution is 2.75. Two synthetic data sets were fitted to using NNLS and Anahess algorithms, one with two components and one with three components. In Fig. 5.21 it can be seen how the synthetic data appear for two components of amplitude 5 each, i.e. rather poor signal to noise ratio.

The results for the fitted intensities/peak areas and (T_1, T_2) are given in Table 5.6 together with the target values. Now, the deviation from the target values is larger due to the poorer signal to noise ratio. Both Anahess and NNLS return fits reasonably consistent with the target values, except for the three component data having the component with short (T_1, T_2). Both algorithms fail to fit well on T_1, but again Anahess appears to perform better with respect to reproduction of the quantitative information (Anahess deviates by 12 % while NNLS deviates by 36 % from the target value) (Figs. 5.22 and 5.23).

Another feature that appears for the low resolution signal is the appearance of peaks of low resolution in the NNLS produced distribution. These peaks are artefacts due to the noise within the dataset, and would vanish if the regularisation

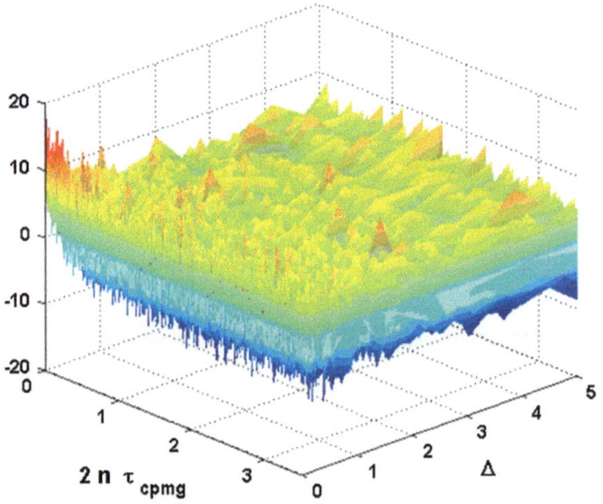

Fig. 5.21 Attenuation of a synthetic signal containing two components of initial amplitude equal to 5 ("2 Comp." in Table 5.6). The half width of the Gaussian noise distribution is 2.75

Table 5.6 Fitted I_0, T_1 and T_2 using Anahess and NNLS, and the corresponding target values, including deviations from the target values

Run	Anahess I_0	T_1 (ms)	T_2 (ms)	NNLS I_0	T_1 (ms)	T_2 (ms)	Target I_0	T_1 (ms)	T_2 (ms)
2	5.5	816.9	214.3	5.2	817	210	5	689.8	193.1
Comp.	4.5	617	643	4.8	622.7	626.6	5	689.8	621.8
3	4.7	621.1	636.2	4.6	627.4	637.4	5	689.8	621.8
Comp.	5.4	804.3	207.9	5.5	804	211.7	5	689.8	193.1
	4.4	145.5	10.2	3.2	174.8	11	5	116	10.4

Run	Anahess ΔI_0 (%)	ΔT_1 (%)	ΔT_2 (%)	NNLS ΔI_0 (%)	ΔT_1 (%)	ΔT_2 (%)	Target I_0	T_1 (ms)	T_2 (ms)
2	−10.0	−18.4	−11.0	−4.0	−18.4	−8.8	5	689.8	193.1
Comp.	10.0	10.6	−3.4	4.0	9.7	−0.8	5	689.8	621.8
3	6.0	10.0	−2.4	8.0	9.1	−2.6	5	689.8	621.8
Comp.	−8.0	−16.6	−7.7	−10.0	−16.5	−9.7	5	689.8	193.1
	12.0	−25.4	−2.0	36.0	−50.7	−10.0	5	116	10.4

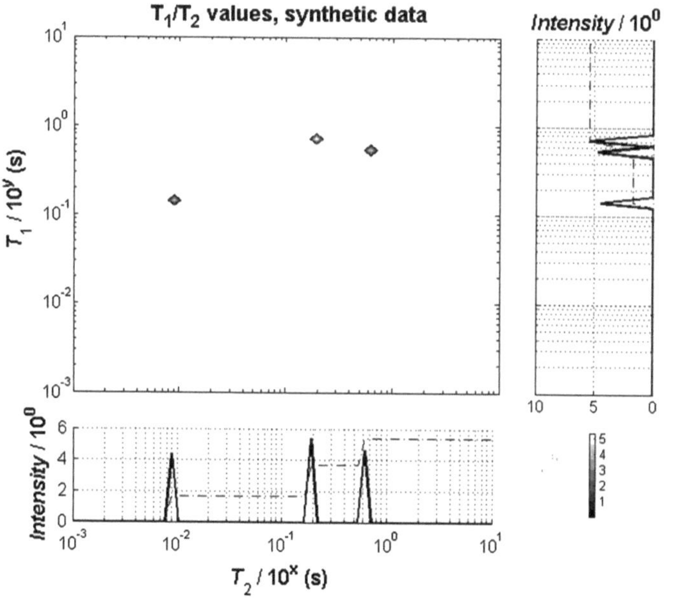

Fig. 5.22 The Anahess result for Comp. 3 in Table 5.6, represented as singular peaks in a log-plot

parameter was reduced. However, if α is reduced, it becomes difficult to reproduce a result close to the target values. The peaks were broadened due to a smaller value of α and the shortest component in particular became very difficult to fit to obtain a reasonable value for the peak area. Thus, a trade-off situation was faced

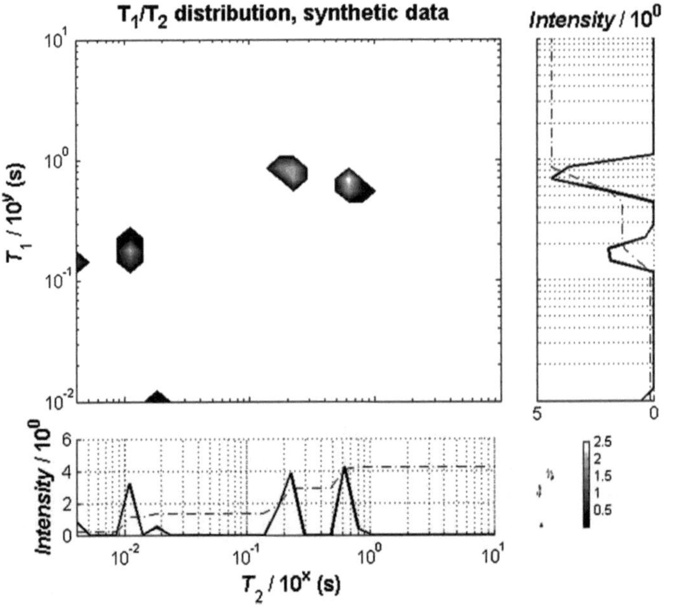

Fig. 5.23 The NNLS results for Comp. 3 in Table 5.6

where the low resolution peaks in the calculation had to be neglected. Again, this was not an issue for the Anahess algorithm, as the fit terminated at three components due to a minimum reached for the BIC and AIC numbers.

5.7 On the Analysis of Two-Dimensional Data Sets from PFG-NMR

The performance of the NNLS and Anahess algorithms has been presented on sets of synthetic data either being singular peaks (discrete components) or simple distributions. The underlying motivation has been to see how able the routines are to quantitatively and qualitatively reconstruct the synthetic distributions. The results from the fit to the synthetic data given in Tables 5.2, 5.3, 5.4, 5.5 and 5.6 speak for themselves; both methods have different strengths and weaknesses, depending on the type of data presented. Anahess is strong in finding the values of the intensities and pairs of (T_1, T_2) with small deviation from the target values, but is not good at determining whether the key values represent a distribution or singular points. Even for a distribution of two peaks represented by 26 non-zero grid points, Anahess only required 3 points to find good values for the peak areas and their average (T_1, T_2). Only for the second peak it indicated that there might be a distribution present as it fitted that region to two points very close in (T_1, T_2)-values. The NNLS is strong at finding distributions, and reconstructs the target values reasonably well. The exception is when the pairs of (T_1, T_2) are in close proximity and there is one peak with higher signal amplitude by one order of magnitude. There are some fundamental differences between the two approaches

- Anahess fits the data to the minimum number of components, but the (T_1, T_2) may take on any value in the fit, i.e. a moving grid approximation.
- NNSL fits the data to a distribution, also for singular peaks, and the (T_1, T_2) are fixed once the grid is set, i.e. a fixed grid approximation.
- Anahess requires the minimum input from the user, once the right kernel that describes the attenuating signal is set.
- NNLS requires important input from the user; including a value for the regularization parameter α, the region of interest (covering the region where the (T_1, T_2)'s are supposed to appear) and the number of grid points (which fixes the (T_1, T_2)-values that the data are fitted to.

The comparison that has been conducted in this section is not a complete one, but it has been performed on a set of data that describes a wide range of typical NMR data. As it already has been concluded that the NNLS does not reproduce well defined symmetric peak(s) in a distribution, there is no new information to be extracted by analysing the more complicated shapes of the peaks.

This chapter will end with an example from a quality control application, which illustrates why the Anahess algorithm was developed: At a dairy company, instabilities were experienced using the spoiler recovery approach for measuring (T_1, T_2). Ten repeated measurements of (T_1, T_2) on a low fat cheese, measurement lasting only 15 min were performed in order to investigate these instabilities. Even though the measurement temperature was set to 5 °C, the cheese melted during the 2.5 h during which the measurements were made (this could be seen by visual inspection). The reason for this heating was found to be due to heating from the RF-electronics when running the high RF duty cycle CPMG-part of the experiment. In the spoiler recovery experiment, this duty cycle can be set short, 0.5 s in this case. Then there is little time for the thermal energy to dissipate between each scan, and a heating of the sample occurs, followed by a melting of the cheese and the associated structural changes within it. When analysing the data using NNLS no significant effect was seen from the melting process, with two main peaks occurring approximately at the same region of (T_1, T_2). In addition, some low-resolution peaks moved more or less randomly. However, when performing the analysis using Anahess, in addition to the components also found by NNLS, a small component was found (slightly less than 2 %) that as a function of time drifts towards higher values of (T_1, T_2). When the cheese had melted, the final (T_1, T_2)-value was reached, from acquisition #4 and upwards. The Anahess fits are superimposed in Fig. 5.24, and demonstrate the stability of the main peaks, while the smallest component that apparently is sensitive to structural changes within the cheese, moves to higher (T_1, T_2) as the cheese melts. By acquisition #4 it has reached a steady state. What is important to note is that at the time of the analysis no particular behaviour was anticipated, but the robustness of Anahess produced important information on the behaviour of the system, as initially it was not considered that RF-heating would take place.

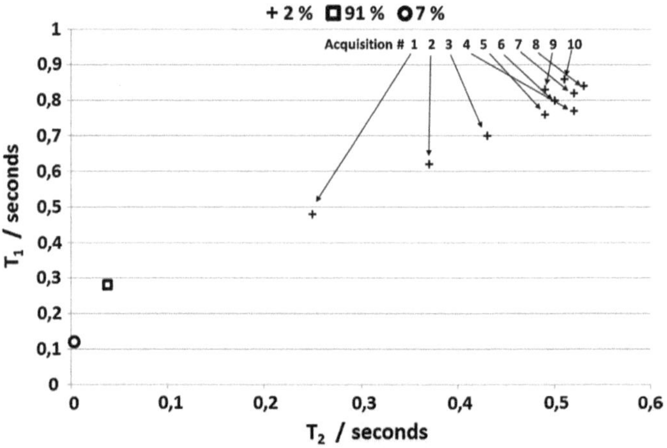

Fig. 5.24 10 repeat measurements on a cheese sample tempered at 5 °C. The markers represented with a thick line for the main components (91 and 7 %) indicates a small variation of the (T_1, T_2) determined from Anahess. The acquisition number provides the chronological order of the acquisitions for the component ('+') that drifts in. (T_1, T_2) due to melting of the cheese

The problem was rectified by reducing the RF duty cycle in order to keep the sample at a steady temperature and avoid melting. It was not possible to conclude with any evolvement of (T_1, T_2) from its fitted distributions using the NNLS algorithm.

5.8 Concluding Remarks

In this section, we have only scratched the surface of data analysis on PFG NMR data. For example, we have not looked at the processing of two-dimensional data where there is an exchange between two or more phases of different values on the dynamic parameters [27, 28]. This may lead to off diagonal components of negative intensity in the two dimensional distribution [29, 30]. Both NNSL and Anahess assume positive amplitudes, so this in the case of such possibilities both algorithms will struggle. However, the tools that are presented in this chapter can be used to analyse some of the most commonly encountered PFG NMR data, and the choice of method will of course depend on what information is required. For example, if accurately determined diffusivities as a function of observation time are required, it is important to keep things as simple as possible, and focus on the region where the measurement is the most accurate. This is almost without exception the initial decay of the PFG NMR experiment, where the deviation from a Gaussian approximation is the least and the slope of the initial decay returns the average diffusion coefficient. If a qualitative picture of the mobility is required, one can apply either the Inverse Laplace Routine (ILT), which produces smoothed distributions in one or two dimensions, or the Anahess algorithm, which produces a set of discrete points as a solution. The Anahess algorithm has at least proven to be well competitive to the ILT approach.

Appendix

In the following a Matlab program is provided that perform a weighted linear fit on a data set.

```
---------------------MAIN PROGRAM--------------------------

% load into Matlab ascii file called data. This containing one row, the
% measured intensities as a function of incrementing gradient strength

nbl     = size(data);
I       = data(1:nbl);

%I      =[1.6044e+09 1.4715e+09 1.3535e+09 1.1901e+09 9.9993e+08 8.1347e+08
%6.3767e+08 4.8984e+08 3.6859e+08 2.7482e+08 2.0493e+08 1.7096e+08 1.3426e+08
%9.9259e+07 8.1729e+07 6.9994e+07] ';

gam     = 7.16e+8;    % the square of the gyro magnetic ratio for the proton

G       =input('What is the gradient strength in G/cm A ? ');                  %G=25;
Istep   =input('What is the minimum current on your first gradient pulse ? ');  %Istep = 2;
Step    =input('What is the step current in the experiment ? ');               %step= 2;
a1      =input('What is the Z-storage time in milliseconds ? ');               %a1=250;
a2      =input('What is the Tau value in milliseconds ? ');                    %a2 = 4;
delta   =input('What is the length of the gradient pulse in milliseconds ? ');  %delta = 3;

for I=1:nbl
        x1(i) = (i-1)*step+Istep; % calculating the applied gradient strengths
end

x       =x1';
a1      = a1*1e-3;
a2      = a2*1e-3;
delta   = delta*1e-3;

B= -(a1+1.5*a2-delta/6)*(x).^2*(G).^2*gam*4*delta.^2

figure(1)
plot(x.^2,I,'r+');
hold on;

Y       = log(I);             % Linearization to the  model Y = A*B
A       = [ones(nbl,1) ,B];

%Weighting of the data

%W      = ones(size(Y));       %equal weight
W       = I.^3;               %weight proportional to the some power of the data-intensities
%Call of the function mlinreg

[b,E,Stdb,Stdfit]=mlinreg(Y,A,W);
```

```
Y_modell = A*b;

I_0= exp(b(1));                     % Fitted initial intensity

Diffusion_coefficient = (b(2))      % Fitted diffusion coefficient

I_modell = exp(Y_modell);           % Plot of exponential attenuation
plot(x.^2,I_modell,'b-');
hold off;

figure(2)
plot(x.^2,log(I/I_null),'r+')       % Plot of linear attenuation
hold on
plot(x.^2,log(I_modell/I_null),'b-')
hold off


-------------------mlinreg.m---------------------

function [b,E,Stdb,Stdfit] = mlinreg(y,Vin,W)

% Weighted multilinear regression
% y    = Vin  * b    + E
% (Jx1) (JxP) (Px1) (Jx1)
%
% where W (Jx1) is the weight of each data point,
% ordinary least squares: use W = ones(size(y));
%
% b      (Px1) = fitting coefficients
% E      (Jx1) = absolute error in each fitting point
% Stdb   (Px1) = standard deviation in percent of fitting coefficients
% Stdfit (1x1) = standard deviation in percent of fit
% beware that this uses the relative

 J          = length(y);
 P          = length(Vin(1,:));

 V          = (W*ones(1,P) ).*Vin;
 VTV        = inv(V'*V);
 b          = VTV*V'*(W.* y );
 E          = y - Vin*b;
 RelE       = E./y;
 Var        = (E'*E)/(J-P);
 Stdb       = 100.*sqrt(Var.*diag(VTV))./abs(b);
 Stdfit     = 100.*sqrt((RelE'*RelE)./(J-P));

-------------------------------------------------------------------
```

When running the program above with a weighting of each data point as the intensity to a power of 4, I^4, the fit will look like the one shown in Fig. 5.1.

References

1. R.M. Cotts et al., Pulsed field gradient stimulated echo methods for improved NMR diffusion measurements in heterogeneous systems. J. Magn. Reson. **83**(2), 252–266 (1969). 1989
2. R.F. Jr Karlicek, I.J. Lowe, A modified pulsed gradient technique for measuring diffusion in the presence of large background gradients. J. Magn. Reson. **37**(1), 75–91 (1980). 1980
3. P.P. Mitra, B.I. Halperin, Effects of finite gradient-pulse widths in pulsed-field-gradient diffusion measurements. J. Magn. Reson. Ser. A **113**(1), 94–101 (1995)
4. L.L. Latour, L.M. Li, C.H. Sotak, Improved PFG stimulated-echo method for the measurement of diffusion in inhomogeneous fields. J. Magn. Reson., Ser. B **101**(1), 72–77 (1993)
5. G.H. Sørland, B. Hafskjold, O. Herstad, A stimulated-echo method for diffusion measurements in heterogeneous media using pulsed field gradients. J. Magn. Reson. **124**(1), 172–176 (1997)
6. G.H. Sorland, D. Aksnes, L. Gjerdaker, A pulsed field gradient spin-echo method for diffusion measurements in the presence of internal gradients. J. Magn. Reson. **137**(2), 397–401 (1999)
7. W.H. Press, *Numerical Recipes in C++: The Art of Scientific Computing* (Cambridge University Press, Cambridge, 2002)
8. J. Stepišnik, Validity limits of Gaussian approximation in cumulant expansion for diffusion attenuation of spin echo. Phys. B **270**(1–2), 110–117 (1999)
9. P.P. Mitra, P.N. Sen, L.M. Schwartz, Short-time behavior of the diffusion coefficient as a geometrical probe of porous media. Phys. Rev. B **47**(14), 8565–8574 (1993)
10. W.H. Press, *Numerical Recipes 3rd Edition: The Art of Scientific Computing* (Cambridge University Press, Cambridge, 2007)
11. F.S. Acton, *Numerical Methods that Work* (Mathematical Association of America, Washington, 1990)
12. C.L. Lawson, R.J. Hanson, *Solving Least Squares Problems* (Society for Industrial and Applied Mathematics, Philadelphia, Pennsylvania, 1995)
13. A.N. Tichonov, A.S. Leonov, in *Ill-Posed Problems in Natural Sciences: Proceedings of the International Conference Held in Moscow*, 19–25 Aug 1991. VSP (1992)
14. A.N. Tikhonov, *Numerical Methods for the Solution of Ill-Posed Problems* (Kluwer Academic Publishers, Berlin, 1995)
15. S.W. Provencher, CONTIN: a general purpose constrained regularization program for inverting noisy linear algebraic and integral equations. Comput. Phys. Commun. **27**(3), 229–242 (1982)
16. S.W. Provencher, A constrained regularization method for inverting data represented by linear algebraic or integral equations. Comput. Phys. Commun. **27**(3), 213–227 (1982)
17. S.W. Provencher, Inverse problems in polymer characterization: direct analysis of polydispersity with photon correlation spectroscopy. Die Makromolekulare Chemie **180**(1), 201–209 (1979)
18. J. Butler, J. Reeds, S. Dawson, Estimating solutions of first kind integral equations with non-negative constraints and optimal smoothing. SIAM J. Numer. Anal. **18**(3), 381–397 (1981)
19. P. Berman et al., Laplace inversion of low-resolution NMR relaxometry data using sparse representation methods. Concepts Magn. Reson. Part A **42**(3), 72–88 (2013)
20. Y.Q. Song et al., T1–T2 correlation spectra obtained using a fast two-dimensional laplace inversion. J. Magn. Reson. **154**(2), 261–268 (2002)
21. R. Fletcher, M.J.D. Powell, On the modification of LDL^T factorizations. Math. Comput. **28**(128), 1067–1087 (1974)
22. B. Borchers, MATLAB routines for square root free Cholesky factorizations. http://infohost.nmt.edu/~borchers/ldlt.html
23. S. Gideon, Estimating the dimension of a model. Ann. Stat. **6**(2), 461–464 (1978)
24. H. Akaike, A new look at the statistical model identification. IEEE Trans. Autom. Control **19**(6), 716–723 (1974)

25. N. Sugiura, Further analysts of the data by akaike's information criterion and the finite corrections. Commun. Stat. Theory Meth. **7**(1), 13–26 (1978)
26. N. van der Tuuk Opedal, G.S. Johan Sjöblom, in *Methods for Droplet Size Distribution Determination of Water-in-oil Emulsions using Low-Field NMR*. Diffusion Fundamentals, vol. 7, (diffusion-fundamentals.org 9): pp. 1–29 (2009)
27. P.T. Callaghan, I. Furó, Diffusion-diffusion correlation and exchange as a signature for local order and dynamics. J. Chem. Phys. **120**(8), 4032–4038 (2004)
28. K.E. Washburn, C.H. Arns, P.T. Callaghan, Pore characterization through propagator-resolved transverse relaxation exchange. Phys. Rev. E **77**(5), 051203 (2008)
29. Y.-Q. Song, L. Zielinski, S. Ryu, Two-dimensional NMR of diffusion systems. Phys. Rev. Lett. **100**(24), 248002 (2008)
30. S. Rodts, D. Bytchenkoff, Structural properties of 2D NMR relaxation spectra of diffusive systems. J. Magn. Reson. **205**(2), 315–318 (2010)

Chapter 6
The Spoiler Recovery Approach (SR)

Abstract Instead of letting the magnetization reach thermal equilibrium before each scan, a pulse sequence is presented, which applies a combination of RF-pulses and magnetic field gradients to spoil any magnetization in any direction of a sample located in an external magnetic field to produce a net nuclear magnetization, the spoiler recovery (SR) sequence. The basic assumption made is that the state of the nuclear spins at the end of the sequence is comparable to the state without any external applied magnetic field, i.e. the directions of the nuclear spins are randomly oriented in space. Due to the presence of the external magnetic field, the net nuclear magnetization will re-equilibrate along the external magnetic field due to T_1 relaxation processes. Using a wait delay equal to T_1 after the spoil process, already 63 % of the magnetization present at thermal equilibrium is regained. The wait time between each scan is then reduced to practically zero, as the ordinary 5 times T_1 recycle delay is no longer necessary to achieve identical conditions between the accumulating scans. Using a spoiler recovery delay equal to T_1, the total experimental time is reduced by 80 % without any major loss of signal to noise. Likewise, the use of the spoiler recovery approach can reduce the acquisition time of two dimensional experiments, as Diffusion-T_2 or T_1-T_2, from the order of hours to the order of minutes. The SR approach is verified using external magnetic fields ranging from 0.047 to 14.1 T. At the highest fields it is also verified that the SR approach can be used to circumvent effects due to radiation damping. A set of applications using the spoiler recovery approach to reduce the acquisition time will be presented, designed either for low or for high external magnetic field.

The determination of the apparent diffusion coefficient, as measured by PFG NMR, has become an important parameter for research, industrial applications and diagnostics. After more than four decades of development of the PFG NMR technique, new methods and applications within PFG NMR are still being developed, which reflects the flexibility and yet unrevealed potential of this technique [1, 2]. One of the major drawbacks of the PFG NMR has been the settling time needed for the system to regain thermal equilibrium of the net nuclear magnetization along the external magnetic field (~5 times T_1) between each scan. This settling time puts for example a limit to how many gradient values or relaxation time values one can apply

G.H. Sørland, *Dynamic Pulsed-Field-Gradient NMR*,
Springer Series in Chemical Physics 110, DOI 10.1007/978-3-662-44500-6_6

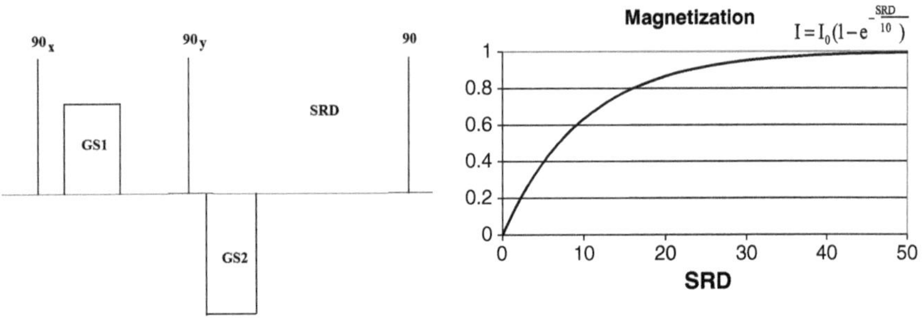

Fig. 6.1 The spoiler recovery sequence and a plot of the magnetization following a variable delay SRD. The second 90° RF-pulse has its phase orthogonal to the phase of the first 90° RF-pulse

before the time of the experiment becomes longer than the stability of the system or what is feasible for example in a quality control application. Methods have been proposed to reduce the experimental time by saturating the magnetization before recording of the actual NMR experiment [3–5] or by performing the diffusion experiment as a single shot experiment [6, 7]. A significant drawback in using some of these approaches is the loss in signal to noise ratio of the experiment and the fact unwanted coherence transfer pathways result in a multi exponential decay from system that should produce a mono exponential decay due to bulk diffusion [3].

Instead of letting the magnetization reach thermal equilibrium before each scan, a pulse sequence that applies a combination of RF-pulses and magnetic field gradients can be used to spoil any magnetization in any direction of a sample located in an external magnetic field to produce a net nuclear magnetization. The basic assumption made is that the state of the nuclear spins at the end of the sequence is comparable to the state without any external applied magnetic field, i.e. the directions of the nuclear spins are randomly oriented in space. Due to the presence of the external magnetic field, the net nuclear magnetization will re-equilibrate along the external magnetic field due to T_1 relaxation processes. Using a wait delay equal to T_1 after the spoil process, defined as SRD in Fig. 6.1, already 63 % of the magnetization present at thermal equilibrium is regained. The wait time between each scan is then reduced to practically zero, as the ordinary 5 times T_1 recycle delay is no longer necessary to achieve identical conditions between the accumulating scans. Using a spoiler recovery delay equal to T_1, the total experimental time is reduced by 80 % without any substantial loss in signal to noise, as 63 % of the signal is regained during the SRD. Likewise, the use of the spoiler recovery approach can reduce the acquisition time of two dimensional experiments, as Diffusion-T_2 or T_1-T_2, from the order of hours to the order of minutes. Furthermore, the rapid T_1-T_2 is quantitatively correct which is not feasible using the saturation approaches [3–5]. In the following sections, the spoiler recovery (SR) approach will be verified using external magnetic fields ranging from 0.047 to 14.1 T. At the highest fields it will also be verified that the SR approach can be used to circumvent effects due to radiation damping [8]. A set of applications using the SR approach to reduce the acquisition time will be presented, designed either for low or for high external magnetic field.

6.1 Verification

The conventional approach for measuring diffusion and or relaxation is to let the magnetization return to thermal equilibrium after application of a set of RF- and gradient pulses. Improvements have been made to reduce this acquisition time, and are based on saturating the magnetization to a steady state before applying the pulse sequences [3–5].The approach shown in Fig. 6.1 is different as initially no assumption is made on any particular state for the net nuclear magnetization. Regardless of the initial state two 90° RF pulses are applied in combination with a bipolar pair of magnetic field gradients of arbitrary shape. This we refer to as a spoiler recovery (SR) sequence or approach.

Figure 6.1 illustrates the timing diagram of the spoiler recovery pulse sequence and the return to thermal equilibrium for the magnetization. The durations and strengths of GS1 and GS2 are chosen such that any coherence transfer pathway arising from the two first 90° RF-pulses does not contribute to any significant signal after application of the third 90° RF-pulse, which is the excitation pulse. It is therefore critical that the phases of the two spoiler gradient pulses are unequal as provided by 6.1

$$\gamma \mathrm{GS1} \int_{0}^{\delta_{\mathrm{GS1}}} z_1(\mathrm{t})\mathrm{dt} \neq \gamma \mathrm{GS2} \int_{0}^{\delta_{\mathrm{GS2}}} z_2(\mathrm{t})\mathrm{dt} \tag{6.1}$$

δ_{G1} and δ_{G2} are the durations of the spoiler gradient pulses, $Z_1(t)$ and $Z_2(t)$ are the positions of the nuclear spins at the time of the two gradient pulses, and γ is the gyromagnetic ratio. To produce a significant difference in the two accumulated phases, the gradients are chosen to be bipolar, of different duration, and the integral of GS2 (see 6.1) to be larger than the integral of GS1 in absolute values. If there initially is any net magnetization along any direction, it will be dephased by the application of the spoiler recovery sequence. To acquire noise only during the SRD interval it is important to choose the GS1 and GS2 values as described above. Question is to what extent unwanted coherence affect the wanted signal, as it evidently does using the approach where a train of RF-pulses is applied to saturate the magnetization instead [3–5]. As the resolution is improved for higher applied magnetic fields, the effect of unwanted coherences should especially be checked for when using high external magnetic fields (>10 T) as well as low external magnetic field (<1 T). Even though a method for reducing the acquisition time substantially using low field NMR instrumentation, as a saturation recovery approach, it does not necessarily apply at higher fields. The reason is that the saturation recovery, which is frequently used in magnetic resonance imaging [9], does not provide spoiler gradients as shown in Fig. 6.1. Consequently there will be unwanted coherences from stimulated echoes, which will interfere after the application of the third 90° excitation pulse [10], and the higher signal to noise ratio, the more the interference from these unwanted coherences will become apparent [3].

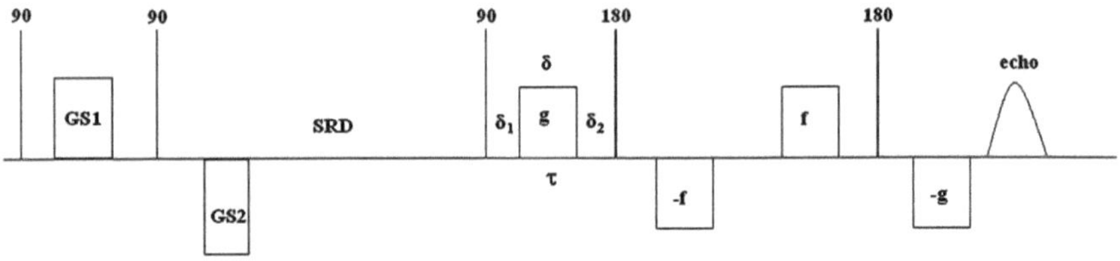

Fig. 6.2 The combined spoiler recovery 11-interval PFGSE sequence

In order to verify the performance of the diffusion experiments on bulk samples when using the SR approach, a linear decay of the logarithm of the echo attenuation in a PFG NMR experiment should be observed. If this is the case, then it is a strong indication that interferences from unwanted coherences are not present. In Fig. 6.2 the spoiler recovery sequence is placed in front of the 11-interval PFGSE and the sequence was tested on pure water on a 0.25 T permanent magnet system with SRD equal to 5 times T_1 (the ordinary approach) and then with a SRD of 0.5 s. The number of accumulating scans was 4 plus one dummy scan, and the gradient strength was ramped 20 times from 0.5 Gauss/cm up to 120 Gauss/cm.

As can be seen from Fig. 6.3 there is no discrepancy between the two methods for acquiring the attenuation due to diffusion, and the logarithm of the decay is linear down to the noise value. At low field there seems to be no interference from unwanted coherences in the diffusion experiment. In the next section, it will be verified that the spoiler recovery approach removes unwanted coherences regardless of recycle delay and the strength of the external magnetic field, at least up to 14.1 T.

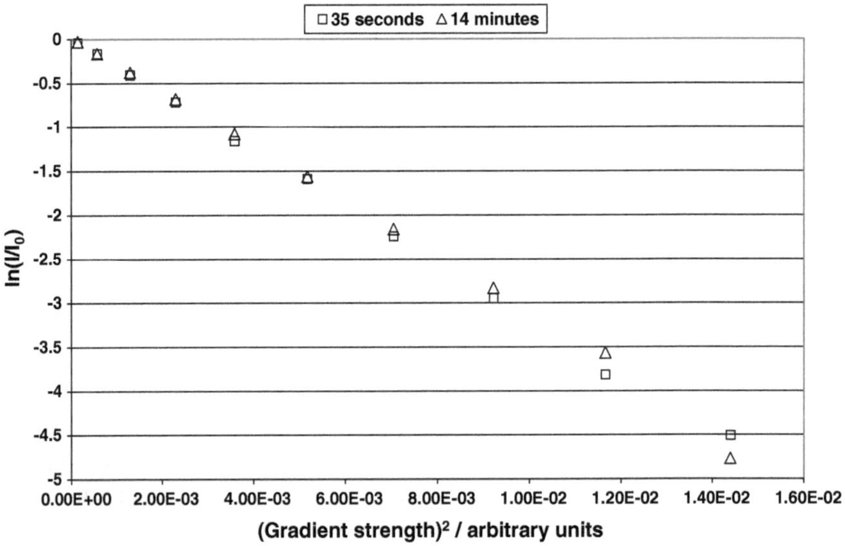

Fig. 6.3 Comparison of the 11-interval PFGSE with (35 s) and without (14 min) the spoiler recovery on pure water

In addition, it will be investigated whether radiation damping may have any influence on the results at higher magnetic field [11]. Radiation damping is an effect that distorts the phase and amplitude of the NMR signal at higher field, due to a coupling between the magnetization of the sample and the RF-coil (see Sect. 6.1.3). The experimental data in the remainder of this section is therefore devoted to high resolution instrumentation where the external magnetic field is larger than 10 T.

6.1.1 The SR Approach Combined with the 13 Interval Bipolar PFGSTE

In Fig. 6.4 the SR sequence is placed in front of the 13-interval PFGSTE, which was employed to measure diffusion of distilled water slightly doped with Cu_2SO_4 using a 11.7 T Avance III instrument equipped with a diffusion probe. As can be seen in Fig. 6.5 the logarithm of the echo attenuation is linear down to −6,

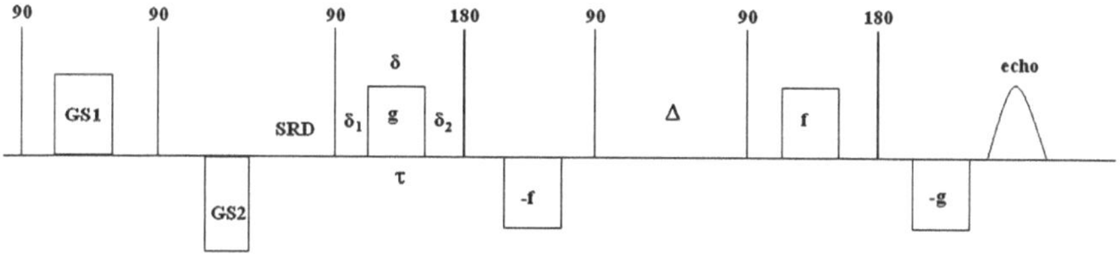

Fig. 6.4 The combined spoiler recovery 13-interval PFGSTE sequence

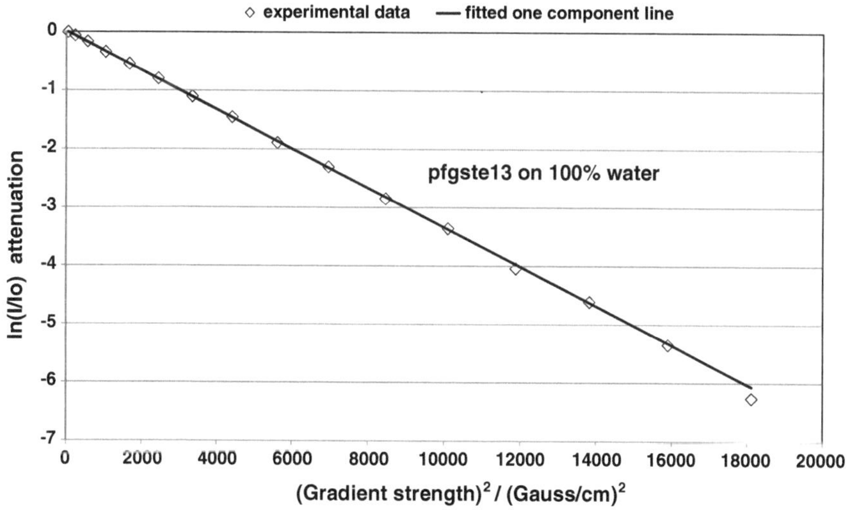

Fig. 6.5 Application of the combined spoiler recovery 13-interval PFGSTE on pure water at 11.7 T

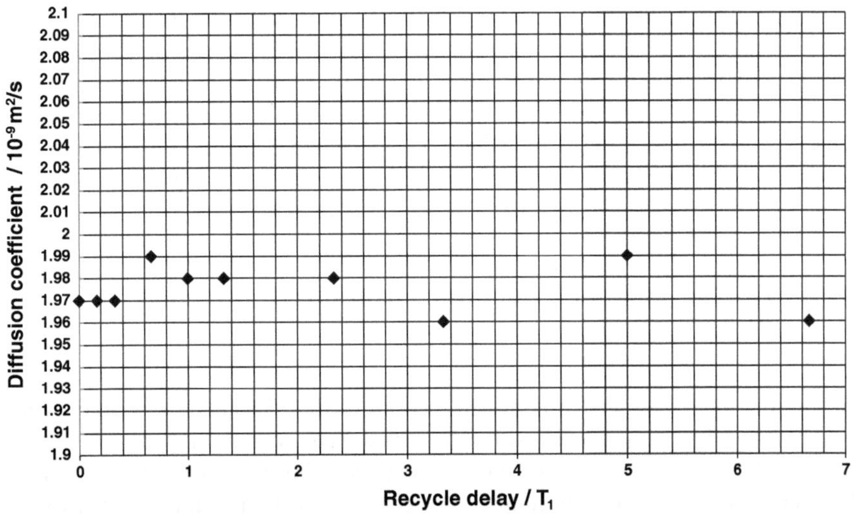

Fig. 6.6 Varying the recycle delay on a doped water sample (T_1 ~ 300 ms) using the spoiler recovery 13-interval PFGSTE. The relative error for fitted diffusion coefficient is less than 2 % and the values are randomly distributed. The resolution of the measured diffusion coefficients is 0.01

where the noise starts to affect the echo signal substantially. When using the saturation approach to reduce recycle delay, it is shown that the non-linearity of the logarithm of the echo attenuation and thus the measured diffusion coefficient is strongly dependent on the recycle delay divided by T_1 [3]. This is not the case for the SR approach, as shown in Fig. 6.6. Here it is confirmed that the measured diffusion coefficient in distilled water is constant regardless of recycle delay between each scan, the linearity is as shown in Fig. 6.5 for all acquisitions, i.e. down to −6 on a logarithmic scale. The conclusion is that the measured diffusion coefficient reflects the real diffusivity of the sample and there are no indications that the use of the spoiler recovery sequence at higher applied magnetic fields should not work as expected. Even though the (recycle delay)/T_1 parameter may be reduced down to practically zero, a finite duration is required to run the pulse sequence and record the echo signal or an FID. Thus, in practice the time between each scan (including the pulse sequence) will not be much shorter than 0.1 s.

The spoiler recovery was tested on a well know system for verifying the SR approach, a solution of 2 mM lysozymes in H2O/D2O. Figure 6.7 displays the attenuation from the highest peak at 0.68 ppm, and the diffusion coefficient from the aliphatic region at 0.68 ppm was found to be 1.11×10^{-10} m^2/s. This is comparable to the value found by others [12]. A 500 MHz Bruker Avance III instrument was used for acquiring the data set.

Another important parameter the spoiler recovery approach introduces is the delay between the end of the spoiler sequence and the 90° excitation pulse, the spoiler recovery delay (SRD). After applying the 90° excitation pulse there are at least two important effects that may interfere with the acquired FID signal, and can affect the outcome of the measurements.

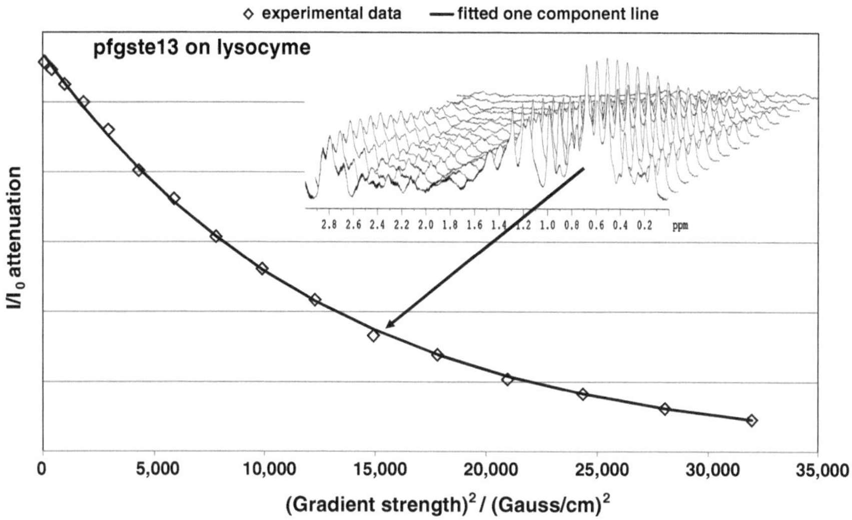

Fig. 6.7 The spoiler recovery 13-interval PFGSTE applied on lysozyme sample

6.1.2 Eddy Current Transients from the Spoiler Recovery at Shorter SRD

When the SRD is made short, it is important to be in control of the eddy current transients following the pair of gradient pulses of opposite polarity, GS1 and GS2. If the transients following the two gradient pulses do not cancel, this can be observed from the lock signal that is usually available on high-resolution instruments. In the presence of such transients, the phase and the baseline of the spectrum of the FID may be distorted. Thus, it is not only 6.1 than must be satisfied when adjusting GS1 and GS2, but also the eddy current transients must be minimized. In Fig. 6.8, the lock signal is visualized with and without a proper adjustment of the amplitudes and durations of GS1 and GS2.

6.1.3 Phase and Baseline Distortions Due to Radiation Damping at Longer SRD

When increasing the SRD to larger values there is the possibility of seeing effects from radiation damping in the acquired spectra. Radiation damping appear as a significant effect at higher field for the following reason: When a 90° excitation pulse is imposed on a sample containing large amounts of protons, a net nuclear magnetisation will be seen from the RF-coil as an oscillating magnetic field. This induces an oscillating current in the RF-coil of the same frequency as found for the nuclear spins. The induced current then produces an oscillating field in the

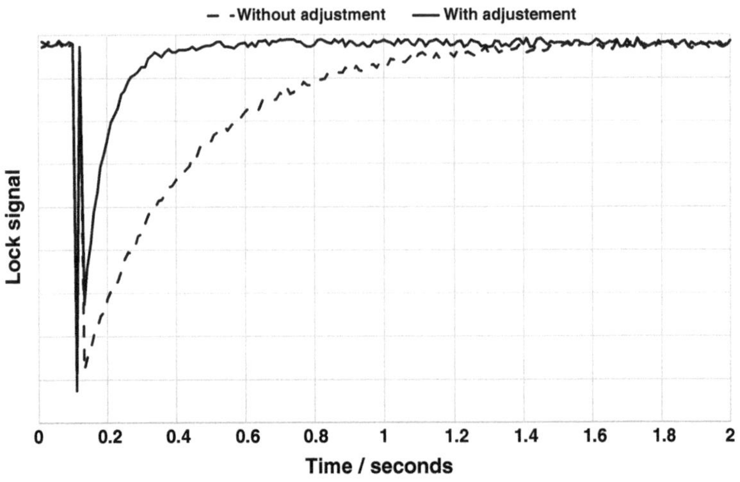

Fig. 6.8 A visualization of how the spin lock signal may be affected by the presence of GS1 and GS2, and how a careful adjustment of the two pulses may suppress the impact of eddy current transients on the lock signal

transverse plane, ideally shifted 90° as compared to the initial excitation pulse. This will act on the nuclear spins and rotate it towards its equilibrium, which is to be aligned with the direction of the applied field. Such a rotation towards the equilibrium may be much faster than the ordinary transverse and longitudinal relaxation times [8], and will especially lead to a significant line broadening of the spectrum.

Figure 6.9 shows a set of FID's and corresponding spectra for increasing value of the SRD. The signal arises from pure water placed in a 14.1 T magnet with a cryogenic probe. The effect from radiation damping can be readily observed, as the line broadening (T_2^*) increases with increasing value of the SRD. Consequently, it is important to keep the SRD-value short enough to avoid this effect.

When acquiring the recovery signal to find a value for T_1, the SRD may have to be set to a long value where radiation damping may interfere as shown in Fig. 6.9. Then a method for reducing the signal intensity at long SRD's must be employed. Previous work has suggested numerous sophisticated set-ups involving hardware modifications [13, 14], or modified pulse sequence and data acquisition schemes [15]. Probably the most effective way and less complicated is the one proposed by Khitrin and Jerschow [16], where the idea is initially to apply a modified pulsed field gradient spin echo sequence. The modification is to impose is a slightly mismatch between the motional encoding and decoding gradients. This will lead to a loss of signal that depends on the mismatch of the gradients. As the fractional loss will be independent on the SRD delay, spoiler recovery measurements may then be applied even at 14.1 T to measure T_1.

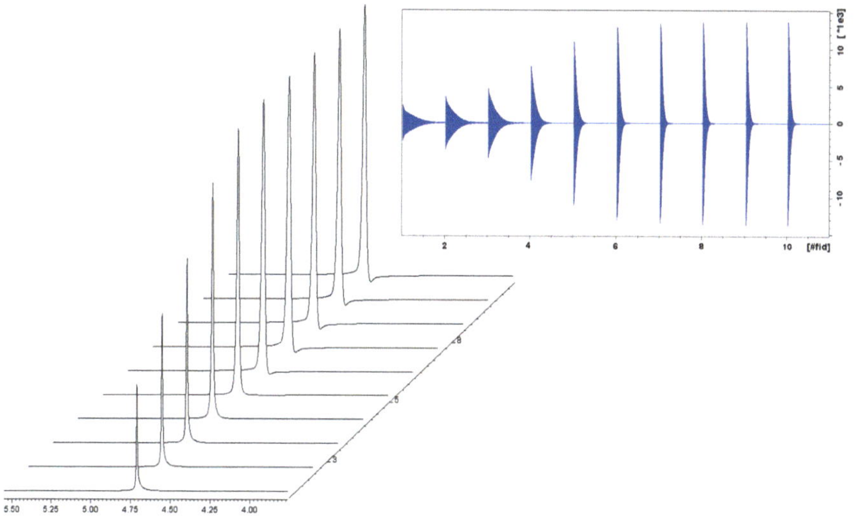

Fig. 6.9 The effects on the FID's (*right part*) and their correspond FT-spectra (*left part*) on pure water using the spoiler recovery sequence shown in Fig. 6.1. The SRD varies from 0.4 to 60 s. Diagram courtesy of Dr. Klaus Zick

6.2 SR Applied at Low Field

The spoiler recovery approach was initially developed for reducing the experimental acquisition time considerably, while preserving quantitative diffusion information. In the following sections, examples of applications using low-resolution instrumentation are presented. The examples show how the SR approach makes it possible to study the dynamics of unstable systems due to the substantial reduction in acquisition time that is achieved.

6.2.1 Characterization of Unstable Emulsions

The SR approach was initially developed due to the need for a rapid method for acquiring droplet size distributions in water in oil emulsions. In the oil processing industry there is a requirement to perform phase separation of these emulsions in order to produce the crude oil prior to refining. So-called demulsifiers or chemicals are added to the emulsion and the surface tension of the interface between the crude oil and brine is altered so that the brine droplets coalesce, sediment and form a bulk fraction at the bottom. The pulse sequence shown in Fig. 6.10 can be used to measure the profile of the average brine droplet size. The crude oil signal is suppressed due to a significant difference in relaxation times, leaving only the brine signal in the profile. A diffusion weighting can then be used to provide a

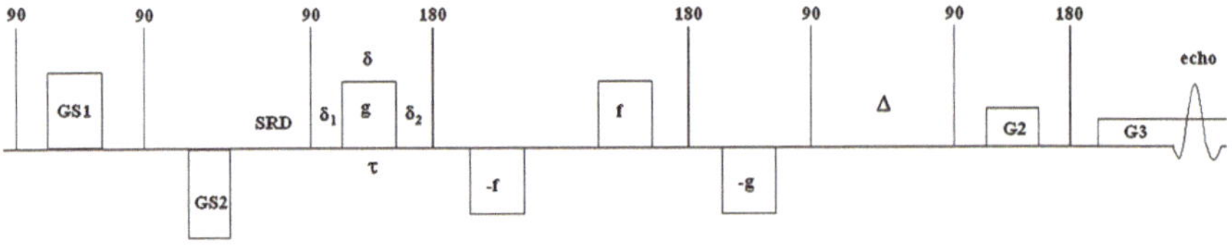

Fig. 6.10 The combined SRD-11PFGSE-profile sequence for determining profile of the average droplet size in a water in oil emulsion. The SRD provides a short recycle delay when choosing SRD = T_1, the 11PFGSE provides a diffusion weighted signal, the z-storage delay resolves the brine signal from the crude oil, and the profile experiment provides a one-dimensional image of the average droplet size

diffusion profile, which subsequently can be transformed to an average (S/V) profile by knowing the observation time and employing for example the short observation time approach by Mitra et al. [17]. Finally, this can be transformed to a profile of average droplet sizes assuming spherical droplets. The methods for characterisation the emulsions will be addressed further in Chap. 7.

Using the SR approach the recording of one droplet size profile takes about 30 s, and this can be repeated instantaneously to produce a large data set of profiles. Any evolution of the emulsion may then be traced on a relatively short time-scale. Figure 6.11 shows a contour plot of the one dimensional profile of the average droplet size in an unstable emulsion of water cut 30 %. A demulsifier has been added to the sample, and the droplets start to coalesce and sediment at

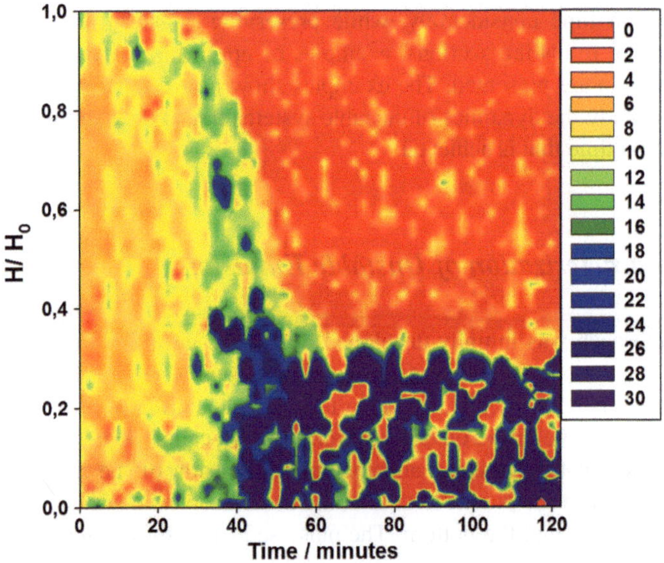

Fig. 6.11 The $(S/V)^{-1}$ contour plots of brine being separated due to the addition of a demulsifier to an emulsion

approximately 20 min after addition of the chemical. After 60 min a bulk water phase is established at the bottom of the sample tube.

Depending on the demulsifier used, the onset of production of water at the bottom may change, and the PFG NMR instrumentation provides a tool that can be used for optimisation of the type of demulsifier and the necessary concentration to be added. In fact, the requirement for a fast method for determining (S/V) of water in an emulsion led to the development of the SR-approach [18].

6.2.2 *Observation Time Dependent Diffusion Measurements*

An important area of PFG NMR is to probe the time dependency of the restricted apparent diffusion. The usual way to conduct the experiment is for a defined observation time Δ, to increment the applied gradient strength and extract the apparent diffusion coefficient from the fitted slope. Usually more than 10 gradient strengths are used to fit the slope, and weight it to the regime where the second cumulant approximation applies [19]. Using well-functioning PFG NMR instrumentation it may not be necessary to apply as numerous gradient values to extract the apparent diffusion. In Fig. 6.12, the logarithm of the echo attenuation is shown for water diffusion amongst densely packed mono sized spheres. Instead of applying 10 incrementing gradients to fit the slope of the attenuation, a minimum of two gradients can be used to determine the apparent diffusion coefficient. Both approaches for determining the apparent diffusion coefficient were repeated 25 times, and the standard deviation was 1 % for both. Thus there is practically

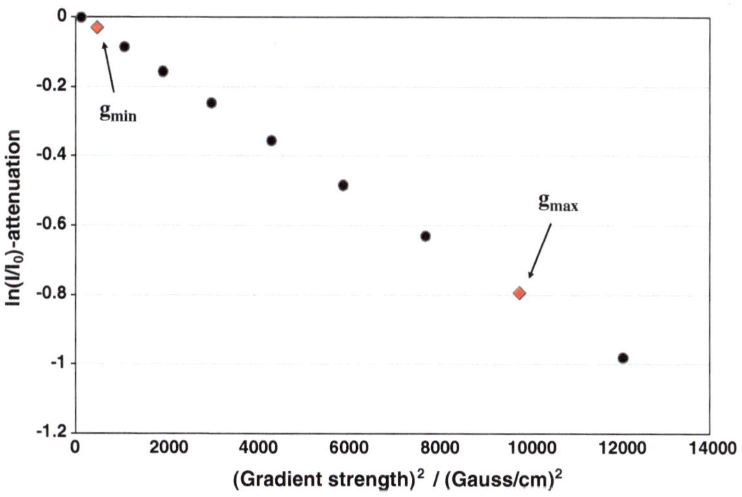

Fig. 6.12 The logarithm of the echo attenuation for incrementing gradients. The *red markers* indicate the gradient strengths applied in the two-gradient fit for finding the slope of the attenuation

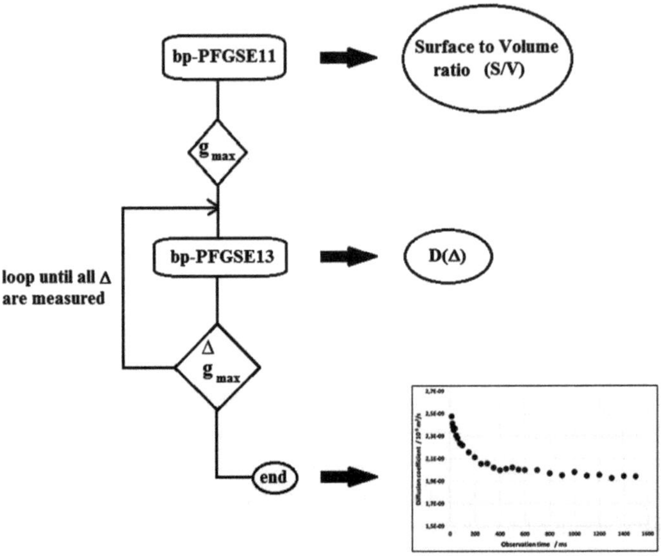

Fig. 6.13 The setup for rapid acquisition of the observation time dependent apparent diffusion coefficients

no benefit from applying the 8 extra gradient values, provided the two gradients, g_{min} and g_{max}, are within the linear regime of the logarithm of the attenuation, i.e. within the second cumulant approximation.

Figure 6.13 shows how a set-up can be run where it is ensured that g_{max} is located within the linear regime of the logarithm of the attenuation. Initially the 11-interval PFGSE (Fig. 6.2) is run at the shortest observation times, and g_{max} is adjusted to be within the linear regime. The measured diffusion coefficient can then be fitted to the short observation time expansion to provide a value for the surface to volume ratio of the cavities in which the water is diffusing. Depending on the measured diffusion coefficient and the observation time in the first 13-interval PFGSTE (Fig. 6.4), an estimate on the g_{max}, the maximum gradient strength shown in Fig. 6.12, necessary to attenuate the logarithm of the NMR signal down to −0.8 can be found and put into the sequence. The output is another apparent diffusion coefficient at an observation time Δ. Δ is then increased to another value and g_{max} is recalculated based on the previous apparent diffusion coefficient and observation time.

$$g\text{max} = \sqrt{\frac{0.8}{\gamma^2 D_{app}^{previous} \delta^2 \left(\Delta^{previous} + \frac{3\tau}{2} - \frac{\delta}{6} \right)}} \tag{6.2}$$

This procedure is repeated for all values of Δ to produce a set of observation time dependent apparent diffusion coefficients, also shown in Fig. 6.13. When

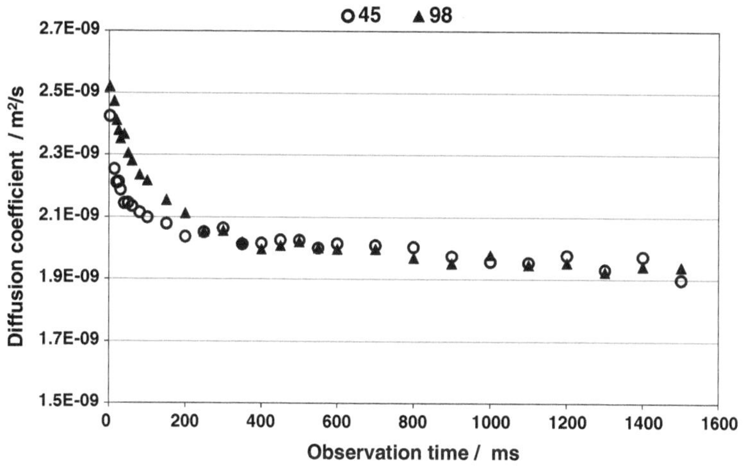

Fig. 6.14 The rapid measurements of observation time dependent diffusion coefficients of water amongst compact mono sized plastic spheres. The measurements were conducted at 35 °C on 20 MHz PFG NMR instrumentation

combining the SR approach with the two-gradient determination of the apparent diffusion coefficient, the experimental time required for producing the data-sets shown in Fig. 6.14 is reduced by more than 80 %, without any substantial loss in signal to noise or accuracy of the determined diffusion coefficient. The data sets in Fig. 6.14 shows the time dependent diffusion of water amongst compact mono sized spheres of diameter 45 and 98 μm, and the 30 observation time dependent diffusion coefficients where recorded within 20 min. Conventionally, with 10 incrementing gradients and a recycle delay of 7 s (~5 times T_1), it would have taken approximately 4 h.

6.2.3 Two-dimensional Applications

Evidently, the application of the SR-approach will reduce the experimental time substantially for numerous PFG NMR experiments that conventionally are quite time consuming. In particular, two-dimensional T_1-T_2, T_2-D or T_1-D experiments will have the experimental time reduced by implementing the SRD into the two-dimensional sequences. The most time consuming two-dimensional experiment is the T_1-D. Implementation of the SRD as shown in Fig. 6.1 will reduce the experimental time so much that it will be the duty cycle of the gradient coils that will put restrictions on how fast the two-dimensional data set can be recorded. As the gradient coils are being heated during the PFG NMR, the rate of heating must not exceed the rate of heat dissipation from the coil. Instead of using a recycle delay of practically zero, it may therefore be necessary to have a recycle delay of ~1 s.

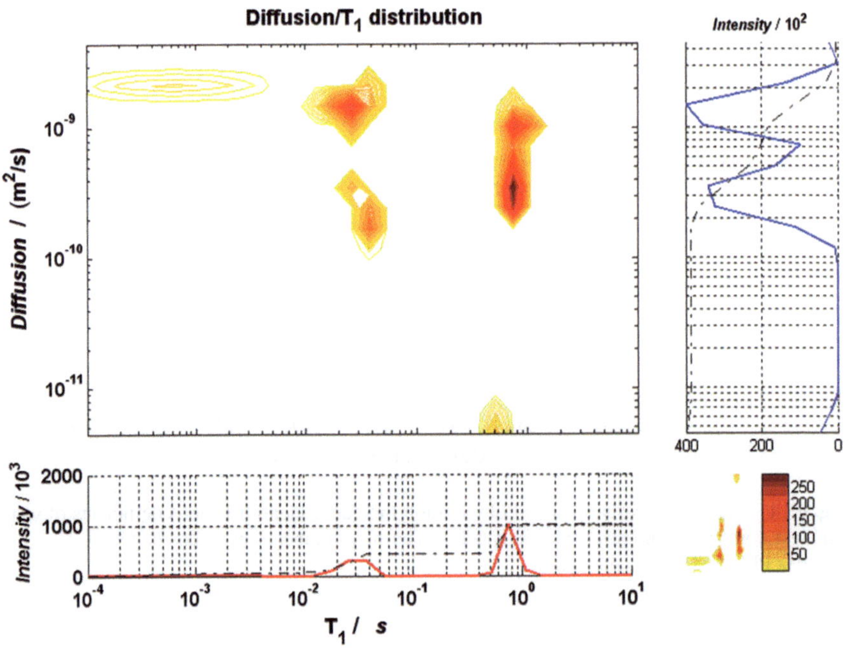

Fig. 6.15 The T_1-diffusion distribution arising from a 1.5″ rock core plug in its native state. The measurements were conducted at 30 °C on 12 MHz PFG NMR instrumentation

This recycle delay may be reduced as the SRD increases in the T_1-D experiment (Fig. 6.15).

In principle, it would be straightforward to replace the acquisition of the echo signal in the T_1-D sequence shown in Fig. 6.1 with a CPMG echo train. This transforms the two-dimensional experiment to a three dimensional T_1-T_2-Diffusion experiment. Although it may be straightforward to add a CPMG echo train into the sequence, it can be difficult to suppress the eddy current transients to achieve a recording of an undistorted CPMG echo train. Preparatory gradients prior to the 90° excitation pulse may be needed in order to improve the performance, which prevents the use of the shortest SRD's (~1 ms). Depending on the number of preparatory gradient pulses necessary to produce a steady state of the eddy current transients, the shortest SRD will be of the order of 10 ms. In addition, it may be required to add a z-storage delay before acquiring the CPMG-attenuation, in order to let the residual eddy current transients settle after the application of the bipolar pairs of gradient pulses. This eddy current settling delay should not have to be more than ~10 ms when the duration of the gradient pulses is ~1 ms (see Chap. 3). Altogether, the preparation parts within the pulse sequence prevent a proper characterization of components of $T_1 < 10$ ms. but lead to significant improvements in the performance of the experiment and return data undistorted by artefacts.

6.3 SR Applied at High Field

The application of the SR approach in high-resolution experiments provides the possibility of acquiring data from components of low-abundance within a reasonable measurement time. Because the recycle delay is practically zero and the SR can be set to T_1, several more scans may be accumulated within the same acquisition time as the traditional method of applying a recycle delay of 5 times T_1. Consequently, an enhanced signal to noise ratio will be achieved. In the following sections, the focus will be on describing the experimental set-up needed for acquiring a rapid DOSY experiment and a quantitative T_1 experiment with spectral resolution (Q-ROSY). After performing the ordinary Fourier transform of the FID's an Inverse Laplace transform is applied on each frequency component in the spectra as a function of gradient strength (DOSY) or recovery delay (Q-ROSY). Thus a two dimensional distribution is produced which provides a distribution of diffusion coefficients or T_1's in one direction whilst having spectral resolution along the other direction. Using the proposed sequences and processing tool a rapid experimental set-up for studying mixtures of molecules has become accessible, i.e. identifying resonances based on their chemical shift as well as their diffusion and relaxation behavior. This includes studies of dimer formation and ligand—protein interactions. One important result of the proposed sequences is that whilst the DOSY returns the same diffusion coefficient from protons sitting on a macromolecule, the Q-ROSY may return different values of T_1 from the same protons. This can be used as a probe to identify the shape of macromolecules in solution, and for example studying the folding of macromolecules [20, 21].

The high-resolution measurements on a 5 mm sample containing 2 mM Lysozyme in H2O/D2O 90/10 were performed using a Bruker 500 MHz Avance III instrument equipped with a 5 mm diffusion probe that can produce a magnetic field gradient up to 1,800 Gauss/cm.

6.3.1 *Rapid DOSY*

Diffusion ordered spectroscopy, DOSY, provides a tool that can correlate the resonance frequency with the molecular mobility of the sample [22, 23] (Fig. 6.16).

Figure 6.1 shows the rapid DOSY sequence, where the attenuation of the Fourier transformed FID for a multi exponential system at a given frequency,ν is written as

$$I(\nu) = I_0(\nu) \sum_i \rho_i (1 - \exp^{\frac{SRD^i}{T_1^i}}) \exp^{-\gamma^2 4 g^2 \delta^2 D^i (\frac{3}{2}\tau - \frac{\delta}{6}) - \frac{4\tau}{T_2^i}} \qquad (6.3)$$

γ is the gyromagnetic ratio, ρ_i is the weighting factor of region *i*, D^i is the diffusion coefficient, T_1^i is the longitudinal relaxation time, *g* is the applied gradient strength, *GS1* and *GS2* are the spoiler gradient pulses, δ_1, δ_2, δ and τ are time

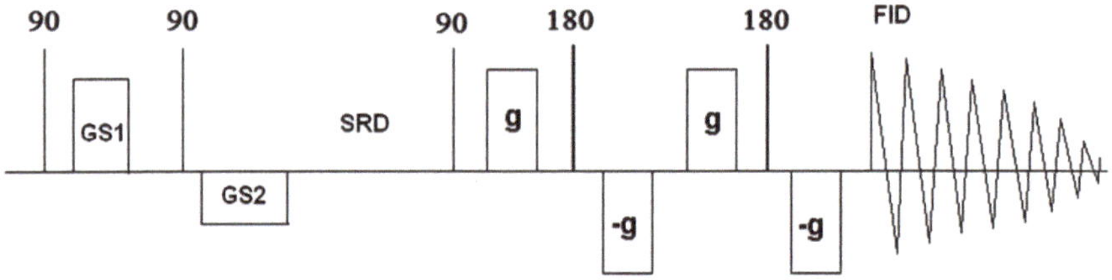

Fig. 6.16 The rapid DOSY experiment

duration parameters, and the SRD is the spoiler recovery delay. By fixing SRD and incrementing the gradient strength only, a set of FID's as a function of gradient strength is produced. After a Fourier transform (FT) of the FID, the spectra is subjected to a one dimensional inverse Laplace routine [24], resulting in a distribution of diffusion coefficients for each frequency point in the spectra. From 6.1 it is seen that a change in SRD will lead to different contributions from components with different T_1 relaxation time. This can be used for solvent suppression, as one usually wants to study macromolecules with short T_1's dissolved in a solvent having a much longer T_1.

Figure 6.17 shows the DOSY Fourier—Inverse—Laplace—Transformed (FILT) spectra arising from the lysozyme sample. As 16 values of the gradient strengths are applied, the maximum number of components fitted along the diffusion direction is set to 16. The grid in the ILT procedure is set to be between

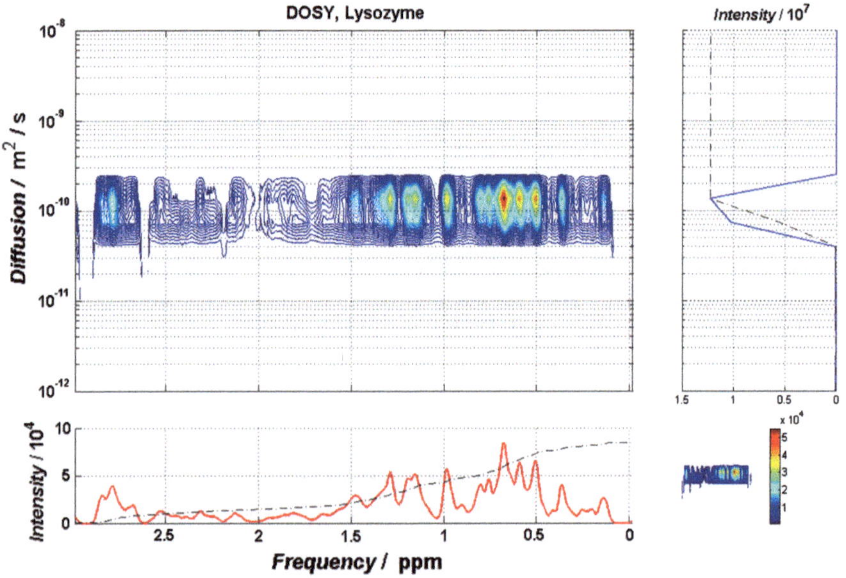

Fig. 6.17 Rapid DOSY applied on a Lysozyme sample

10^{-9} and 10^{-11} m^2/s. Thus, the resolution is such that it is not easy to achieve a mono exponential decay presented as a sharp peak (as a single data point) along the *D*-dimension. Instead, a weighting between two points in the *D*-dimension is more likely to appear, and this will represent the mono exponential decay due to the single component's diffusion. This explains why the diffusion distribution of the rapid DOSY of the lysozyme is rather broad compared to the signal along the frequency axis. The resolution along the *D*-dimension can be improved by increasing the number of measurement points at the expense of increased experimental time. This may also increase the time it takes to process the data matrix. For the data presented here 8192*16 data points are processed into a FT-1DILT in less than 1 s using Intel Core i7 930 @ 4 GHz. The weighting/smoothing of the data set for the 1D-ILT procedure is then fixed.

6.3.2 Rapid Q-ROSY

When using the saturating RF-train to reduce the recycle delay [3–5], relaxation ordered spectroscopy is not necessarily quantitative. The saturation levels of the different components within the sample is likely to be dependent on the individual relaxation times when applying the RF-train prior to the recovery sequence. When combining the spoiler recovery with the acquisition of an FID, a quantitative method for determining T_1-frequency correlated distributions is available (Fig. 6.18).

In Fig. 6.2 the simplest sequence using the spoiler recovery sequence is shown, only an extra 90° RF-pulse followed by acquisition of the FID. Again having Fourier transformed the FID of the mono exponential system, the attenuation at a given frequency,ν is written as

$$I(\nu) = I_0(\nu) \sum_i \rho_i (1 - \exp^{\frac{SRD}{T_1^i}}) \tag{6.4}$$

Compared to the standard inversion recovery sequence there are some important advantages using the spoiler recovery sequence. Again, the 5 times T_1 recycle delay is avoided and the acquisition time is for each observation time (SRD) the number of scans times (SRD + acquisition time of FID). Thus, the sequences with shortest SRD are recorded in a second or so while the only time consuming

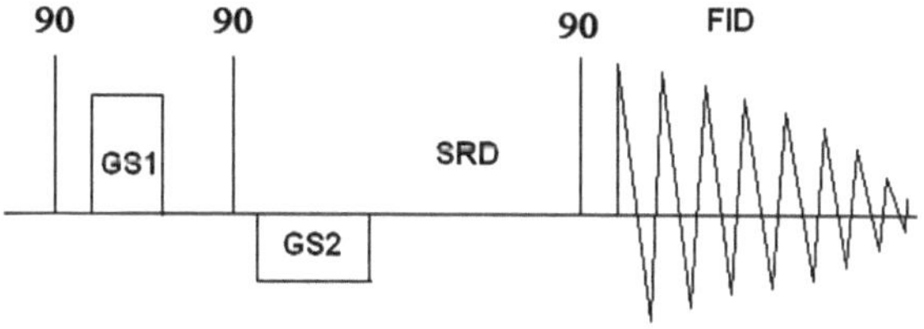

Fig. 6.18 The Q-ROSY experiment

acquisitions are the ones when SRD approaches 5 times T_1. Secondly, the signal does not start at a negative value, passing through the 0 or baseline where the signal is within the noise, and ending up at thermal equilibrium. In the spoiler recovery sequence, the signal starts at zero with the shortest SRD time and grows up to thermal equilibrium as the SRD time increases. This makes it easier to analyse using 1D-ILT as the data set will not be as sensitive to possible offset values from the baseline.

Figure 6.19 shows the Q-ROSY Fourier—Inverse—Laplace—Transformed (FILT) spectra arising from the lysozyme sample. As 16 SRD values were used, the maximum number of components along the T_1 direction is 16. The grid in the ILT procedure is set to be between 10^1 s and 10^{-3} s. Thus, the resolution is such that it again is not easy to achieve a mono exponential decay presented as a sharp peak along the T_1 direction. While all lysozyme peaks show the same diffusion coefficient in Fig. 6.17, the region between 2.5 and 3.0 ppm has a much shorter spin lattice relaxation time T_1. This must be due to the environment this group of protons are experiencing in the lysozyme molecule. The peak at ~0.6 ppm seems to have a broader T_1 distribution than the adjacent peaks. Maybe a larger number of data points (SRD delays) could have revealed a bimodal structure for some of the peaks in the lysozyme spectra.

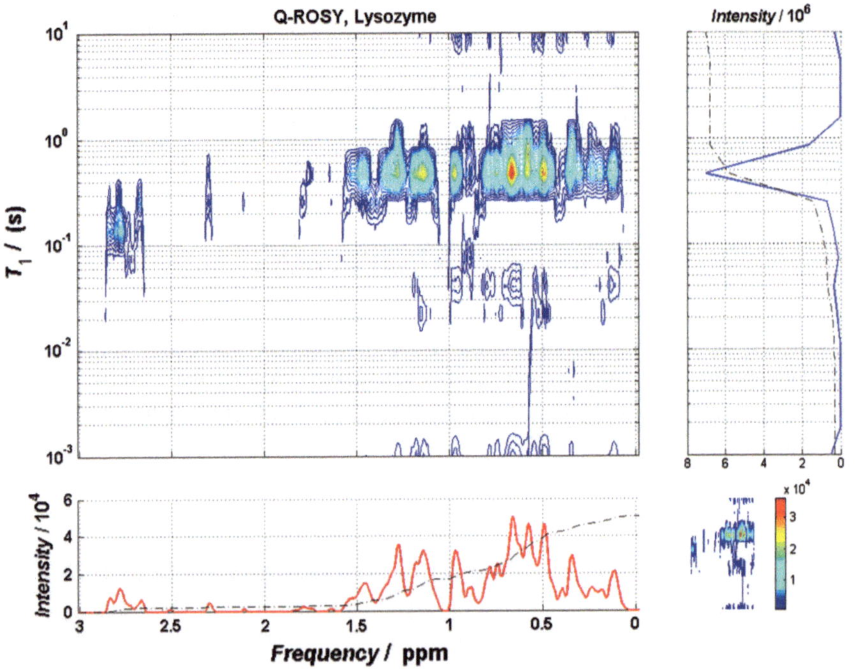

Fig. 6.19 Rapid Q-ROSY applied on a Lysozyme sample

6.3.3 *Applications of the Rapid DOSY*

The data acquired with the rapid DOSY and Q-ROSY methods also requires a software for fast processing of the data. In Fig. 6.20 a snapshot of the software in which the method applied by Anthonsen et al. [25] is implemented. After reading the data in an appropriate format, field of view for the ILT may be specified together with the smoothing/regularisation parameter. Thereafter the FT-ILT distribution is produced in just a few seconds.

For demonstration of the performance of the rapid DOSY on real samples, asphaltene solutions were prepared by dissolving asphaltenes in toluene-D_8 (99.6 atom %D, Aldrich) at a concentration of 1 wt% and then magnetically stirred overnight. This solution was then diluted by toluene-D_8 to produce an asphaltene solution of low abundance (~0.01 wt%). Due to the poor signal to noise ratio the latter solution required a relatively large number of scans to achieve an acceptable signal to noise ratio. Using the conventional approach (recycle delay of 5 times T_1), it would take 10 h to run the diffusion experiment on the low abundance sample. Alternatively, with the spoiler recovery method, the DOSY data set was acquired in approximately 1 h, without any loss in signal to noise ratio. The instrument used was a 14.1 T with a cryogenic probe, capable of delivering gradient strengths up to 50 Gauss/cm, and the temperature was set to 25 °C.

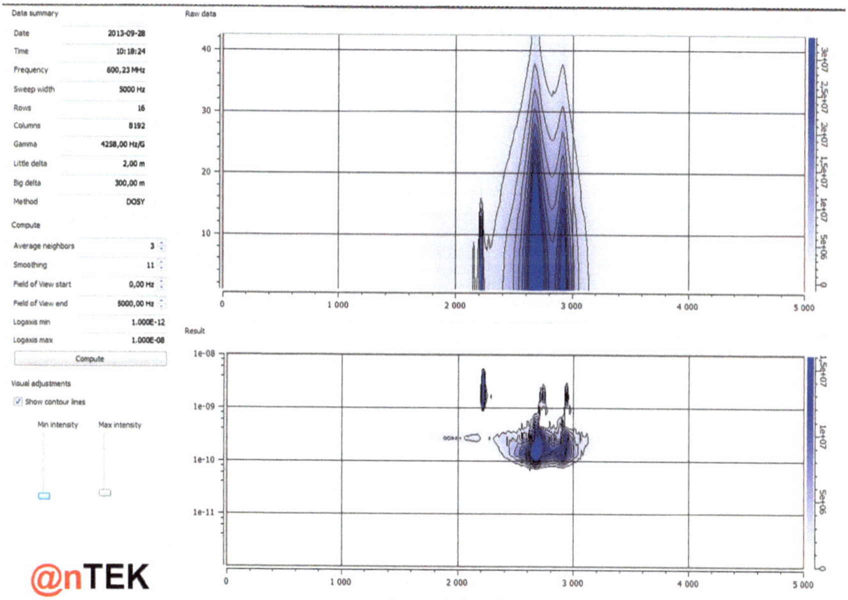

Fig. 6.20 A snap-shot of the fast software for performing an FT-ILT. It displays the DOSY applied on a sample of asphaltene solution, raw data in the *upper graph* and the FT-1DILT *below* (www.antek.no)

The DOSY distribution of asphaltene in Fig. 6.21 shows two major peaks around 0.9 and 1.4 ppm. A broad band covers the whole frequency range from 0.5 to 3.0 ppm, but only the range 0.5–2 ppm is shown to avoid the inference from the protons present in the toluene-D_8. Compared to the lysozyme, the diffusion distributions vary more, over one order of magnitude. Whilst the lysozyme exhibit no significant variation in diffusion coefficient as function of frequency a variation in diffusion coefficients as a function of frequency is seen in the asphaltene FT-1DILT. A major cluster is observed in the region 0.7–1.6 ppm with diffusion coefficients in the range $(1-3) \times 10^{-10}$ m^2/s. Interestingly, two regions around 0.9 and 1.3 ppm indicate narrow peaks in the frequency domain and a diffusivity of the order of $(1-2) \times 10^{-9}$ m^2/s. It is tempting to assign these peaks to the presence of rather well defined fractions of smaller molecules in the asphaltene solution. However, the value of the diffusion coefficient also fits well to the value found for the toluene-D_8 at 2 ppm, and measurements on pure toluene-D_8 indicate that impurities are present that may interfere in this region of frequencies. Figure 6.22 shows the identical measurement on the asphaltene solution of low abundance, but with 20 times more scans. At 2 ppm the shoulder of the proton toluene-D_8 signal is now resolved and returns a diffusion coefficient of $(1–3) \times 10^{-9}$ m^2/s. The narrow peak at 1.3 ppm has vanished, but the one at 0.9 ppm is still present and increased in signal intensity as compared to the cluster of asphaltene signal around 1.3 ppm. Unfortunately, it cannot be ruled out that this

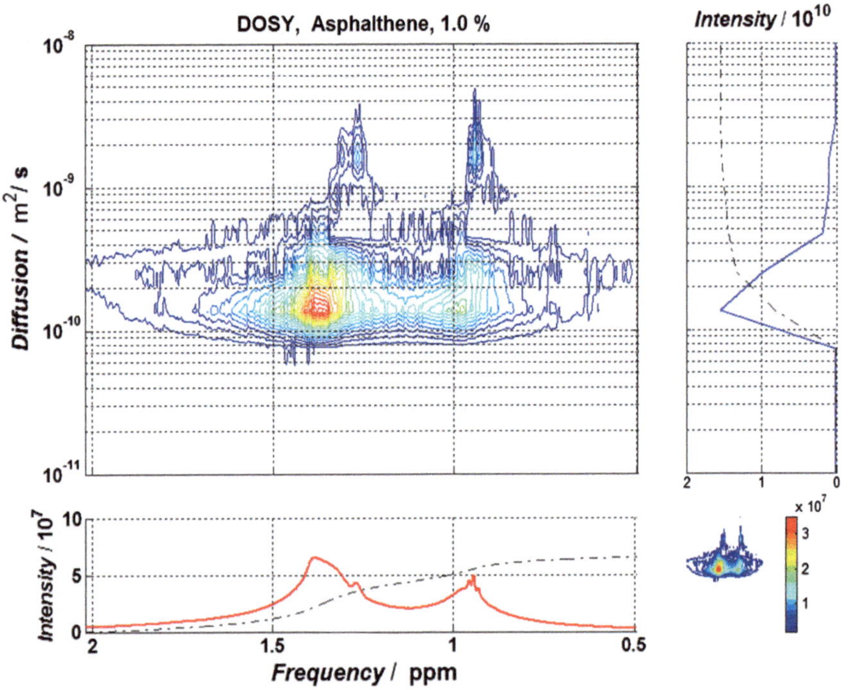

Fig. 6.21 Rapid DOSY applied on the asphaltene solution of 1.0 wt%

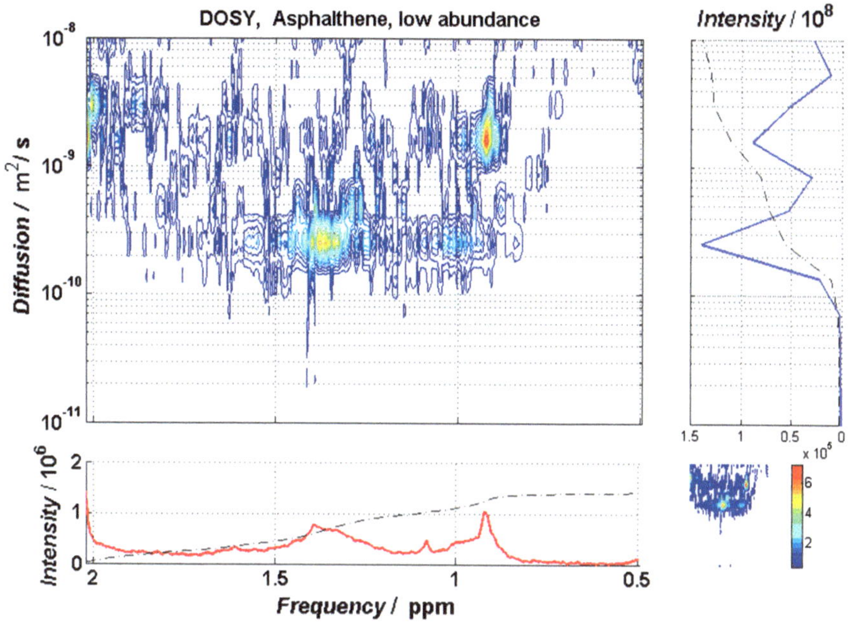

Fig. 6.22 Rapid DOSY applied on the asphaltene solution of ~0.01 wt%

signal at least partially originates from the impurities in the toluene-D. The main peak of the cluster shows a shift in diffusion coefficient towards higher values, and there is a region of less intense signal at diffusion values of the order of 10^{-9} m^2/s. Despite the presence of the interfering impurities of the toluene-D_8, it is therefore likely that some of the largest asphaltene compounds have segregated.

The fast DOSY and Q-ROSY experiments provide a tool for measuring spectrally resolved diffusion coefficients in solutions of low abundance. Water-soluble surfactants in emulsions may be studied, either at the interface (low mobility) or at present in the oil or water phase (high mobility). Drug release may be monitored from a water in oil in water (w/o/w) double emulsion, as the drug initially present in a water droplet will have a lower mobility than the drug that has moved through the droplet into the bulk water phase [26]. Repeating the DOSY-experiment at various times after production of the double emulsion, initially with the entire drug confined in the water droplets, may then reveal drug release rate as the high mobile fraction of the drug will evolve.

References

1. P.T. Callaghan, *Translational Dynamics and Magnetic Resonance* (Oxford University Press, Oxford, 2011), p. 547
2. W.S. Price, *NMR Studies of Translational Motion Principles and Applications* (Cambridge University Press, Cambridge, 2010)

3. T. Stait-Gardner, P.G. Anil Kumar, W.S. Price, Steady state effects in PGSE NMR diffusion experiments. Chem. Phys. Lett. **462**(4–6), 331–336 (2008)
4. J. Mitchell, M.D. Hürlimann, E.J. Fordham, A rapid measurement of T1/T2: the DECPMG sequence. J. Magn. Reson. **200**(2), 198–206 (2009)
5. L. Venturi, K. Wright, B. Hills, Ultrafast T1–T2 relaxometry using FLOP sequences. J. Magn. Reson. **205**(2), 224–234 (2010)
6. M.D. Pelta et al., A one-shot sequence for high-resolution diffusion-ordered spectroscopy. Magn. Reson. Chem. **40**(13), S147–S152 (2002)
7. Y.-Q. Song, X. Tang, A one-shot method for measurement of diffusion. J. Magn. Reson. **170**(1), 136–148 (2004)
8. V.V. Krishnan, N. Murali, Radiation damping in modern NMR experiments: progress and challenges. Prog. Nucl. Magn. Reson. Spectrosc. **68**, 41–57 (2013)
9. R.H. Hashemi, W.G. Bradley, C.J. Lisanti, *MRI: The Basics* (Wolters Kluwer Health, The Netherlands, 2012)
10. J.M. Fauth et al., Elimination of unwanted echoes and reduction of dead time in three-pulse electron spin-echo spectroscopy. J. Mag. Reson. **66**(1), 74–85 (1986)
11. N. Bloembergen, R.V. Pound, Radiation damping in magnetic resonance experiments. Phys. Rev. **95**(1), 8–12 (1954)
12. W.S. Price, F. Tsuchiya, Y. Arata, Lysozyme aggregation and solution properties studied using PGSE NMR diffusion measurements. J. Am. Chem. Soc. **121**(49), 11503–11512 (1999)
13. P. Broekaert, J. Jeener, Suppression of radiation damping in NMR in liquids by active electronic feedback. J. Magn. Reson. Ser. A **113**(1), 60–64 (1995)
14. W.E. Maas, F.H. Laukien, D.G. Cory, Suppression of radiation damping by Q switching during acquisition. J. Magn. Reson. Ser. A **113**(2), 274–277 (1995)
15. H. Barjat, D.L. Mattiello, R. Freeman, Suppression of radiation damping in high-resolution NMR. J. Magn. Reson. **136**(1), 114–117 (1999)
16. A.K. Khitrin, A. Jerschow, Simple suppression of radiation damping. J. Magn. Reson. **225**, 14–16 (2012)
17. P.P. Mitra, P.N. Sen, L.M. Schwartz, Short-time behavior of the diffusion coefficient as a geometrical probe of porous media. Phys. Rev. B **47**(14), 8565–8574 (1993)
18. G.H. Sørland, Characterization of emulsions by PFG-NMR. AIP Conf. Proc. **1330**(1), 27–30 (2011)
19. J. Stepišnik, Validity limits of Gaussian approximation in cumulant expansion for diffusion attenuation of spin echo. Phys. B **270**(1–2), 110–117 (1999)
20. G. Grasso, J.J. Titman, Chain folding and diffusion in monodisperse long n-alkanes by solid-state NMR. Macromolecules **42**(12), 4175–4180 (2009)
21. A.G. Palmer, D.J. Patel, Kurt Wüthrich and NMR of biological macromolecules. Structure **10**(12), 1603–1604 (2002)
22. K.F. Morris, C.S. Johnson, Diffusion-ordered two-dimensional nuclear magnetic resonance spectroscopy. J. Am. Chem. Soc. **114**(8), 3139–3141 (1992)
23. C.S. Johnson Jr, Diffusion ordered nuclear magnetic resonance spectroscopy: principles and applications. Prog. Nucl. Magn. Reson. Spectrosc. **34**(3–4), 203–256 (1999)
24. S.W. Provencher, A constrained regularization method for inverting data represented by linear algebraic or integral equations. Comput. Phys. Commun. **27**(3), 213–227 (1982)
25. H.W. Anthonsen, G.H. Sørland, K. Zick, J. Sjoblom, Quantitative recovery ordered (Q-ROSY) and diffusion ordered spectroscopy using the spoiler recovery approach. Diffus. Fundam. **16**, 12 (2012)
26. F. Leal-Calderon, J. Bibette, V. Schmitt, *Double Emulsions*, in *Emulsion Science* (Springer, New York), pp. 173–199

Chapter 7
Emulsion Characterisation

Abstract Emulsions are dispersions of oil and water classified according to the nature of the continuous phase. The simplest classification is water-in-oil emulsions and oil-in-water emulsion. A variety of pulsed field gradient nuclear magnetic resonance (PFG NMR) experiments will be presented, providing data that can be converted into droplet size distributions (DSD), water profiles or surface to volume profiles. PFG NMR is a versatile technique, and often there are several possible methods to determine the DSD. How to choose the appropriate method is discussed, as it sometimes may be a challenge to select the best sequence for a given emulsion system. As the key to successful droplet size measurements is proper resolution of the water signal from the crude oil, this will also be given a thorough treatment, along with artefacts that may interfere with this process.

The 20 MHz NMR instrumentation dedicated for characterizing emulsions

"The low field NMR instrumentation dedicated for characterizing emulsions by PFG-NMR. It features pulsed field gradients up to 400 Gauss/cm *on sample tubes of* 20 m *diameter, and the sample may be tempered by air within the range* −50 to +100 °C. *The gradient coils are mounted on the pole base to reduce the impact of knocking of the gradients on the PFG-NMR experiment. Eddy current dead times are down to* ~0.4 ms."

G.H. Sørland, *Dynamic Pulsed-Field-Gradient NMR*,
Springer Series in Chemical Physics 110, DOI 10.1007/978-3-662-44500-6_7

Emulsions are dispersions of oil and water classified according to the nature of the continuous phase. The simplest classification is water-in-oil (w/o) and oil-in-water (o/w) [1]. Nevertheless, more complex systems, such as double emulsions, in which small droplets are dispersed into droplets, can be formed for a period depending on the system prepared. Double emulsions can be oil-in-water-in-oil (o/w/o) or water-in-oil-in-water (w/o/w) [2]. As the emulsions are thermodynamically unstable, they do not form spontaneously. However, the presence of surface active agents (surfactants) can enhance the formation of the emulsions [3] and can stabilize them once they have been formed. Emulsions play a central role in our daily life since they are present in food [4], petroleum [5], pharmaceutical and cosmetic industries [6]. A proper characterization of emulsions, either being water in oil or oil in water, is important for various purposes. Droplet size is a key factor to control the flavor of a food, as the size distribution of fat droplets may affect how the food is experienced when being tasted. In the oil separation industry, it could be the characterization of a small but unwanted brine fraction present within crude oil that is critical as here small sized brine droplets may be difficult to separate from the heaviest crude oils [7–9]. In the pharmaceutical industry, double emulsions have particular importance. Drugs are dissolved in small water droplets that are confined in larger oil droplets which finally are suspended in a water solution [2]. When the double emulsion is injected in vivo, the rate at which the drug is released depends on several factors, including the concentration of the drug, the droplet size of the oil and on the droplet size and quantity of water droplets inside the oil droplets. All these parameters can in principle be determined by PFG NMR prior to injection.

The mechanisms of emulsion formation and stability are more or less similar for all industries being involved with emulsions. In order to describe the features of an emulsion and being able to predict behaviors as sedimentation, coalescence or flocculation, a complete characterization would be of interest. Important features of an emulsion are the average droplet size, the distribution function of the sizes of the droplets (droplet size distribution), the quantity of oil and water and their interaction at the boundaries, and the time stability of the emulsion. The droplet size distribution (DSD) influences many characteristics, for instance the rheology [10, 11], the stability of an emulsion [12] and emulsion liquid membrane performance [13]. There are presently several different techniques available to obtain the DSD. The suitability of the techniques, the quality and reliability of the measurements, and the ease at which to use them depend on the system to be studied. Some of the techniques require a sample preparation that may alter the characteristics of the emulsion. Other techniques consider only a small portion of the sample, making the results less representative. Some require the sample to have particular properties, for example the success of light scattering depends on whether or not the sample is transparent to visible light. This is often not the case for petroleum emulsions [14]. Thus, if any droplets are measured at all, only the droplets in vicinity of the container wall are included in the measurement. This technique is also not ideal for concentrated emulsions. In addition, light scattering does not discriminate between single droplets and clusters [15]. The optical microscopy is another popular technique used for droplet size characterization, but it often requires dilution of the sample. In addition to being potentially tedious and labor intensive, there are wall effects to consider when the emulsion droplets are placed between microscope slides, and only a small part of the sample is analyzed [16]. Compared to the abovementioned techniques PFG NMR has several advantages in obtaining the DSD. Firstly, the entire sample is considered, no sample preparation or dilution is required and the measurements can be relatively fast. Secondly, the non-perturbing handling of the sample means that the same sample can be analyzed several times. Thirdly, the technique is particularly useful for opaque samples.

In this chapter a variety of experiments will be presented, providing PFG NMR data that can be converted into DSD's, water profiles or $(S/V)^{-1}$ profiles. PFG NMR is a versatile technique, and often there are several possible methods to determine the DSD. How to choose the appropriate method will also be discussed, as it sometimes may be a challenge to select the best sequence for a given emulsion system. As the key to successful droplet size measurements is proper resolution of the water signal from the crude oil, this will also be given a thorough treatment, along with artefacts that may interfere with this process.

7.1 Separating the Oil and Water Signals

In order to achieve a proper characterization of the emulsions by PFG NMR, it is crucial to resolve the signal arising from the two main components present, the oil and the water phase. When it is not possible to separate the oil and water signals by their chemical shifts in a Fourier transformed spectrum, other characteristics of the two components must be applied to separate their contribution to the PFG NMR experiments. When working with low resolution NMR-instrumentation, this will be the relaxation times, the difference in temperature dependency on the viscosity, the mobility, or the observation time dependency on the root of the mean squared displacement. Water provides a better probe for the characterization of the emulsion than oil, as the latter may contain hydrocarbon chains of varying molecular weight and thus result in a broad distribution of T_2-values or diffusion coefficients even when being present as a bulk fluid. Hence, it is difficult to assign a distribution of T_2 or Diffusion values from an oil signal to a distribution of droplets. For the reverse situation, bulk water is an excellent probe due to the NMR-signal's single exponential decay for both relaxation and diffusion. Focus will therefore be on separating the water signal from the oil, and using this signal to probe the emulsions regardless of they are water in oil or oil in water emulsions.

7.1.1 The Inversion Recovery Approach

A frequently used approach for suppression of the oil signal is to make use of the fact that the oil and water have significantly different T_1-relaxation times [17, 18]. The usual procedure to resolve water from oil is therefore to use an inversion recovery sequence. This works by initially applying a 180° RF-pulse. Thereafter a wait time is applied, t_{rd}, where the duration is adjusted to when the net nuclear magnetic moment for the oil fraction is zero along the direction of the external field (z-axis) with an apparent single longitudinal relaxation time T_1^{oil}.

$$t_{rd} = \ln(2) \cdot T_1^{oil} \tag{7.1}$$

At t_{rd} the magnetization brought onto the xy-plane by a 90° RF-pulse, and is assumed to arise from the water phase only. Subsequently, the appropriate PFG NMR sequence is applied for extracting the droplet size distribution of the water phase [19, 20].

This method unfortunately only applies when the T_1-relaxation of the oil fraction can be described by a single exponential T_1 decay. When there is a broader range of relaxation times present, the zero net nuclear magnetization along the z-axis of the various components will appear at different durations following the initial 180° RF-pulse. There will always be a well-defined duration, where net nuclear magnetization of the components within the oil adds up to be zero. At the time of application of the 180° RF-pulse, t_{rd}, there will initially be a zero magnetization from the oil phase, but due to the finite duration of the PFG NMR sequence, some of the oil components will relax towards thermal equilibrium faster than others. Consequently, the remaining net nuclear magnetization in the xy-plane of the components within the oil phase will not necessarily add up to zero, and a non-zero nuclear magnetization from the oil phase will appear in the xy-plane, as documented in Fig. 7.1. The amplitude of this magnetization will depend on the characteristics of the oil phase and the duration of the PFG NMR sequence.

In Fig. 7.1 the inversion recovery approach has been applied on a sample containing crude oil only, where the duration between the initial 180° RF-pulse and the PFG NMR sequence was adjusted so that the sum of the net nuclear magnetization was zero at the time of the 90° RF-pulse (IR = 76.8 ms). This is verified by the recording of noise only from a spin echo. The spin echo sequence was then exchanged with an ordinary stimulated echo sequence using a z-storage delay of 250 ms. Even though there apparently was no initial magnetization following the first 90° RF-pulse, a significant signal appeared at the time of the stimulated echo. This signal appears even if a proper phase cycle and crusher gradients are applied, and must therefore be a consequence of the distribution of T_1-values from the oil phase. As expected, the signal intensity is dependent on the duration of the z-storage-interval, as it reaches a maximum at approximately 200 ms, and then starts to decrease. The decrease is due to the relaxation of all

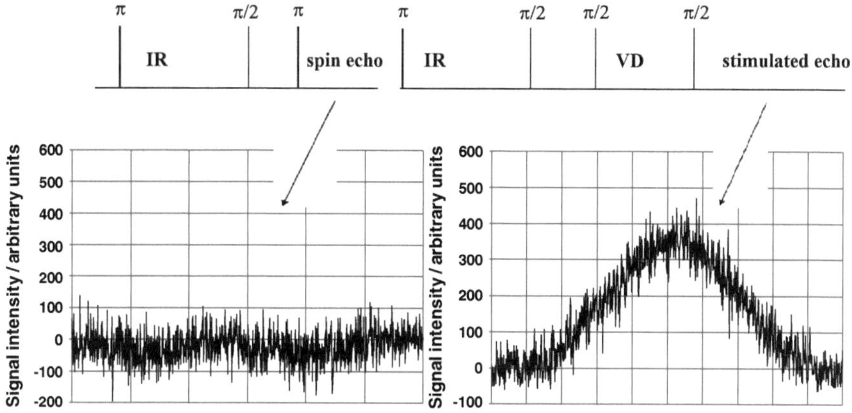

Fig. 7.1 Two inversion recovery experiments: *The left part* is an Inversion Recovery where the delay IR is adjusted (IR = 76.8 ms) such that the following spin echo acquisition only produces noise. *The right part* is an Inversion Recovery with IR = 76.8 ms, followed by an acquisition of stimulated echo where VD, the z-storage delay, is 250 ms

components within the oil phase towards thermal equilibrium, and at some duration, the oil signal no longer contributes to the stimulated echo signal. At this duration, the oil signal is suppressed due to relaxation, which will be discussed in the next two sections.

When the PFG method is introduced after the IR-sequence, it is tempting to state that the unwanted oil signal, which is produced at the time of the stimulated echo, will be crushed by the applied gradient prior to the recording of the stimulated echo, i.e. in the read interval. Obviously there is no signal to crush in the prepare interval, but that is only because the net magnetization is zero from different components within the crude oil. However, all components will undergo motional encoding both in the prepare and read intervals, and will thus contribute to the stimulated echo signal as the wanted water signal. As long as there is a distribution of T_1-values for the oil phase, an oil signal will always contribute to the PFGSTE. That is, when the duration of IR is defined as the time when the signal from the oil phase after the IR is zero. This the traditional way of determine droplet size distributions using the approach by Packer and Rees [17–20].

7.1.2 T_2-Separation

In Fig. 7.2 the ordinary T_2 distribution from a brine in crude oil emulsion is shown at 30 °C. Due to the significant difference in T_2 relaxation times of the brine and the crude oil, it is straightforward to separate the brine from the crude

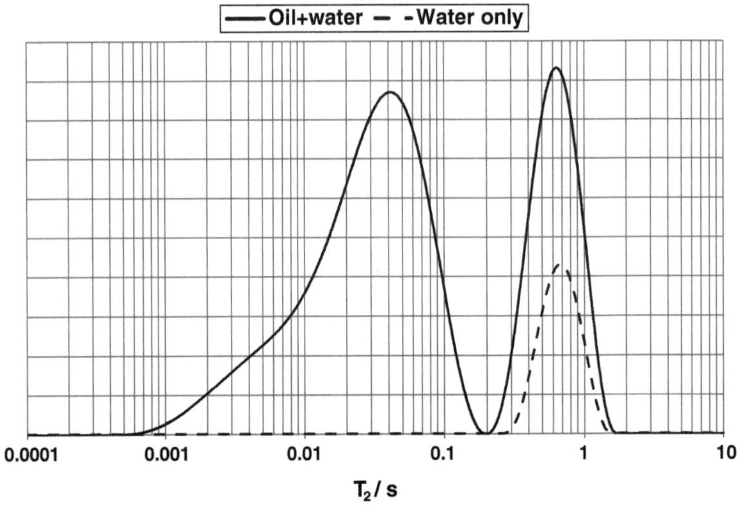

Fig. 7.2 The separation of water signal (*dotted line*) from the total signal (*left part* of the distribution represented by a *solid line*) using CPMG data

oil signal using a T_2-separation procedure. By running an ordinary CPMG sequence, where the first echo signal acquired is the one where the crude oil signal has decayed to an insignificant value, the remaining attenuating signal will have its origin from the water phase only. The dotted line in Fig. 7.2 represents the T_2-distribution processed from the CPMG data where the first echo signal acquired arises from water signal only. This procedure for separation of the water signal from the crude oil signal can be built into the PFG NMR sequence(s) that are used for characterization of the emulsion. The suppression of the oil signal shown in Fig. 7.2 resulted in a loss of approximately 60 % of the water signal as well. As the water content in the emulsion initially is 30 %, this loss will not cause a major decrease in signal to noise ratio. However, if the water fraction was 10 % or less, the loss of signal would require an increased number of accumulating scans, in order to compensate for the signal loss during the suppression of the oil signal. This requirement leads to an increased experiment time.

7.1.3 T_1-Separation

When the T_2-distributions of oil and water overlap significantly, as would be the case for very small water droplets, enhanced surface relaxivity or the presence of internal magnetic field gradients, the separation due to non-overlapping distributions along the T_1-direction may apply instead. The ability to resolve the signals along the T_1 direction will depend on the T_1/T_2 ratios. In

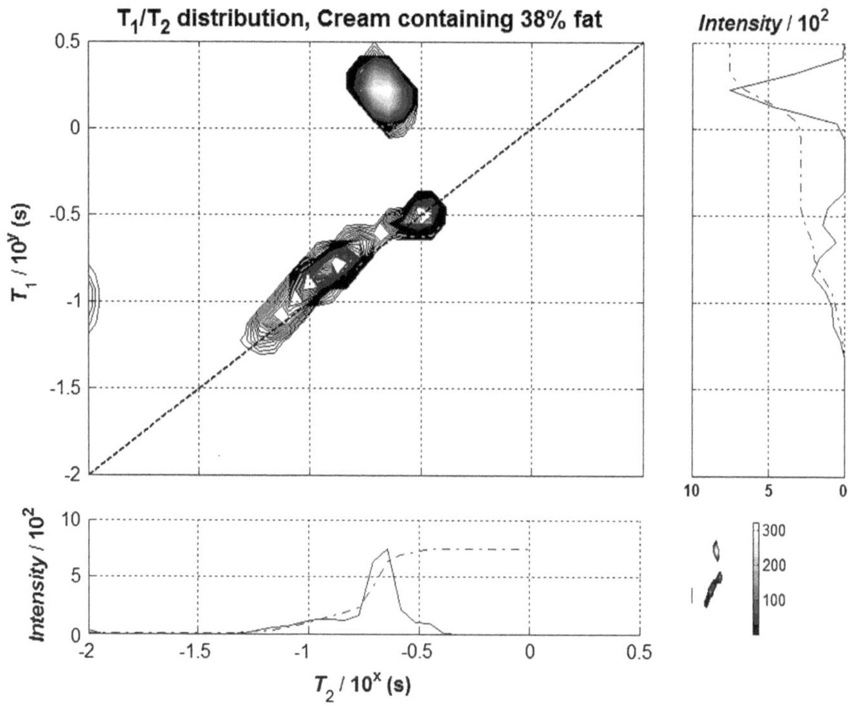

Fig. 7.3 The two-dimensional inverse Laplace transform applied on data arising from a T_1-T_2-correlated experiment on cream. The fat signal is located along the diagonal (*dotted line* where $T_1 = T_2$) while the remaining signal is from water

Fig. 7.3 the 2D-ILT distribution is shown for a sample of cream containing 38 % fat recorded at 35 °C. From the projections to the x-axis (T_2-relxation) and y-axis (T_1-relaxation), no resolved signal is seen in the T_2-regime while the water component is well resolved along the T_1-direction. For the water signal the T_1/T_2 ratio $\gg 1$ while it is approximately 1 for the oil signal found at or just above the diagonal where $T_1 = T_2$. Instead of using an ordinary CPMG-sequence, the stimulated echo sequence must be applied using a long enough z-storage delay, in order to resolve the water component from the crude oil signal by T_1-separation. In Fig. 7.3 the T_2 distributions arising from a combined stimulated echo CPMG experiment are shown with two values of the z-storage delay, $\Delta = 5$ ms (solid line) and $\Delta = 1{,}000$ ms (dotted line). The water in crude oil emulsion used is identical to the one producing the distributions shown in Fig. 7.2. In the version with the shortest Δ both crude oil and water signals are present, while for the longest Δ the T_2-distributions the water signal only remains. The loss of water signal due to the oil suppression delay is 50 %, which is an improvement compared to the loss experienced using the T_2-separation approach.

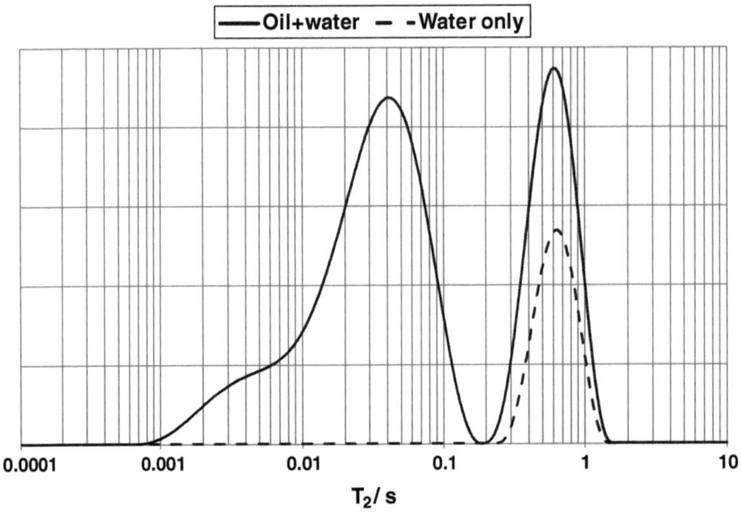

Fig. 7.4 The separation of water signal (*dotted line*) from the total signal (*left part* of the distribution represented by a *solid line*) using combined Stimulated Echo CPMG data

When the water and oil signal is resolved in both the T_1 and T_2 direction, the direction in which most of the water signal is preserved should be chosen. Assuming a mono exponential decay of the oil signal, the suppression time should be 5 times the relaxation time of the oil to ensure a satisfactory suppression of the oil signal. As the oil in general has a distribution of relaxation times, the T_2-distribution at various suppression times should be evaluated. The shortest suppression time can then be found, for which the acquired PFG NMR data do not contain interfering oil signals.

As the T_1/T_2 ratio of the oil is close to unity for many emulsion systems, this should imply the use of oil suppression is preferred along the T_1 direction, as $T_1/T_2 > 1$ for the water. However, in many emulsion systems, the surface relaxivity at the interface between the water and oil is weak, and the T_1/T_2 may be close to unity for the water as well. The results shown in Figs. 7.2 and 7.4 were produced using the same sample, and for this particular sample, suppression along the T_1-direction provided the smallest loss of water signal compared to the initial water signal.

Provided the emulsion returns resolved oil and water T_2-distributions, and depending on the T_1/T_2-ratios of both the oil and the water present in the sample, either a T_1 or T_2 method can be used for suppression of the oil signal. However, the approach using T_1-suppression provides a z-storage delay in the sequence, and this can be made use of when characterizing the emulsion. When determining average droplet size of the emulsion, pulsed field gradients in a diffusion sequences are in use, either before or after the oil suppression sequence. When the oil suppression appears before the diffusion sequence, the preparatory gradients in the diffusion sequence may be placed in the z-storage delay used for oil suppression. This is not possible when using the T_2-suppression method, and thus the flexibility using T_1-suppression is better when preparatory gradients are needed (Chap. 3).

When determining water profiles of unstable emulsions, where sedimentation or coalescence may take place within minutes, the time required to acquire the profile should be minimized. In order to acquire profiles on a timescale much less than the phase separation process, it may be necessary to conduct one scan only. In this case the use of a stimulated echo sequence to suppress the oil signal is not the best choice, as it in principle requires a phase sequence involving at least two scans to get rid of the unwanted coherences. In such a situation, the T_2-suppression approach is a better choice, for example when determining water profiles in unstable emulsions [21].

7.1.4 Temperature Separation

When neither the T_1 nor the T_2 relaxation times resolve the water from the oil/fat signal at a given temperature, there is the possibility of running the PFG NMR measurements at a lower temperature, as the temperature dependency on viscosity is usually stronger for the oil than the water component [22, 23]. A lowering of the temperature will result in a larger shift of the T_2 values for the oil than for the water. In Fig. 7.5 the processed T_2 distributions are shown for an emulsion at various temperatures. While there is a considerable overlap of the distributions at the highest temperature, there are two well-resolved peaks at the lowest temperature (oil peak to the left in Fig. 7.5). When a temperature is found, the T_1 and T_2 distributions from oil and water are well separated, either the T_1 or T_2 method can be applied to resolve the water signal from the oil.

The major drawback with changing the sample temperature is that it may require extra hardware for cooling of the sample, as the permanent magnet usually is tempered to a value above room temperature. In addition, the fixed temperature at which the water and oil is resolved is not necessarily at the desired measurement condition. The characteristics of the emulsion are temperature dependent, the colder the sample is the less sensitive it is to coalescence or sedimentation. Thus, the information gathered at a wrong temperature is not an optimal solution. The separation by the difference in root of the mean squared displacement at longer observation times may then apply instead.

7.1.5 Root of Mean Squared Displacement Separation

When neither the T_1 nor the T_2 relaxation times resolve the water from the oil signal, and reduction in sample temperature is not an option, an important parameter

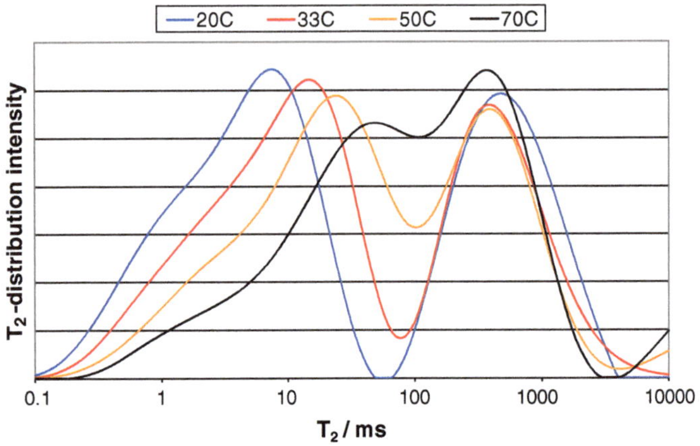

Fig. 7.5 The inverse Laplace transform applied on data arising from a CPMG experiment on a crude oil in water emulsion at different temperatures. The *blue* distribution represents resolved contributions from oil (*left peak*) and water (*right peak*)

to optimize is the difference in observation time dependent restricted mobility of the oil and water phase. Assuming a water in oil emulsion, for the shortest observation times the apparent diffusion coefficient of water deviates slightly from the bulk value, which is of the order of 10^{-9} m^2/s. Depending on the length of the hydrocarbon chains, the apparent diffusion coefficient of the oil phase will vary from 10^{-9} m^2/s and downwards. It will be slightly affected by the presence of the water droplets. While it is possible for the oil to move over distances that exceed the dimensions of the water droplets, the water is confined within the dimension of the droplet. Consequently, the water will be the component with the highest apparent diffusion coefficient at the shortest observation times, but will gradually turn to having the lowest apparent diffusion coefficient at longer observation times. As the water is confined within the droplets, the root of the mean squared displacement as measured from the PFG NMR experiment will not exceed the dimension of the droplet. When the asymptotic limit is reached (Chap. 2), there is no longer an observation time dependency on the attenuation of the PFG NMR experiment. For the attenuation of the oil signal, there is still an observation time dependency present and thus its slope will become steeper as the observation time is being increased in the PFG NMR experiment.

In Fig. 7.6 the attenuations from diffusion experiments performed on a water in oil emulsion are shown using the 13-interval PFGSTE sequence at two different observation times, 35 and 200 ms. At the shortest observation time there is no clear two-fraction model present in the attenuation, and thus it is not possible to resolve either of the components in the PFG NMR experiment. When increasing the observation time to 200 ms, a two-fraction model is seen. The fast attenuating signal originates from the oil phase while the slow attenuating signal arises from water trapped in droplets. As the observation time is increased further the slope of the fast attenuating signal becomes steeper while the slope of the water remains unchanged. Consequently, the

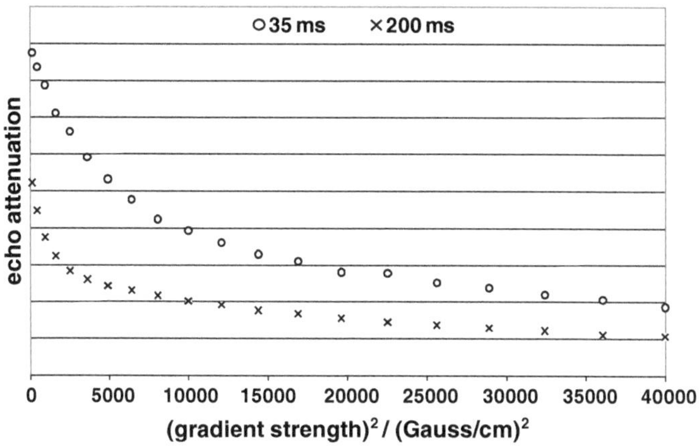

Fig. 7.6 The attenuation of the water in oil emulsion using the 13-interval PFGSTE at two observation times [$\Delta = 35$ ms ('o') and $\Delta = 200$ ms ('x')]

water signal has reached the asymptotic limit, and it is resolved from the crude oil signal at a given gradient strength (~100 Gauss/cm when $\Delta = 200$ ms in Fig. 7.6).

When including this approach for resolving water signal from the oil signal along with the other approaches described earlier, there exist NMR methods to resolve the water signal regardless of the oil viscosity. When the oil viscosity is high (i.e. non-overlapping T_1 and T_2 with water) the T_1- or T_2-methods apply. As the oil viscosity approaches the water viscosity (i.e. overlapping T_1, T_2 and diffusion coefficients), the root of the mean squared displacement method applies.

An important feature of using the root of the mean squared displacement approach is the use of relatively short observation times to suppress the oil signal. As shown in Fig. 7.6, the oil signal is easily suppressed at an observation time of 200 ms, and the remaining fraction of the water signal due to gradient suppression is then approximately 0.9 of the initial signal. Assuming a relaxation time of approximately 600 ms, the total suppression of the water signal is at a fraction of 0.65 relative to the initial signal. Thus, this approach provides the smallest loss in water signal due to suppression of the oil signal, and it will be the best choice of method in the presence of low water content in the emulsion.

7.2 Droplet Size Distributions

Stejskal and Tanner [24–26] pioneered the work on restricted diffusion by studying the diffusion using the pulsed field gradients (PFG NMR) experiment, where they utilized the difference in the relaxation times for oil and water to separate the signals. Since this initial work on restricted diffusion, the method has been further developed to its current status where it may be used to characterize various food and crude oil emulsions [18, 27–30]. As the NMR experiment can be performed in numerous ways, a large variety of approaches has been developed for this purpose. Even though the size determination using different NMR approaches has been compared with other techniques with promising results, many of the methods are based on the work by Packer and Rees [20] with the assumption that the shape of the distribution follows a log-normal distribution [29, 31–33]. Most droplet size distributions do follow a log-normal distribution. However, it might not be the case for all systems. The loss of accuracy in the determined distribution shape can be of importance in order to fully understand the behavior of an emulsion. Peña and Hirasaki [34] included a CPMG sequence to avoid the a priori assumption of the distribution function. Nevertheless, they still applied the same diffusion model as used by Packer and Rees to find the droplet sizes. It should also be noted that these methods assumes a mono exponential decay of the oil signal due to longitudinal relaxation. This is not the case in for example crude oil emulsion, where the relaxation components from the crude oil may vary several orders of magnitude. Aichele et al. [30] presented a technique using PFG NMR with diffusion editing (DE) to quantify brine/crude oil emulsions. This technique made no assumptions regarding the distribution function. However, each measurement was relatively long, (5–7 h), and proved sensitive to coalescence.

In this chapter, two methods which employ a combination of Pulsed Field Gradient (PFG), Stimulated Echo (STE) and CPMG sequences to obtain the droplet size distributions (DSD) for crude oil emulsions within 1–2 min, will be presented. The methods combine either the short diffusion time model developed by Mitra et al. [35] or a modification of the original method presented by Packer and Rees [20]. This modification will be referred to as the asymptotic approach.

7.2.1 The Short Observation Time Approach

As shown in Chap. 2, there is a situation where the surface relaxation term is absent in the solution of the diffusion propagator, i.e. the short observation time expansion. By assuming piecewise smooth and flat surfaces and that only a small fraction of the particles are sensing the restricting geometries, the apparent restricted diffusion coefficient can be written as

$$\frac{D(t)}{D_0} \approx 1 - \frac{4}{9\sqrt{\pi}}\sqrt{D_0 t}\frac{S}{V} + \phi(\rho, R, t) \tag{7.2}$$

D(t) is the time dependent apparent restricted diffusion coefficient, D_0 is the unrestricted diffusion coefficient in bulk fluid, and t is the observation time. The higher order terms in t, $\varphi(\rho, R, t)$ contain the deviation due to finite surface relaxivity and curvature (*R*) of the surfaces. At the shortest observation times, these terms may be neglected such that the deviation from bulk diffusion depends on the surface to volume ratio alone. In a porous system a large span in droplet sizes must be assumed, but (7.2) is also valid for a heterogeneous system. If ξ_i is the volume fraction of the droplets with surface to volume ratio $(S/V)_i$, (7.1) can be expressed as

$$\sum_i \xi_i \frac{D_i}{D_0} \approx \sum_i \xi_i \left[1 - \frac{4}{9\sqrt{\pi}}\sqrt{D_0 t}\left(\frac{S}{V}\right)_i\right] = \left(1 - \frac{4}{9\sqrt{\pi}}\sqrt{D_0 t}\,\overline{\left(\frac{S}{V}\right)}\right) \tag{7.3}$$

Measurements of the early departure from bulk diffusion combined with a linear fit of the experimental data to the square root of time will thus result in an average value for the surface to volume ratio (S/V).

Transforming a T_2 distribution into a droplet size distribution

Assuming that the water molecules probe the droplets within the sample, there is a simple relation between T_2 values and the droplet sizes [36–38]

$$T_2 \approx \frac{V}{S\rho} \tag{7.4}$$

This couples the surface to volume ratio to the surface relaxivity, ρ, and makes it difficult to assign the T_2 distribution to a (V/S) distribution. However, if the assumption made in (7.4) holds for any droplet size, with ξ_i being the volume fraction of droplets with surface to volume ratio $(S/V)_i$ and corresponding relaxation time T_{2i}, following Uh and Watson it may be written [39]

$$\sum_{i=1}^{n} \xi_i \frac{1}{T_{2i}} = \sum_{i=1}^{n} \xi_i \rho_i \left(\frac{S}{V}\right)_i \approx \rho \sum_{i=1}^{n} \xi_i \left(\frac{S}{V}\right)_i = \rho\,\overline{\left(\frac{S}{V}\right)} \tag{7.5}$$

Here the basic assumption that the surface relaxivity ρ is independent of droplet size is made use of. The left hand side of (7.5) is the harmonic mean of the T_2-distribution weighted by the fraction ξ_i of nuclei with relaxation time T_{2i} and n is the number of subdivisions of droplet sizes. This average can be calculated from the T_2-distribution obtained in a CPMG measurement where the magnetization attenuation is transformed into a T_2 distribution by solving an inverse problem using e.g. an Inverse Laplace Transform (ILT) routine. Then the surface relaxivity ρ can be calculated from (7.5) if the average surface to volume ratio is already found from the diffusion experiment. Consequently, the measured T_2-distribution can be transformed into an absolute droplet size distribution (*V/S*) by means of the relationship inherent in (7.4).

In summary, the procedure for deriving absolute droplet size distributions is as follows:

1. The average surface to volume ratio is found from fitting (7.3) to a set of diffusion measurements at short observation times.
2. The average (*S/V*) can be correlated to the average $(1/T_2)$ found from a CPMG experiment. From (7.5), (7.4) can then be written as

$$\overline{\left(\frac{1}{T_2}\right)} \approx \rho\,\overline{\left(\frac{S}{V}\right)} \Rightarrow \rho = \overline{\left(\frac{1}{T_2}\right)} \times \overline{\left(\frac{S}{V}\right)}^{-1} \tag{7.6}$$

   hence the surface relaxivity ρ is found, which then is assumed to be droplet size independent.

3. Under the assumption of the surface relaxivity, ρ, being independent of droplet sizes, the value of ρ can then be used in (7.4) thus resulting in a linear relation between T_2 and the volume to surface ratio, which is a measure of the droplet size. By multiplying the T_2 distribution by the calculated surface relaxivity, the T_2 distribution is scaled to a droplet size distribution in absolute length units.

7.2.2 The Asymptotic Approach

As shown by Packer et al. [20], there is a situation where the surface relaxation term is absent in the solution of the diffusion propagator, i.e. for diffusion within closed cavities and when the diffusing molecules have covered mean free path lengths $\gg$ cavity dimension ($(6\,D_0\,t)^{1/2} \gg R_{cavity}$). In such a situation the attenuation of the NMR signal from diffusion within the closed droplet can be simplified to [5, 20]

$$\frac{I}{I_0} \approx \exp^{-\frac{\gamma^2\delta^2g^2R^2}{5}} \tag{7.7}$$

δ is the gradient pulse length, g is the applied gradient strength and R is the droplet radius. In a heterogeneous system a distribution in droplet sizes must be assumed. As long as ($(6\,D_0\,t)^{1/2} \gg R_{cavity}$) holds for all sizes (7.7) is also valid for a heterogeneous system. If ξ_i is the volume fraction of the droplets with surface to volume ratio $(S/V)_i$, (7.7) can be expressed as

$$\frac{I}{I_0} \approx \sum\nolimits_i \xi_i \exp^{-\frac{\gamma^2\delta^2g^2R_i^2}{5}} \tag{7.8}$$

When the exponent in (7.8) is small for all i, the exponential functions are transformed into a series expansion where two first terms will be applied:

$$\frac{I}{I_0} \approx \left(\sum\nolimits_i \xi_i - \sum\nolimits_i \xi_i \frac{\gamma^2\delta^2g^2R_i^2}{5}\right) = 1 - \frac{\gamma^2\delta^2g^2\overline{R^2}}{5} \tag{7.9}$$

where $\overline{R^2}$ yields the average value of the square of the droplet radius. Measurements of the early departure from I_0 as a function of applied gradient strength may then result in a value for the average square of the droplet radius. This can be used in combination with a T_2 distribution to produce a droplet size distribution as shown in the following.

Transforming a T_2 distribution into a droplet size distribution

Assuming that the water molecules are probing the droplets within the sample, there is a simple relation between T_2 values and the droplet sizes [36–38]

$$T_2 \approx \frac{V}{S\rho} \tag{7.10}$$

This couples the surface to volume ratio to the surface relaxivity, ρ, and makes it difficult to assign the T_2 distribution directly to a (V/S) distribution. However, if the assumption that (7.10) holds for any droplet size applied, with ξ_i being the volume fraction of droplets with surface to volume ratio $(S/V)_i$, and corresponding relaxation time T_{2i}, it may be written

$$\sum_i \xi_i \frac{1}{T_{2i}^2} = \sum_i \xi_i \rho_i^2 \left(\frac{S}{V}\right)_i^2 = 9 \sum_i \xi_i \rho_i^2 \left(\frac{1}{R_i^2}\right) \approx 9\rho^2 \overline{\left(\frac{1}{R^2}\right)} \quad (7.11)$$

With the additional assumption that the surface relaxivity ρ is independent of droplet size, the left hand side of (7.11) now becomes the harmonic mean $\overline{\frac{1}{T_2^2}}$ of the T_2-distribution weighted by the fraction ξ_i of nuclei with relaxation time T_{2i} and n is the number of subdivisions of droplet sizes. This average can be calculated from the T_2-distribution obtained in a CPMG measurement where the magnetization attenuation is transformed into a T_2 distribution by solving an inverse problem using e.g. an Inverse Laplace Transform (ILT) routine. Then the surface relaxivity ρ can be calculated from (7.11)

$$\rho = \frac{1}{3}\sqrt{\overline{\left(\frac{1}{T_2^2}\right)} \times \overline{\left(\frac{1}{R^2}\right)}^{-1}} \quad (7.12)$$

$(\overline{R^2})$ is the quantity that can be measured by PFG NMR according to (7.9). $\left(\overline{\frac{1}{R^2}}\right)^{-1}$ is a parameter that needs to be solved. Again denoting (S/V) as $(3/R)$, as for spherical droplets and assuming the surface relaxivity to be independent of droplet size the following relations can be found

$$\overline{T_2^2} = \frac{\overline{R^2}}{9\rho^2}, \ \overline{\left(\frac{1}{T_2^2}\right)} = 9\rho^2 \overline{\left(\frac{1}{R^2}\right)}, \ \overline{\left(\frac{1}{T_2}\right)} = 3\rho \overline{\left(\frac{1}{R}\right)} \quad (7.13)$$

From these expressions the following may be deduced

$$\overline{\left(\frac{1}{R^2}\right)} = \overline{\left(\frac{1}{T_2^2}\right)} \left(\frac{\overline{T_2^2}}{\overline{R^2}}\right) \quad (7.14)$$

Substituting this in the equation for ρ

$$\rho = \frac{1}{3}\sqrt{\left(\frac{\overline{R^2}}{\overline{T_2^2}}\right)} \quad (7.15)$$

Finally, the measured T_2-distribution can be transformed into an absolute droplet size distribution (*V/S*) by means of the relationship inherent in (7.10). In summary, the procedure for deriving absolute droplet size distributions is as follows:

1. The square of the average droplet radius is found from fitting (7.9) to a diffusion measurement at long observation times.
2. The square of the average droplet radius can be correlated to the average value for $(\overline{T_2^2})$ found from a CPMG experiment. Then an expression for the surface relaxiavity can be found

$$\rho = \frac{1}{3}\sqrt{\left(\frac{\overline{R^2}}{\overline{T_2^2}}\right)}$$

3. Under the assumption of the surface relaxivity, ρ, being independent of droplet sizes, the value of ρ is applied in (7.10) resulting in a linear relationship between T_2 and the volume to surface ratio which is a measure of the droplet size. By multiplying the T_2 distribution with the calculated surface relaxivity the distribution is scaled to a droplet size distribution in absolute length units.

7.2.3 Verification

The theoretical models for determining droplet size distributions were tested on brine in crude oil emulsions. The emulsions were prepared by mixing the crude oil with brine at room temperature (25 °C). The total sample volume of all the emulsions was 15 mL. The mixing was performed by an IKA Ultra Turrax homogenizer (10 mm head) with a stirring speed up to 24,000 rpm for a defined number of minutes. Depending on the mixing speed and duration of the mixing, the droplet size distributions varied. The emulsions were analysed immediately after mixing, and parallel to the PFG NMR measurements, emulsions from the same batch were analysed by microscopy.

The microscope used for an alternative measurement of droplet size distributions consists of a Nikon Eclipse ME 600L Microscope and a CoolSNAP-Pro cf camera with Image-Pro PLUS 5.0 software of Media Cybernetics. By utilizing a custom-made macro that identifies and measures the size and shape of dark objects in a picture, the distribution of the droplets was obtained. The shape measurement was useful to exclude droplets with a non-spherical shape and clusters of droplets. In order to obtain a droplet size distribution by using the microscope technique, the emulsions had to be diluted. The original emulsions were too concentrated for the macro to properly separate and distinguish the droplets. The dilution was performed by adding approximately 0.2 g emulsion drop wise to 2 g of the crude oil, and thereby gently mixing. The droplet size distribution was obtained by placing a small drop of the diluted emulsion on a glass slide. The emulsion was then flattened by placing a smaller glass slide on top of it. Thereafter, several pictures of the emulsion

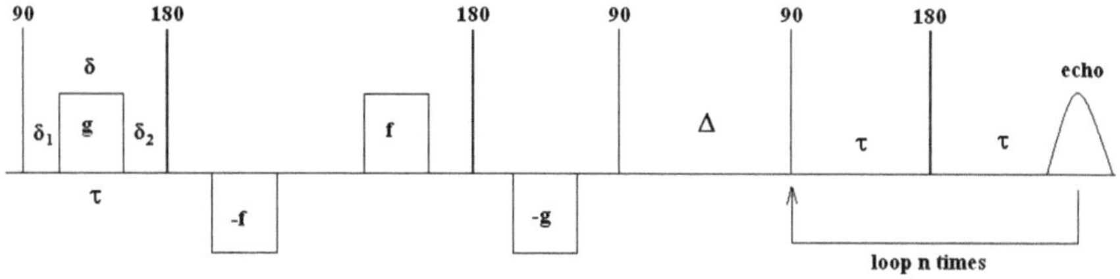

Fig. 7.7 The combined pulsed field gradient-stimulated echo—CPMG sequence

were taken. To get a credible impression of the droplet sizes for each emulsion, more than 800 droplets were counted for the production of the distribution population.

The PFG NMR measurements were performed on a low field Oxford Instruments MARAN Ultra spectrometer, 23 MHz tempered at 40 °C, with the ability of producing shaped gradients up to 400 G/cm. The NMR sample tubes of 18 mm diameter were filled with approximately 3 mL of the emulsion. The PFG NMR sequence as shown in Fig. 7.7 makes use a combination of three different sequences to obtain the droplet size distribution. The first part is the 11-interval PFGSE sequence used to weight the NMR signal with a diffusion dependent term. Originally this sequence was developed for diffusion studies in the presence of internal magnetic field gradients [40]. In an emulsion system, the internal gradient strengths are found to be negligible, but still the 11-interval PFGSE sequence is made use of because it minimizes the effect from eddy current transients. Consequently an eddy current dead time of just 500 μs can be applied without any sophisticated preemphasis adjustment [41]. In the second part of the sequence in Fig. 7.7, the signal is stored along the z-direction for a period of Δ. The duration of Δ is adjusted so that the crude oil signal have relaxed to thermal equilibrium, and the remaining signal arises from the water phase only. In addition, any residual eddy current transients, which the 11-interval PFGSE sequence does not cancel, will vanish during this interval. As the base RF sequence is a stimulated echo sequence, it is appropriate and sufficient to employ the phase sequence by Fauth et al. [42]. Depending on the T_2 values of the crude oil, the water signal was separated from the crude oil signal by setting the Δ to 1.5 s + −0.5 s, depending on the T_1 distribution of the crude oil being investigated.

During the third and last part of the sequence, a CPMG sequences is performed on the remaining water signal. The acquired data is then subjected to an Inverse Laplace routine, which returns a T_2 distribution from brine only. The distribution is diffusion weighted depending on the strength of the applied magnetic field gradient *g* used in the first part of the sequence. When the applied gradient strength is set to zero and Δ is made long enough to suppress the crude oil signal, the T_2 distribution obtained can be used to calculate the volume weighted droplet size distribution of the dispersed droplets. An important feature about this method is that it does not make any assumptions on the distribution shape.

There is a fundamental difference between droplet size distributions obtained from image analysis and PFG NMR. A distribution obtained from image analysis of droplets observed in a microscope will give a size distribution based on number based intensity. The PFG NMR method yields a volume-based size distribution; the droplet dimensions are calculated from the surface to volume ratio. This means that a direct comparison of the results from the two techniques cannot be achieved. A number based size distribution will yield a smaller mode and be differently shaped compared to a volume based size distribution. Figure 7.8 illustrates the difference of weighting by a number- and volume-based distribution obtained by the PFG NMR. In a volume-based distribution, the smallest droplets will be discriminated by the larger ones, and vice versa. In order to compare the two techniques, the volume distribution obtained from the NMR has been changed/transformed to a number based distribution by dividing the volume intensity of each diameter interval by its volume. The opposite could have been done; transforming the number-based distribution of the microscope to a volume-based one. Due to the low amount of data points compared to the NMR data, such a transformation could easily become imperfect. If a data set contains one or two droplets that are significantly larger than the other ones, the transformed distribution would be skewed by these particular droplets. The distributions presented from both techniques are also normalized.

Figure 7.9 displays the size distribution for an emulsion from a crude oil. The correlation between the PFG NMR and the microscope is good. The shapes of the two distributions are comparable, and the registered dimensions of the smallest and largest droplets present in the emulsion are similar for both techniques. Opedal et al. [43] measured on a large set of emulsion system with varying water content, and compared the data acquired with PFG NMR against image analysis from microscope pictures (as shown in Fig. 7.10). The conclusion was that there was no significant discrepancy between the PFG NMR data and the image

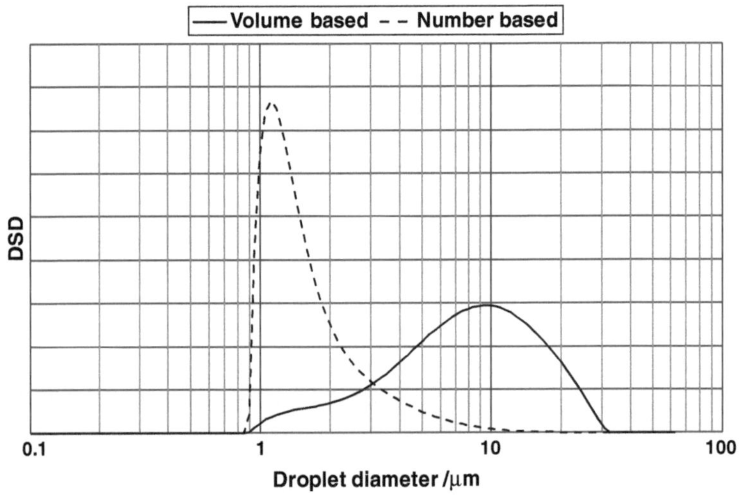

Fig. 7.8 Volume-based and number-based size distribution of crude oil emulsion with 40 % water cut obtained by PFG NMR

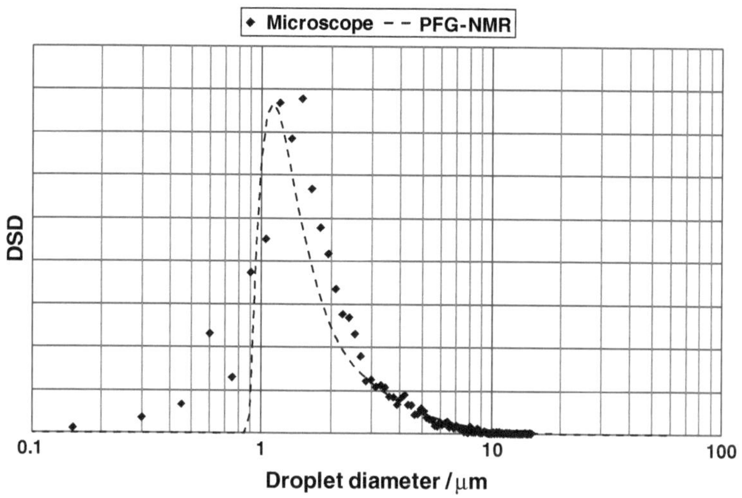

Fig. 7.9 Comparison of the DSD of crude oil from PFG NMR (*dotted line*) and microscopy, 40 % water cut

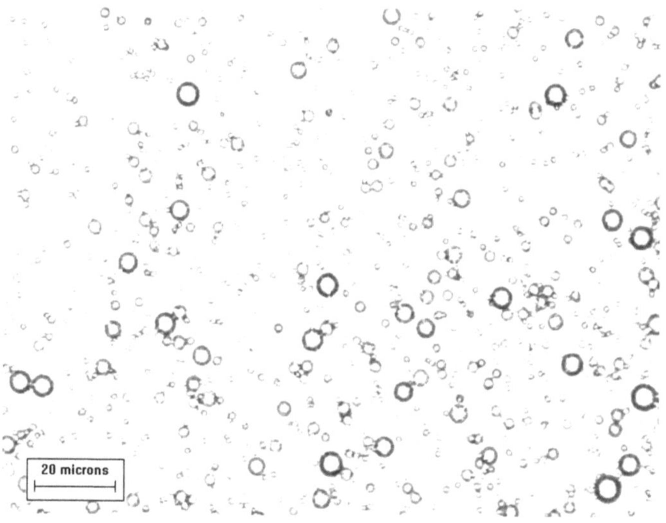

Fig. 7.10 Microscope image of an emulsion of crude oil, 40 % water cut

analysis. For the system studied, the resolution of the microscope was good enough to measure all the droplet sizes, otherwise there would have been a bias towards larger droplet sizes in the distribution.

For the particular emulsion sample shown in Figs. 7.8, 7.9 and 7.10, the volume-based average droplet diameter is ~9 μm, and the short observation time approach can be applied to determine the droplet size distribution. However, for many crude oil emulsions, even the volume average of the droplet diameter may be as small as ~1 μm, and the asymptotic approach has to be employed when determining the droplet size distribution.

The results presented here demonstrate that the PFG NMR approach is a competitive alternative to existing techniques, and eliminates many of the drawbacks associated with microscopy/image analysis methods. The sample does not have to be transparent, as for optical methods, the number of droplets measured is much larger compared to number based counting methods, and it does not assume any particular shape of the distribution a priori. Figure 7.11 shows a more complete study where the same crude oil is used, but with different water cuts. Here it is quite evident that the applied PFG NMR method is sensitive to a variation in the droplet size distribution, which indeed is being determined by the method. As the water content increases the droplet sizes increases, but the distribution function also changes. The distribution function appears bimodal for the lowest water cut with approximately equal volumes of the bigger and smaller droplets, the relative volume of the bigger droplets increases as the water cut is increased.

Using the approaches presented in this section, as much information as possible is utilized in order to extract the droplet size distribution. In addition, the suppression of the oil signal does not assume a mono exponential behaviour. Suppression methods are chosen where the oil signal is reduced to an insignificant amount instead of having it passing a zero-value, which is not necessarily unique for all the components within the oil (documented in Fig. 7.1).

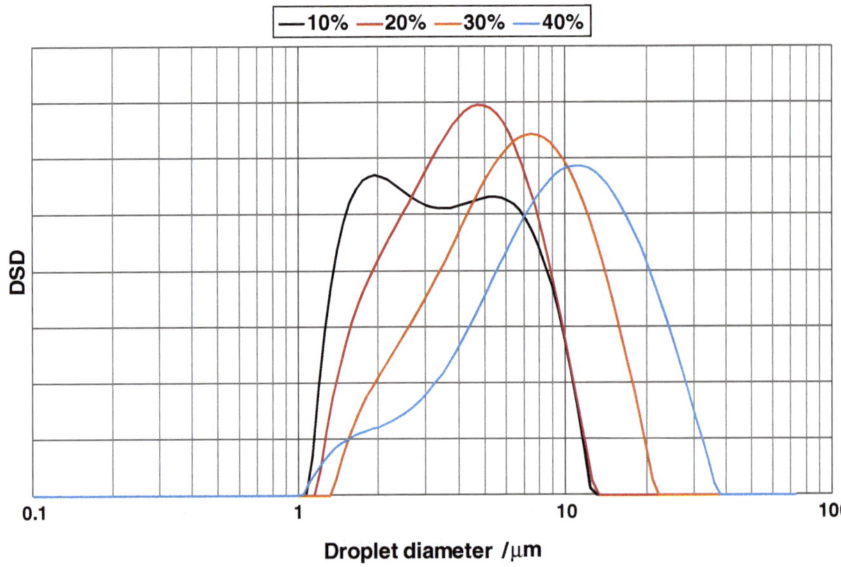

Fig. 7.11 The acquired normalized droplet size distributions at different water cuts; 10, 20, 30, and 40 %

One drawback of the PFG NMR method is the fact that it relies on the use of an Inverse Laplace Transform to produce the distribution function, and thus the width and number of peaks in the distribution is depending on the smoothing parameter (Chap. 5), which is to some degree subjective. For the droplet size distributions to be comparable and consistent, the smoothing parameter is set to the same value in all measurements. As the experimental parameters are fixed, and the noise in the PFG NMR data is approximately the same for all the experiments, the use of the same smoothing parameter is justified.

7.3 Water and Inverse of Surface to Volume, $(S/V)^{-1}$, -Profiles

In addition to the determination of droplet size distributions in w/o emulsion, there are other physical parameters that can be determined, which provides knowledge of the emulsion with respect to homogeneity and stability. In particular, the water content and the average droplet size, which correlate to $(S/V)^{-1}$, can be used to predict emulsion stability.

The one-dimensional image of the water, known as the water profile [21], is produced by combining the CPMG sequence with a profile experiment, as shown in Fig. 7.12. Initially, the NMR signal contains a crude oil and water contribution. When a significantly lower transverse relaxation time for the crude oil is measured, the C1 loop is used to suppress the contribution from the crude oil. The first gradient echo is then from water only, and a Fourier transform of this echo yields a water profile. The second loop is used to measure a gradient echo at an even longer observation time, which may result in another water profile. These two profiles may then be used to correct for transverse relaxation of the water signal, resulting in a third water profile, which is unaffected by transverse relaxation processes. In Fig. 7.13 the procedure for extrapolating back to zero observation time is shown. I_1 corresponds to the signal intensity after the C1 loop at time t_1, and I_2 corresponds to the signal intensity after the C2 loop at time t_2.

The intensities are written when assuming a mono exponential decay

$$\begin{aligned} I_1 &= I_0 \exp^{\left(\frac{-t_1}{T_2}\right)} \\ I_2 &= I_0 \exp^{\left(\frac{-t_2}{T_2}\right)} \end{aligned} \tag{7.16}$$

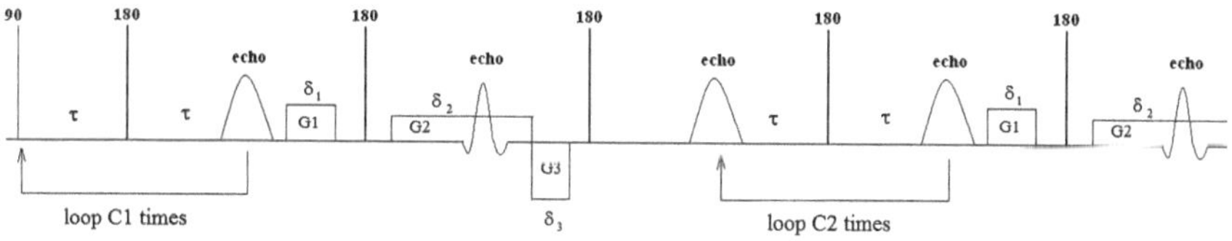

Fig. 7.12 The PFG-NMR pulse sequence for quantitative determination of a water profile

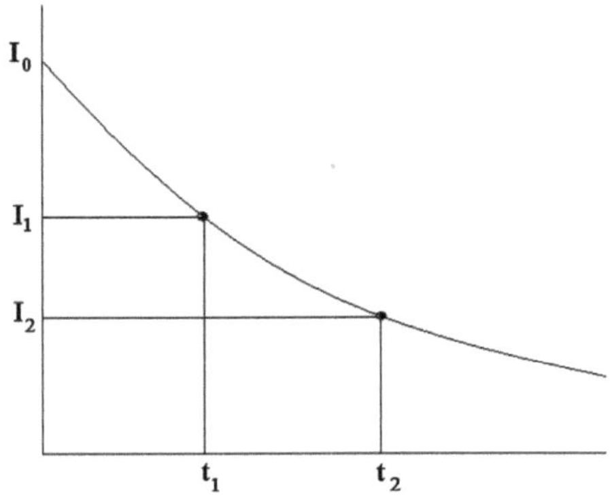

Fig. 7.13 Schematic view on the procedure for achieving a quantitative water profile

By solving out the T_2-relaxation times, the following relation for the initial water signal applies

$$I_0 = \exp^{\left(\frac{\ln(I_2)\cdot t_1 - \ln(I_1)\cdot t_2}{t_1 - t_2}\right)} \tag{7.17}$$

When rescaling the water intensity to I_0 in each pixel to their corresponding I_0, the resulting profile is proportional to the water content along the sample. By calibrating with a sample that contains 100 % water, one can measure the water content along the length of the sample tube. C1 is determined by measuring the signal for pure oil phase and determining at which C1 value the entire oil signal has vanished. When there is a situation where the viscosity of the crude oil is such that the T_2-values of the water and crude oil more or less overlap, the combined CPMG-profile experiment cannot be used. Instead, the best option is to suppress the crude oil signal due to significant differences in root of mean squared displacements, RMD, at longer observation times. Figure 7.14 represents a PFG-NMR sequence, where the 13-interval PFGSTE part is used to suppress the crude oil signal when Δ_i is made sufficiently long. Again, the experiment is performed at two different values of Δ_i, and the same procedure for extrapolation back to zero observation time as used for the combined CPMG-profile experiment can be applied. A negligible relaxation is assumed during the prepare and read intervals as these times are much less than the typical T_2. Likewise, the gradient strength is assumed to be so weak at the value when the crude oil signal is suppressed, that the echo signal of the water confined in the droplets has attenuated insignificantly. If the latter requirement is not fulfilled, the profile could be acquired at two different gradient strengths and an extrapolation to zero gradient strength can be applied as for the extrapolation back to zero observation time.

Figure 7.15 shows the profiles acquired on a 100 % water sample, and two emulsion produced with 30 and 50 % water cut. The method resolution is 0.35 mm, i.e. the distance between the measuring points in the profile. The emulsion with 30 % water content is stable after production, and provides a flat profile around 30 %. On the contrary, the system with 50 % water is not stable after

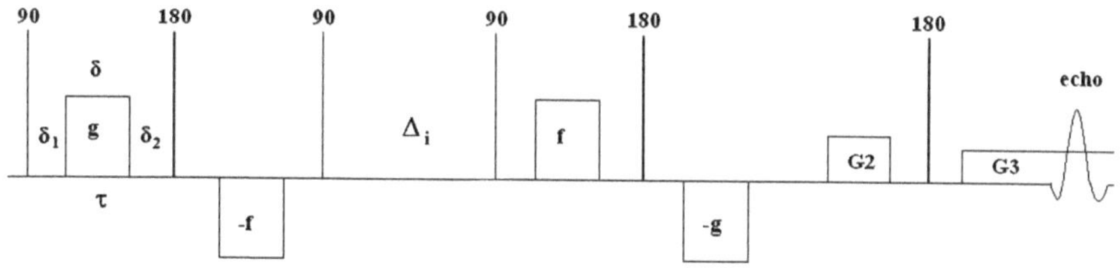

Fig. 7.14 The combined 13-interval PFGSTE profile experiment

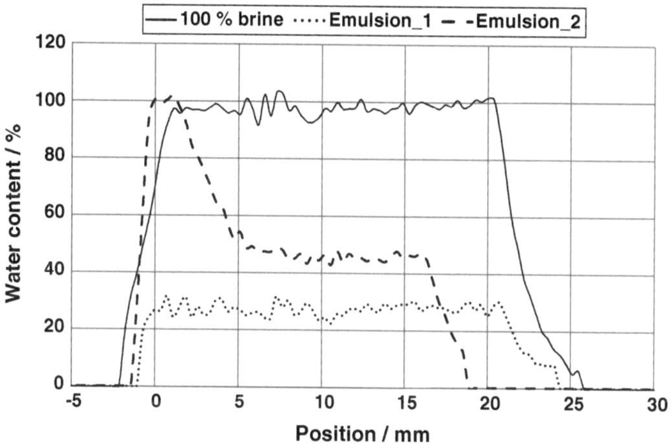

Fig. 7.15 Water profiles acquired from the combined CPMG profile. Emulsion 1 is the sample with 30 % water cut while emulsion 2 is the sample with 50 % water cut

production, and a bulk phase at the bottom is instantly produced and a bulk phase of oil emerges at the top, between 18–25 mm, where no water is found.

In Fig. 7.16, the approach for acquiring water profile in an emulsion is applied, where the crude oil signal is suppressed due to a significant difference in root of mean squared displacement. For this system the relaxation times of the crude oil was significantly different from the water, so both methods for acquiring the water profile are comparable. When comparing the two approaches, it is quite evident that the bulk water phase is also suppressed in the RMD-approach, so the total water content is not measured, only the emulsified water.

The data presented in Figs. 7.15 and 7.16 are rather noisy. This is a consequence of the rather poor signal to noise, as the number of scan was two. As the aim often is to study the separation process, it is important to minimize the experiment time, and for the CPMG-profile sequence an acquisition time of 30 s is achieved by running two scans only. Consequently, every 30 s a water profile can be measured, and separation processes can be studied at the expense of noise in the water profiles.

In the production of crude oil, it is important to remove the emulsified water from the crude oil, where a "dry" oil is defined as a crude oil of water content less than 0.5 %. Although the previous experiments quantify the water content

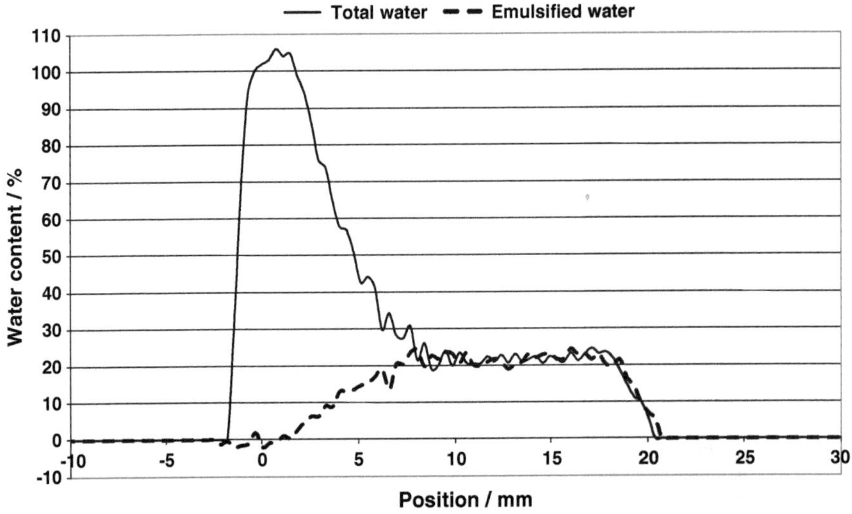

Fig. 7.16 Water profiles acquired from the combined 13-interal PFGSTE profile

properly, the data are too noisy to determine such low amounts of residual water still present in the crude oil after a separation process. The usual solution is simply to increase the number of scans until a satisfactorily signal to noise is achieved. Figure 7.17 shows an acquisition containing 48 scans, which implies increasing the acquisition time to 12 min. The residual water content is high (2–4 %) so that it cannot be defined as a dry oil, and there is clearly a gradient in the water content, which indicates that the process of emulsion separation is not over yet.

When comparing the PFG NMR approach for quantifying residual water within the crude oil against Karl Fisher titration of equivalent samples the correlation is good [44]. Compared to Karl Fisher titration the obvious benefit using the PFG NMR approach is the possibility of measuring profiles in the residual water signal in the crude oil, i.e. gradients in water content, and not just a value for water content as a result from the quantification.

Another profile experiment that is useful when monitoring the stability of an emulsion is the average droplet diameter as a function of position in the sample. In principle, the method is the same as for determining the droplet size distribution, except that the CPMG decay is replaced with a profile experiment, as shown in Fig. 7.18. This sequence produces diffusion weighted profiles, which can be transformed into $(S/V)^{-1}$ profiles by imposing the short observation time model or the asymptotic limit on the apparent diffusion coefficient. It should also be noted that as the aim is to study separation of an emulsion, the spoiler recovery approach is incorporated into the sequence. This reduces the acquisition time of the $(S/V)^{-1}$ profile down to approximately 1 min. Then only two gradient strengths are applied to determine the diffusion weighted profiles, which is the minimum number of gradient values required to produce a value for the apparent diffusion coefficient. In fact the

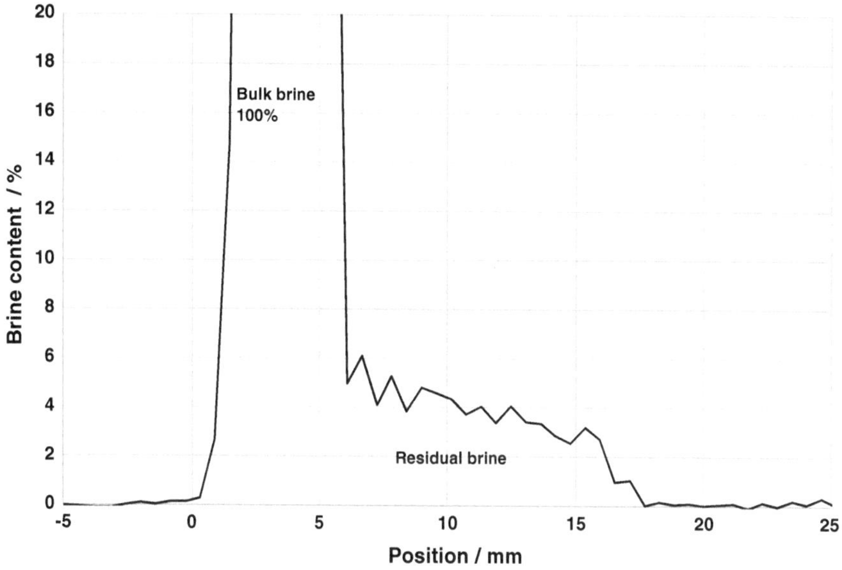

Fig. 7.17 Water profile acquired from the combined CPMG profile, number of scans = 48

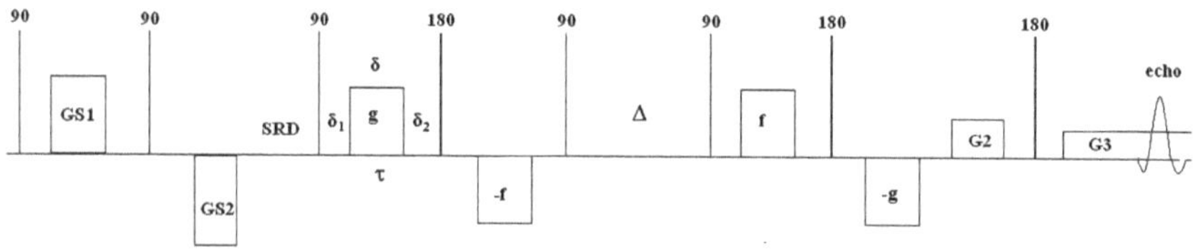

Fig. 7.18 The combined 11-interval PFGSE-STE-profile sequence for determining $(S/V)^{-1}$ profiles

requirement for determining $(S/V)^{-1}$ as fast as possible was the trigger for developing the spoiler recovery method. During the conventional approach that took approximately 5 min, the $(S/V)^{-1}$ arising from an unstable emulsion could change a lot, and therefore return very noisy data that were impossible to interpret. However, as it is not necessary to maintain quantitative information on the emulsified water in order to determine $(S/V)^{-1}$, a method that circumvented the recycle delay by initially spoiling all NMR signal was developed. After spoiling the NMR signal, a duration equal to T_1 to recover 64 % of the signal was applied, followed by the PFG NMR sequence. The procedure could then be repeated using a negligible recycle delay.

Figure 7.19 shows the acquired $(S/V)^{-1}$ profile from a stable emulsion and its corresponding droplet size distribution in the upper right corner. Depending on the water cut of the emulsion, it takes 1–4 min to acquire a $(S/V)^{-1}$ profile with satisfactorily resolution using the spoiler recovery approach.

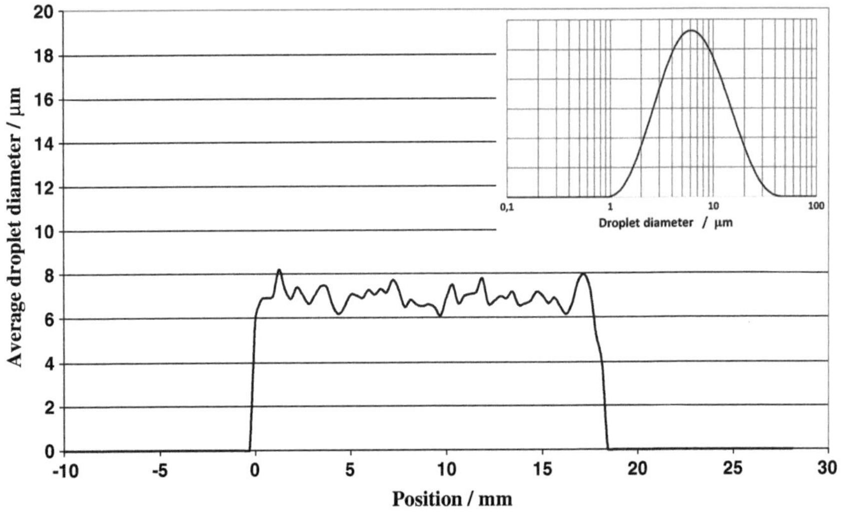

Fig. 7.19 $(S/V)^{-1}$ profile from a stable emulsion and its corresponding droplet size distribution (*upper right corner*)

7.4 The Characterization of Unstable Emulsions

Emulsions are thermodynamically unstable dispersions of oil and water. However, macro emulsions are often kinetically stabilized due to large activation energies associated with flocculation and/or coalescence. Coalescence is the process of fusing of dispersed droplets, and is driven by a free energy reduction associated with removal of interfacial energy. The destabilization of water-in-crude oil emulsions is illustrated in Fig. 7.20, and it includes three primary steps: water droplet flocculation, gravimetric sedimentation and coalescence. Subsequently, a free water layer is formed.

One problem often experienced during exploitation of oil fields, is the occurrence of highly stable water-in-crude oil emulsions. In many fields the co-production of water is often substantial, up to water cuts of 50–70 % [45]. These emulsions are detrimental for the industry because they cause flow assurance problems due to high viscosity of the emulsion, and significant effort is required to separate the emulsified water from the crude oil, which impacts on the reservoir economics. Techniques for enhancing separation efficiency include the use of electro coalescence [46] and addition of chemicals, so-called demulsifiers. Demulsifiers induce sedimentation or coalescence of the emulsified water droplets by affecting the strong interfacial films which acts as barriers towards coalescence. Electrostatic and chemical destabilization can also be used in combination. The addition of chemicals however brings extra costs and may lead to environmental consequences when the water phase is disposed. Hence, there is a need for more knowledge of the stabilization mechanism of heavy crude oil emulsions to increase separation efficiency, production rates and reduce environmental impact.

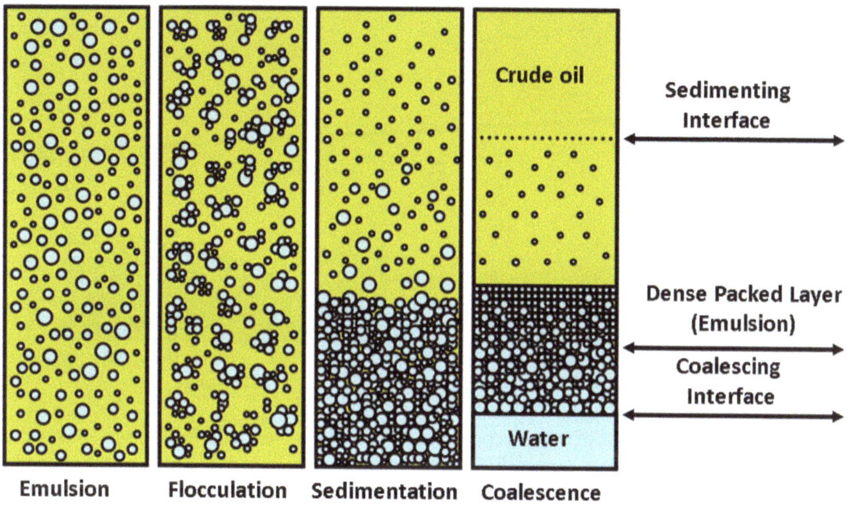

Fig. 7.20 Destabilization of crude oil emulsions (w/o)

In order to study the separation of emulsions, a set of w/o emulsions with 40 % water cut were prepared and mixed with different amounts and/or types of demulsifiers. An Ultra Turrax T25 (Ika®-Werke Co., Germany) with a 10 mm stirring head was used to emulsify the samples (4 min at 24,000 rpm), and thereafter the demulsifier was added into the emulsion. The emulsion sample was shaken for 15 s in order to homogenise the distribution of the demulsifier before the PFG NMR measurements were made at a sample temperature of 40 °C. The height of the samples was 20 mm.

In Fig. 7.21 the brine profiles are recorded and plotted as a contour plot as a function of observation time. The amount and type of demulsifier is unknown. Every minute a brine profiles was recorded and the entire contour plot therefore consists of 120 brine profiles. As can be seen at the shortest observation times, the water cut is 40 % along the whole sample but just after 10 min a bulk water fraction evolve at the bottom of the sample (left plot in Fig. 7.21). The bulk fraction continues to grow throughout the measurements that lasts for 2 h. At the end of the measurements, the bulk fraction seems to approach an asymptotic value, which is approximately 20 % of the sample (left plot in Fig. 7.21). Consequently, there is still 20 % left of the water suspended as droplets in the crude oil. No significant fraction of bulk oil is observed, and the conclusion is therefore that the separation of the emulsion due to addition of a demulsifier is far from being complete. Indeed coalescence takes place in the separation process, but also sedimentation as a densely packed layer of water droplets is observed just above the bulk water layer. This layer of approximately 70 % water cut is the broadest up to 40 min.

In the right plot of Fig. 7.21, the situation is different, as the bulk fraction evolves much faster. After 2 h, the bulk fraction is slightly above 30 %, indicating a remaining 10 % of water as droplets in the crude oil. Now, there is a buildup

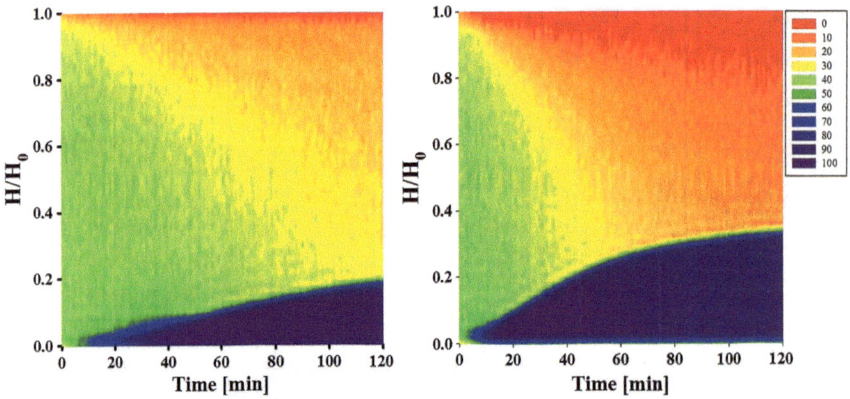

Fig. 7.21 The contour plot of water separation from an emulsion as a function of observation time. The amount and type of demulsifier added is unknown. The x-axis provides the time after starting the water profile measurements (time evolution is then recorded). The y-axis provides the brine content as a function of position in the sample; 0 is at the *bottom* and 1.0 is at the *top* of the sample which is 20 mm long. The *color bar* provides the brine content. For no brine (or less than 5 %) the color is *red* and for bulk brine the color is *dark blue*

of a bulk oil phase at the top of the sample, whilst there is an intermediate emulsion layer of approximately 20 % water cut. As there is water still present in the crude oil, the separation process is not complete for this system either. However, the sensitivity of the rapid water profile determination is such that any residual water content (<5 %) is not accounted for. What may seem as bulk oil in the plot, with zero water content, may therefore very well contain significant amounts of residual water. In order to quantify the residual water content, the more time consuming method with increased number of scans must be applied [44].

For the plots shown in Fig. 7.22 the bulk water fraction builds up rapidly to almost the total water content within the 2 h measuring time. After 2 h there are only isolated regions where the water cut is 10 %, and it is tempting to conclude that the greater the rate of production of the bulk phase, the better is the separation of the water from the crude oil. However, in the right plot Fig. 7.22, the bulk water fraction evolves much faster than all the others do. It also has the highest demulsifier concentration; nevertheless, it only reaches a content of 35 %. Just above the layer of bulk water there is apparently a region of crude oil and possible residual water, to which the method is not sensitive. Higher up in the sample a region of water cut 10 % is found. For the water from this region to enter into the bulk water layer, it must therefore traverse a region of lower water concentration. This process appears to be not very energetic favorable, and it may take time to complete the separation. The conclusion from these experiments is that the most efficient way of adding demulsifier for separation of the emulsion, is not to make the concentration as high as possible. The left plot in Fig. 7.22, having the second highest demulsifier concentration, provides the most efficient way of separating the water from the crude oil of the four samples.

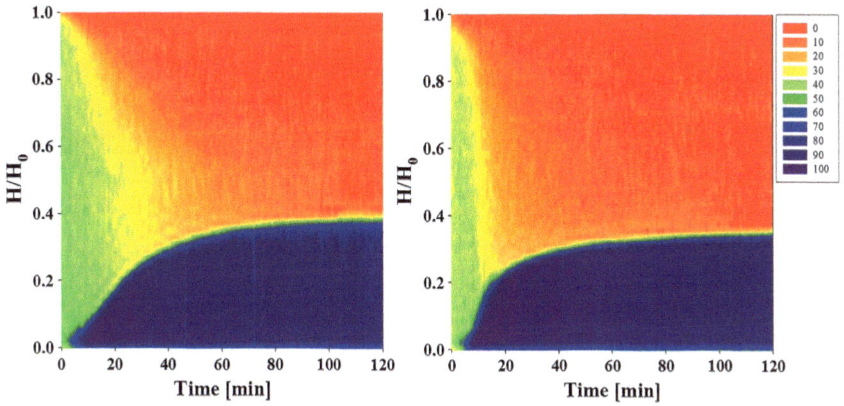

Fig. 7.22 The contour plot of water separation from an emulsion as a function of observation time. The amount and type of demulsifier is unknown

All the separations shown in Figs. 7.21 and 7.22 produce a bulk fraction of water at the bottom of the sample, and both sedimentation and coalescence processes take place prior to the formation of the bulk water layer. However, the brine profiles do not return information on how the droplet sizes change during the separation process. This additional information can be obtained via the use of the fast methods for determining droplet size distributions and/or $(S/V)^{-1}$ profiles described earlier. A separation of the water from the oil phase involves a flow of water droplets towards the bottom of the sample tube. The measured droplets sizes can be determined either from diffusion measurements at short observation times or at the asymptotic limit. In order to conduct reliable measurements, convection compensation is required in the PFG NMR sequences to reveal true values for the $(S/V)^{-1}$ or the average droplet radius. Assuming that the convection compensating sequences provide reliable apparent diffusion coefficients, the method for determining droplet size distributions is a bulk method while the acquisition of the $(S/V)^{-1}$ profile may provide information on possible differences as a function of position in the sample. As a fraction of bulk water will develop during the separation process, a pulse sequence that suppresses any bulk water signal present must be employed, i.e. the application of the RMD-approach.

The left plot in Fig. 7.23 shows the contours of the $(S/V)^{-1}$ profiles for an emulsion of 30 % water cut and a demulsifier of unknown concentration added to the sample. In the right plot in Fig. 7.23, the corresponding brine contour plot is shown for another sample that is prepared in an identical way. The timing of the $(S/V)^{-1}$ and brine contour plots is not identical as the plots were not recorded simultaneously, and it is seen that the brine profile starts to evolve approximately 20 min later than the $(S/V)^{-1}$ profile.

A profile and droplet size distribution were measured every 2 min and the resolution in the profile was 0.35 mm. For the first 30 min, the profile is more or less stable and produces a droplet diameter of approximately 10 μm. This droplet

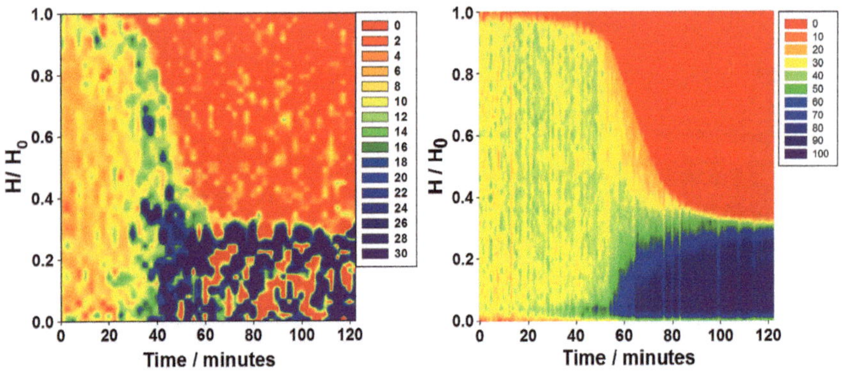

Fig. 7.23 The brine (*right*) and $(S/V)^{-1}$ (*left*) contour plots of water separation due to the addition of a demulsifier to an emulsion

diameter is confirmed by the bulk droplet size measurements. After 30 min an evolution in $(S/V)^{-1}$ and droplet sizes take place. During the following 20 min the $(S/V)^{-1}$ starts to increase, which indicates coalescence taking place, and thereafter a buildup of larger droplets occur at the bottom and hardly anything from $H/H_0 > 0{,}3$ (H/H_0 is the position in the sample normalized against the height of the sample H_0). 30 μm is a limit, which defines when the water is no longer present as droplets but as bulk water. As there is noise in the measured diffusion coefficient, a droplet size of 30 μm is a value that may be measured on a sample of bulk brine due to experimental noise in the measured diffusion coefficient, but this limit can be altered by using other experimental settings. The accuracy when measuring larger droplets can be improved for example by increasing the observation time in the diffusion experiment or an increase in the number of accumulating scans. For the given experimental settings, a measured droplet diameter larger than 30 μm in the $(S/V)^{-1}$ indicates that a bulk water fraction is present, which is confirmed by the corresponding brine profile. Here the rapid buildup of a bulk water fraction occurs at the time when the coalescence starts. Figure 7.24 confirms the coalescence and loss of emulsified water by a shift towards larger droplet sizes and a loss in signal intensity from the water present as droplets as the time after adding the demulsifier increases.

The reason for the isolated red regions in the $(S/V)^{-1}$ contour plot (instead of a continuous blue colour as in the brine profile), is due to the diffusion measurements that are the basis of achieving the $(S/V)^{-1}$ profile. When a bulk fraction of water is present at the bottom of the sample, the measured apparent diffusion will not be identical to the bulk diffusion coefficient, but a value slightly above or below due to noise in the measurement. If the measured apparent diffusion coefficient is higher than the bulk, it will report negative values for the $(S/V)^{-1}$, and the procedure for generating the contour plot marks this value as 0. Consequently, the isolated regions are merely a manifestation of the presence of a bulk water phase. This information can be applied to discriminate between regions of large droplets; at an observation time

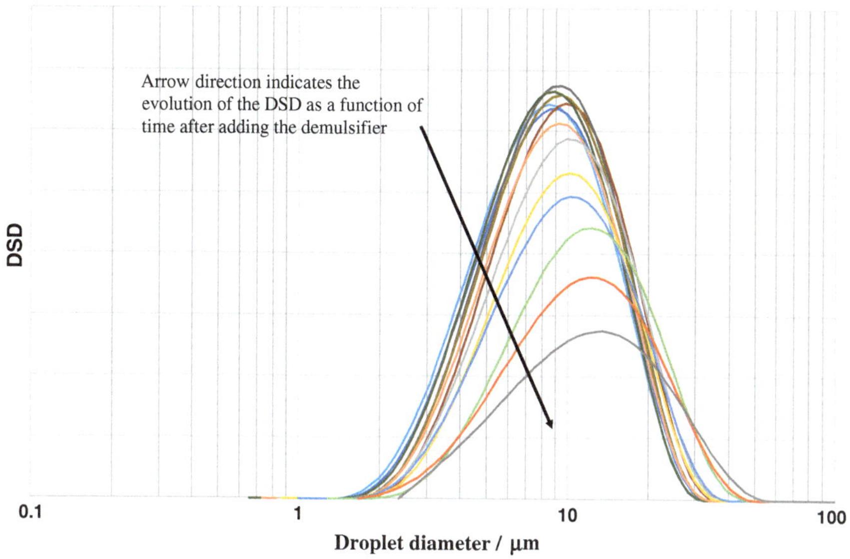

Fig. 7.24 The acquired droplet size distributions at different timings after adding the demulsifier to the emulsion

of approximately 40 min in Fig. 7.23 where there are no red isolated regions present in the lower portion of the sample. Consequently, there are larger droplets present in a sediment phase at this time. From 50 min and onwards red isolated regions start to appear in this area, which confirms that the bulk phase of water is present.

Figure 7.25 shows another system with a demulsifier added, where the evolution during the measurement time is significantly different. To the right the brine contour plot shows the evolution of a crude oil phase in the upper layer (residual brine content of less than 5 % disregarded). After approximately 1 h, a region of high brine content starts to evolve, but there is no clear evidence of a 100 % bulk brine phase at the bottom when removing the sample from the NMR instrument after two hours. When looking at the $(S/V)^{-1}$ contour plot (left plot Fig. 7.25) the average droplet sizes increase, except for in the upper layer where there is no significant water signal to measure on. Sedimentation and coalescence is therefore taking place and densely packed layer of approximately 4 mm is present at the bottom 2 h after having added the demulsifier.

By combining the measurements of brine- and $(S/V)^{-1}$-profiles and droplet size distributions, the separation process of an emulsion is satisfactorily characterized as one may distinguish between coalescence and sedimentation. Consequently, in work with optimizing the separation process the PFG NMR is an appropriate and valuable tool for assessing the efficiency of the demulsifier and assisting in choice of the right concentration to use for a particular crude oil and initial droplet size distribution. In addition other vital operating conditions as the effect of temperature can be investigated [47].

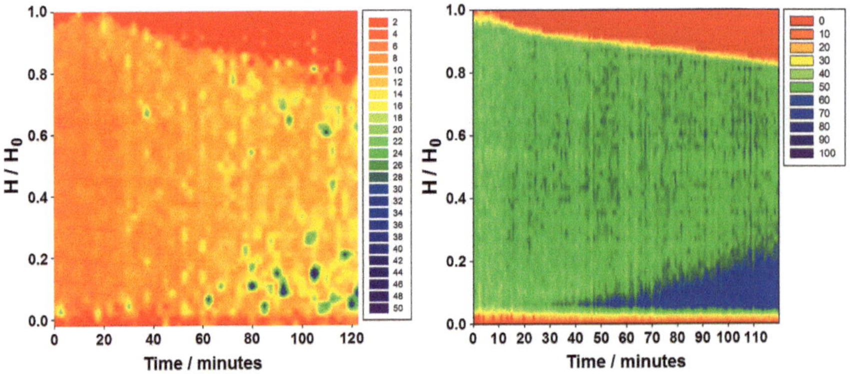

Fig. 7.25 The brine (*right*) and $(S/V)^{-1}$ (*left*) contour plots of water separation due to the addition of a demulsifier to an emulsion

7.5 Emulsion Inversion

The presence of surface active agents (surfactants) enhances the formation of emulsions [48] and can stabilize them once they have been formed. Because of the instability of the emulsion, the nature of the dispersed phase in the continuous phase may change from w/o to o/w in an emulsion or vice versa. This process is called inversion [49]. Inversion can happen due to a change in water-oil ratio (WOR) or physicochemical properties. Variation in formulation occurs due to changes on hydrophilic-lipophilic interactions that promote changes in the surfactant affinity towards oil and water. The hydrophobic lipophilic deviation (HLD) is defined as a formulation variable that includes several physicochemical parameters such as, salinity, temperature and characteristic parameters of surfactants that affect phase behavior of the emulsion. This variable provides information on the deviation of a system from a balanced formulation (HLD = 0). HLD also allows the prediction of the emulsion type [50].

The low field NMR method for the study of the inversion of an emulsion is based on how water molecules are affected by turbulence applied to a system. For a w/o emulsion, water molecules are contained into moving droplets due to the stirring applied on the system which has a relatively high viscosity. The self-diffusion of these water molecules is limited by their confinement within the droplets. At the inversion point, the water becomes the bulk phase in a turbulent regime under continuous stirring, and the system now has a relatively low viscosity. This change, from being confined within brine droplets with a restricted self-diffusion to a turbulent flow amongst oil droplets, is traceable by means of NMR through the T_2^* measurements as shown below.

When assuming a Gaussian diffusion propagator and a mono-exponential attenuation of the NMR signal due to relaxation processes, Karlicek et al. [51] has shown that the solution to the transverse magnetization is

$$I = I_0 e^{-\frac{t_1}{T_2}} e^{-\frac{t_2}{T_1}} e^{-\gamma^2 g^2 D \int_0^t \left(\int_0^{t'} g(t'')dt''\right)^2 dt'} e^{i\gamma v \cdot \int_0^t \int_0^{t'} g(t'')dt''dt'} \tag{7.18}$$

In the following section focus will be on the combined STE-CPMG sequence shown in Fig. 7.26 with the presence of a background gradient G_0, identified as g(t) in (7.18). The parameters used in this sequence are: $\tau = 150\ \mu s$, $\Delta = 1$ s. This background gradient could either be made by passing a constant electric current through the gradient coils, or one may just apply the small background gradient that is present due to a small inhomogeneity in the external magnetic field. The echo attenuation for the even and odd echoes in this sequence is found by evaluation of the integrals in (7.18) [52]

$$\begin{aligned} I_{odd} &= textI_0 e^{-\frac{2n\tau}{T_2}} e^{-n\gamma^2 G_0^2 D \frac{3\tau^3}{2}} e^{in\gamma G_0 v \tau^2} \\ I_{even} &= I_0 e^{-\frac{2n\tau}{T_2}} e^{-n\gamma^2 G_0^2 D \frac{3\tau^3}{2}} \end{aligned} \tag{7.19}$$

Equation (7.19) is valid for laminar flow, v, and it can be seen that the even echoes are unaffected by a constant velocity. When the velocity is not constant, as in turbulent flow, also even echoes will be affected. Collecting the exponential terms into an apparent T_2^*, the attenuation of the even echoes is written

$$I = I_0 e^{-n\tau\left(\frac{2}{T_2} + \gamma^2 G_0^2 D \frac{3\tau^2}{2}\right)} = I_0 e^{-\frac{n\tau}{T_2^*}} \tag{7.20}$$

If there is no turbulence and τ is kept short, the apparent T_2^* should coincide with the true T_2. Otherwise the apparent T_2^* will be shortened to a value depending on the τ-value, the apparent diffusion coefficient, and the degree of turbulence. Figure 7.27 demonstrates the effect of turbulence on the CPMG decay of the brine signal present in a w/o emulsion (left part) and as bulk brine (right part).

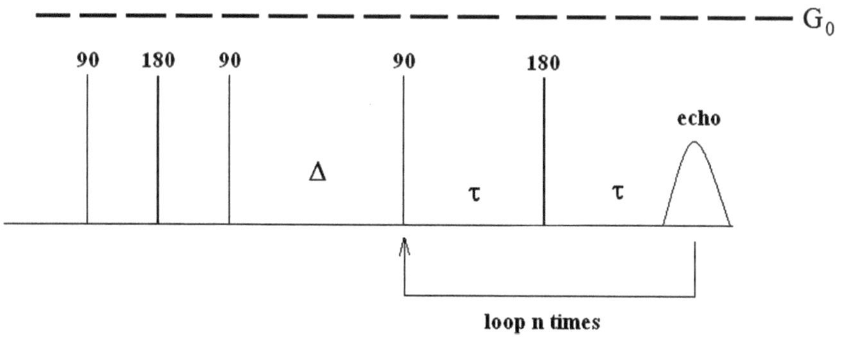

Fig. 7.26 The combined STE-CPMG sequence with a background gradient G_0 present. The STE-part of the sequence is used for suppression of the oil phase due to significant difference in T_1 between the water and oil phase

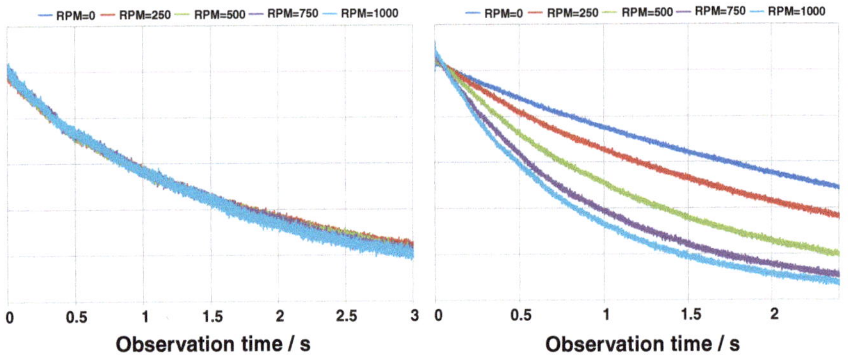

Fig. 7.27 The decay of the echo signals from the CPMG sequence as a function of stirring rate (rpm) for 30 % (v/v) brine (3.5 % NaCl) in a w/o emulsion containing 1 % Span® 80 in Primol as the oil phase (*left part*) and pure brine (3.5 % NaCl) (*right part*)

The signal from the oil phase of the emulsion is suppressed as it has a T_1 that is approximately 10 times shorter than the T_1 of the brine. When including a z-storage delay as shown in Fig. 7.26, the Δ-delay may then be chosen such that the oil signal does not contribute to the CPMG part of the sequence. As the stirring rate is increased, so is the decay of the apparent T_2^* for the bulk brine, while the brine within droplets is hardly effected by the stirring rate. In other words, by continuously monitoring the apparent T_2^* the point of inversion may be monitored when titrating an oil emulsion with brine. At some concentration of brine, the emulsion breaks and undergoes an inversion with oil droplets surrounded by a bulk brine phase. The method only records the initial decay because it is sufficient to determine the average T_2 from the water phase. In this range the signal yields a mono exponential decay and reflects the average T_2.

A mixture of 9:1 mass ratio of Primol 352 and xylene was used as oil phase. Primol 352 is a medical grade mineral oil provided by Exxon Mobil, USA. Analytical grade xylene and sorbitan oleate (Span® 80, HLB = 4.3) were received from Sigma-Aldrich, 1 % (w/v) of Span® 80 in oil phase was used as model oil. The synthetic brine containing 3.5 % (w/v) of NaCl and 4 % (w/v) of sorbitan monooleate (Tween® 80, HLB = 15) was used as the aqueous phase. Tween® 80 was provided by Sigma-Aldrich.

The initial volume of the oil sample was 1.5 mL and the addition rate of the brine to the oil was varied from 0.05 to 0.45 mL/min in the measurements. Each addition was injected into the sample every 2 min. A mixer was used to stir the oil phase and the added brine in the NMR tube during the measurements. Stirring rates were 500, 750 and 1,000 rpm during the measurements. In all the experiments, the aqueous phase was added to the oil phase. The inversion was induced by increasing the water-oil ratio (WOR) and decreasing the hydrophobic lipophilic deviation (HLD) of the system by addition of Tween® 80. Consequently, the emulsions showed transitions from water-in-oil to oil-in-water. A low field NMR with resonance frequency of 21 MHz (Advanced Magnetic Resonance Ltd.) was used for conducting the NMR experiments

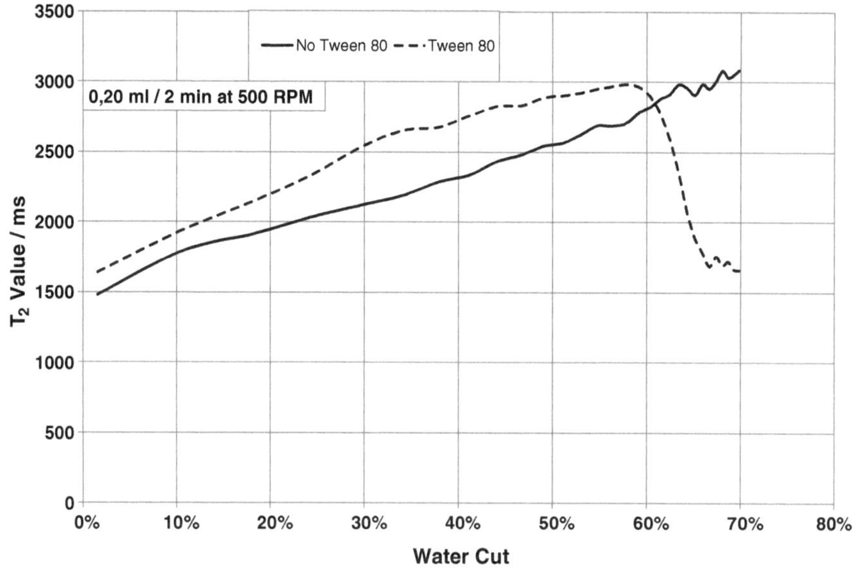

Fig. 7.28 The measured average T_2^* for the water as a function of the added brine with (*solid line*) and without (*dotted line*) Tween® 80. (Stirring rate = 500 rpm, Addition rate = 0.20 mL every 2 min)

The fundamental principle of this method is based on the state of mobility for the water phase. A continuous stirring regime is needed to be able to observe a change in the system. Prior to the inversion point, water molecules are confined within droplets in a high viscous system, which is not significantly affected by turbulence. After the phase inversion occurs, the water phase changes to bulk characteristics being a low viscous system, which is significantly affected by turbulence. Thus, a sudden change in the T_2^* value is expected as confirmed in Fig. 7.28.

Figure 7.28 represents the T_2^* value as a function of the aqueous volume fraction in two systems with and without surfactant added. The NMR experiment was performed at 500 rpm and the addition rate was 0.20 mL of brine every 2 min. The initial volume of the oil phase was 1.5 mL. The evolution of T_2^* clearly differs for two systems. For the system without the surfactant Tween® 80 in the aqueous phase the T_2^* increases continuously as the water content is increased while a discontinuity is observed at ~60 % water cut for the system containing the surfactant Tween® 80 in the aqueous phase. The explanation is that inversion takes place when the brine fraction reaches 60 % for the Tween® 80 system while no inversion takes place for the no Tween® 80 system. Under the assumption of droplet size independency of the surface relaxivity, the average $(S/V)^{-1}$ correlates to the average $(1/T_2)$ found from the CPMG experiment. Therefore, the increasing trend in the T_2^* within the regime where T_2^* changes continuously, is primarily due to an increase in the droplet sizes for the emulsion as the water content increases. The sudden drop in the T_2^* value in Fig. 7.28 is interpreted as an inversion from w/o to o/w emulsion. More specifically, the inversion point is most likely the brine fraction at which the T_2^* value is the maximum before suddenly dropping. The drop is due to the introduction of turbulence on the T_2^* value which occurs when brine becomes the continuous phase. Electrical conductivity measurements confirm that the inversion takes place at the corresponding water cut [50].

Microscopy pictures were compared against the NMR data to confirm that inversion takes place. Figure 7.29 displays three distinct stages for the T_2^* value as a function of the brine fraction. The first stage lasts until the T_2^* value drops. Microscope picture shown in Fig. 7.30a indicates that at a water fraction equal to 10 %, a single w/o emulsion is formed. During the first stage the water-in-oil emulsion evolves from being dispersed to aggregated as it can be seen at a brine fraction of 25 % (Fig. 7.29b), which is consistent with the tendency of Span® 80 to form aggregates [53]. Moreover, an increase in the droplet size is observed, which is consistent with the increase of T_2^*.

As the brine fraction approached the inversion point, where T_2^* dropped (around 60 % brine fraction), double emulsion formation was observed. This is documented in a picture taken at 60 % brine fraction (Fig. 7.29c), where larger oil droplets (50 μm) with several inner water droplets are observed. The presence of the double emulsions is established by observing the motion of the brine droplets inside the larger oil droplets in the microscope. An increase of the T_2^* is primarily due to the increase of the droplet size, also confirmed by microscopy images. Just

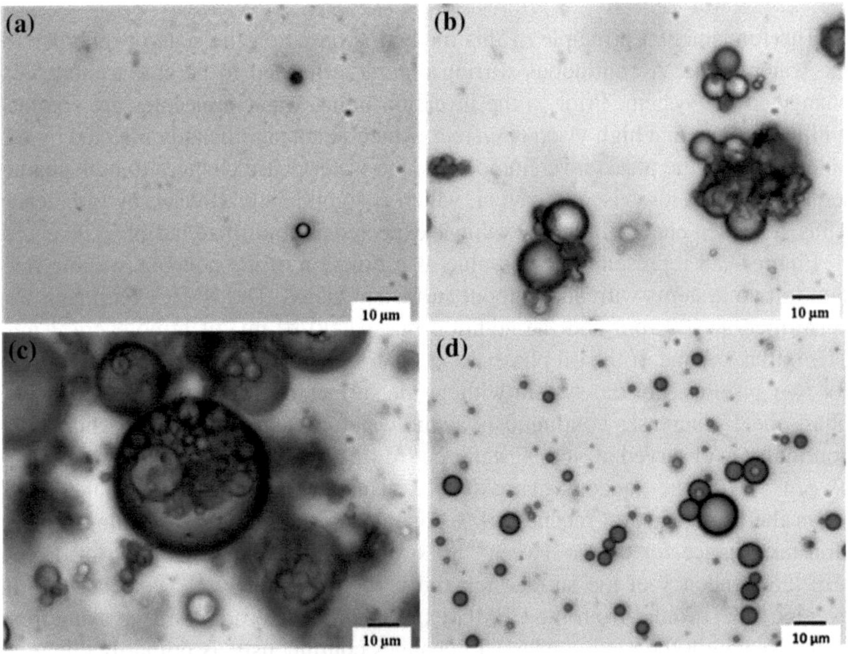

Fig. 7.29 Microscopic images of the oil phase contained 1 % (w/v) of Span® 80 at different amount of the brine contained 4 % (w/v) Tween® 80; **a** Brine content = 15 %, **b** Brine content = 45 %, **c** Brine content = 60 %, **d** Brine content = 70 % (Scale bar corresponds to 10 μm)

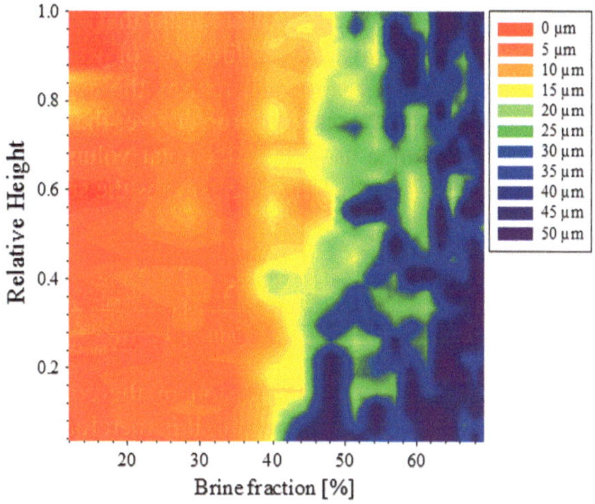

Fig. 7.30 A contour plot for the droplet size of the oil phase contained 1 % (w/v) Span® 80 measured using the NMR as a function of the added brine contained 4 % (w/v) of Tween® 80

after the drop of the measured T_2^*, the emulsion turns into a single o/w emulsion, as can be seen in Fig. 7.29d.

Figure 7.30 depicts the average droplet diameter measured as a function of the position in the sample tube. In order to measure the $(S/V)^{-1}$ profile the stirring was stopped during the PFG NMR experiment. Results showed that while increasing the amount of brine at the lower region of water cuts, the average droplet size remained around 10–12 μm. There may have been an increase in droplet sizes as the brine content increased, as one should expect from the increasing T_2^*. However, due to the noise in the measurements, this behavior was not confirmed for the lowest water cuts. The mean droplet size started to increase significantly from a brine content of approximately 40 %. From this point and up to ~58 %, the determined average droplet diameters increase continuously from 12 up to 50 μm, which was the limit diameter determined by the experimental parameters in the PFG NMR. Above 58 % water cut, the inversion took place (Fig. 7.28).

While the w/o emulsion is characterized by restricted diffusion of water confined in droplets, the situation is more complicated for double emulsions and oil-in-water (o/w) emulsions. For the o/w emulsion to be characterized as w/o emulsion, a well-defined synthetic oil of low viscosity is required. A crude oil would not suit as a probe for the oil droplet sizes, because there is already a distribution of diffusivities and relaxation times due to a relatively large span in molecular sizes. To distinguish between the restricted diffusion and distribution of relaxation times due to the coupling between surface relaxivity and $(S/V)^{-1}$ would then be difficult, at the least. The only possible situation that can be exploited is when the asymptotic limit of the diffusion measurements is applied, but in this case the relaxation times of the crude oils are so short and the diffusion coefficient so low, that the signal to noise would be approaching the noise level before the asymptotic limit is reached. Alternatively, the water can be used as a probe for determining the average oil droplet in the o/w system. This is done by suppressing the crude oil signal by difference in relaxation times, to create a situation where the water is effectively diffusing amongst spheres (oil droplets) that do not contribute with a signal. Then the situation is equivalent to diffusion of water amongst compact spheres [54], where the short time approach returns the average (S/V) of the water phase. To transfer this to an average oil droplet size, the volume fraction of the water phase must be determined as well as the total volume. Acknowledging that the surface area in the (S/V) for the oil is the same as the surface area for the water, the $(S/V)_{oil}$ is

$$\frac{S}{V_{oil}} = \frac{S}{V_{total} - V_{brine}} = \frac{S}{V_{brine}} \left[\frac{\frac{V_{brine}}{V_{total}}}{1 - \frac{V_{brine}}{V_{total}}} \right] = \frac{S}{V_{brine}} \left(\frac{Frac_{brine}}{1 - Frac_{brine}} \right) \quad (7.21)$$

By probing the restricted diffusion of the continuous phase in the oil in water emulsion and quantify its fraction, it is therefore possible to determine the average droplet diameter of the oil droplets. It is not as straight forward to determine a distribution of crude oil droplets in a o/w system, as there is no easy way to transform the distribution of relaxation times of the water phase to a distribution of $(S/V)_{oil}$.

The intermediate situation during the inversion process, the double emulsion, can be characterized by probing the water phase, both as water droplets inside larger oil droplets, and as the continuous phase.

7.6 Double Emulsions

Multiple emulsions, also called "emulsions of emulsions", can be described as emulsions where the droplets contain smaller droplets dispersed within the same liquid. These double emulsions can be classified as two types, oil-in-water-in-oil (o/w/o) and water-in-oil-in-water (w/o/w), and they are thermodynamically unstable systems [55, 56]. Formation of double emulsions has been observed as a precursor to phase inversion [50, 57, 58], such as the example given in the previous section. There are two ways of preparing the double emulsion system, the one-step and the two-step method. The one-step preparation uses on set of emulsifiers, but suffers from poor stability because of the fast migration of the emulsifiers between the phases [5]. In order to have a good control in the double emulsion preparation and to obtain an emulsion of reasonable stability, a two-step emulsification process is followed. This method uses a set of two different emulsifiers: a hydrophobic surfactant (used in the water-in-oil emulsion) and a hydrophilic (used in the oil-in-water emulsion). In the first step, the water-in-oil (w/o) emulsion is prepared under a high stirring rate. In the second step the previous w/o emulsion is gently mixed with an aqueous phase, which produces the double emulsion (w/o/w) [59]. Figure 7.31 illustrates the two-step emulsification process for double emulsion formation.

In order to investigate the potential of PFG NMR measurements for double emulsion characterization, a 50 % v/v mixture of MQ-water and the oil phase, composed by a 5 % w/w solution of Span®80 (HLB 4.3, Sigma-Aldrich, Germany) dissolved in Mineral Oil (Sigma-Aldrich, Germany) was prepared.

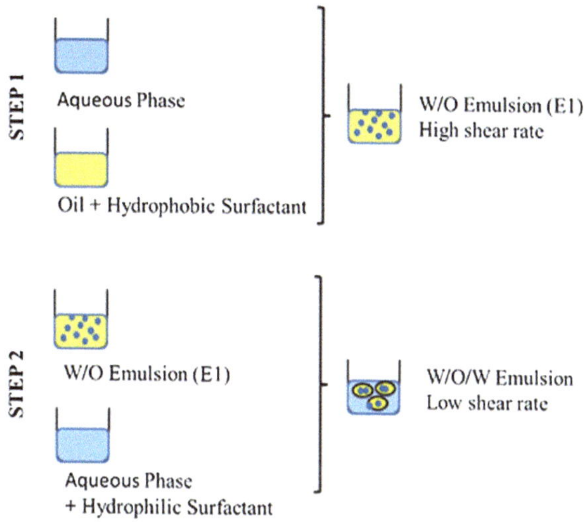

Fig. 7.31 Illustration of the two-step process for double emulsion formation [59]

The mixture was stirred for 3 min at a shear rate of 24,000 rpm by using an UltraTurrax® (IKA®-Werke Co., Germany) at room temperature. As a result, a w/o emulsion (E1 in Fig. 7.32) was obtained. In the second step, 50 % v/v of the emulsion E1 and a solution of 2 % w/w of Tween®80 (Sigma-Aldrich, Germany) in MQ-water was stirred for 1 min at 250 rpm by using a mixer IKA®-Werke Eurostar Digital (IKA®-Werke Co., Germany).

Figure 7.32 shows an image of the produced double emulsion. When observing the oil droplets in the microscope, water droplets are observed to move around within the oil droplet but do not move outside. Consequently, they are confined within the oil droplet and form a double emulsion with water as the continuous phase.

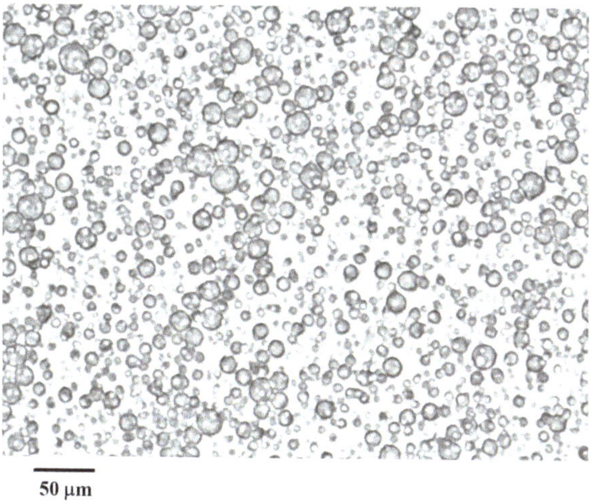

Fig. 7.32 Verification of the formation of double emulsions. Oil droplets are observed with brine droplets inside

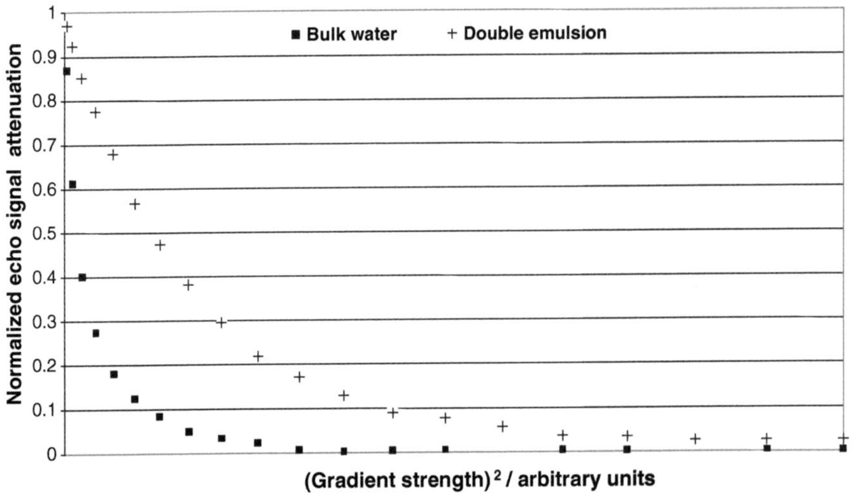

Fig. 7.33 The attenuations of signal from a double emulsion and bulk water using the 13 interval PFGSTE sequence where $\Delta = 1$ s

In order to characterize the droplet size distribution of the water phase, it is necessary to simultaneously suppress two components, the oil phase and the continuous water phase. The oil phase can be suppressed using the T_1-approach, but this method cannot be applied for the continuous water phase, as this component will have the approximately the same T_1 and T_2 as the emulsified water component, or even longer. The only way to distinguish the emulsified water and the water in the continuous phase is to distinguish them due to a significant difference in the measured root of the mean squared displacement (RMD) at longer observation times. The duration of this observation time is dependent on the water droplet sizes, and in Fig. 7.33 the echo attenuation from a 13 interval PFGSTE sequence is shown where the observation time is 1 s. At this Δ-value the oil signal is suppressed, leaving the water signal as the only contributor to the attenuating signal. Also shown in the plot is the attenuation of bulk water acquired under the same experimental conditions. While the bulk water rapidly approaches the noise level for the signal, the initial slope of the double emulsion is less and there is a small component (~10 % of the total signal) which attenuates much more slowly with increasing gradient strength. If the oil phase was not satisfactorily suppressed, such a component could have been attributed to a residual oil signal. However, when running a CPMG after suppression of the continuous water phase, the T_2-distribution peaks up around 1 s. As the T_2 of the oil phase is 0.2 s, the slow diffusing component must have its origin from water. By combining the RMD-approach with the CPMG, the existence of double emulsion is verified by PFG NMR, as well as from observing restricted movement of droplets inside oil droplets in the microscope.

Having established the existence of double emulsions, the profile of the emulsified water is recorded using the RMD-approach. To make it quantitative the profile was recorded at two observation times and two gradient strengths in the RMD

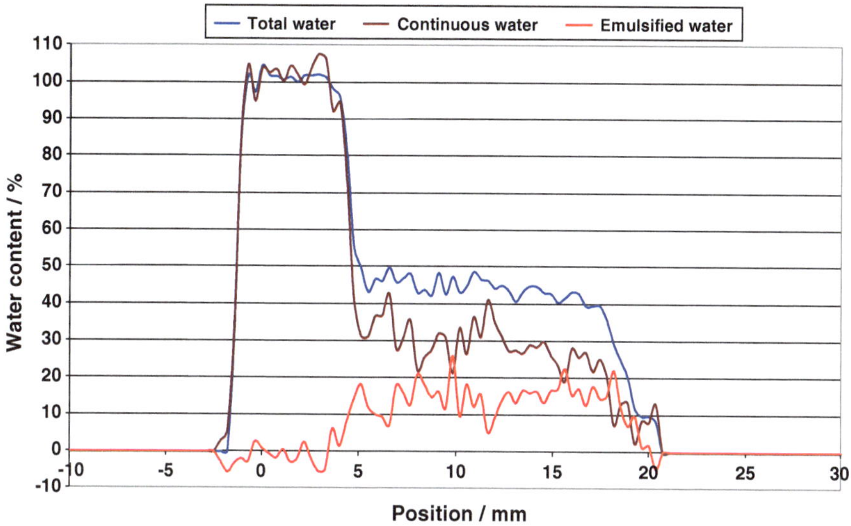

Fig. 7.34 Water profiles from a double emulsion. The *blue* profile corresponds to the profile of the total water, the *red* profile to the emulsified water and the *brown* profile to the water continuous phase

part of the sequence. The total water profile was recorded using the T_2-approach, and subtraction of the emulsified water from the total water then provides the profile of the water continuous phase. As seen in Fig. 7.34 the double emulsion is not a stable emulsion. A bulk phase emerged only minutes after production of the emulsion, and it increased with time. Nevertheless, even in the presence of the bulk phase of water, it was possible to identify the double emulsion. As can be seen in the water continuous profile there is a weak gradient, which confirms the instability of the system. Regarding the oil droplets, images taken on microscope confirmed that creaming took place, i.e. a densely packed layer of oil droplets containing water droplets emerged at the upper part of the sample.

Figure 7.35 shows the number based droplet size distribution and a corresponding microscope image of the double emulsion. As there is no other simple way to measure the droplet size distribution of the brine droplets confined in oil droplets it was only confirmed visually that there was a reasonably good agreement between the NMR droplet size distribution and the actual sizes as seen in the microscope.

In principle the average diameter of the oil droplets could have been determined by measuring the apparent diffusion of the water continuous phase within the double emulsion, quantify the fraction of the water continuous phase, and thereby use (7.21) to calculate $(S/V)_{oil}$. However, for the current sample, the presence of the bulk water phase alters the apparent diffusion such that there is no straightforward relation between the apparent diffusion and the (S/V) of the water continuous phase within the double emulsion. The apparent diffusion coefficient of the water continuous phase could have been determined from the data in Fig. 7.33, but the data are weighted between the bulk water diffusion coefficient and the desired apparent

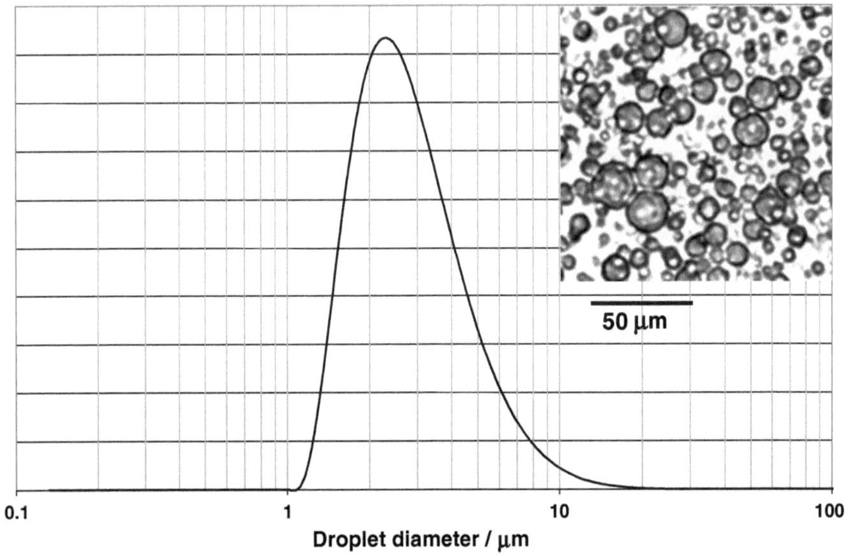

Fig. 7.35 The number based droplet size distribution of water in a w/o/w double emulsion, and a corresponding microscopy image of a double emulsion

diffusion coefficient from the water continuous phase within the double emulsion. For the separation of the two water continuous phases, either diffusion weighted profiles must be recorded or a slice selection using shaped RF-pulses is required to exclude the region containing the bulk phase in the PFG NMR experiment.

7.7 On the Use of PFG-NMR for Characterising Emulsions

The possibility of using of PFG NMR to determine droplet sizes or distribution of sizes has been known since the 1960s, where the most popular method has been the one presented by Packer and Rees [20]. In this chapter several different approaches have been applied, which combine the diffusion experiments with relaxation time measurements aiming at making a fast and accurate method for characterizing emulsions. With these combinations there are no assumptions made regarding the droplet size distribution function, except for the assumption that the decay of the signal from a relaxation time experiment can be represented by multi exponential decay, which subsequently can be subjected to an Inverse Laplace Transform. Another important difference compared to the more traditional PFG NMR method for determining the droplet sizes [20], is the use of the various methods for resolving the water from the oil signal. In work with crude oils and synthetic oils the viscosity of the oil and thereby the molecular mobility and its characteristic relaxation times may change

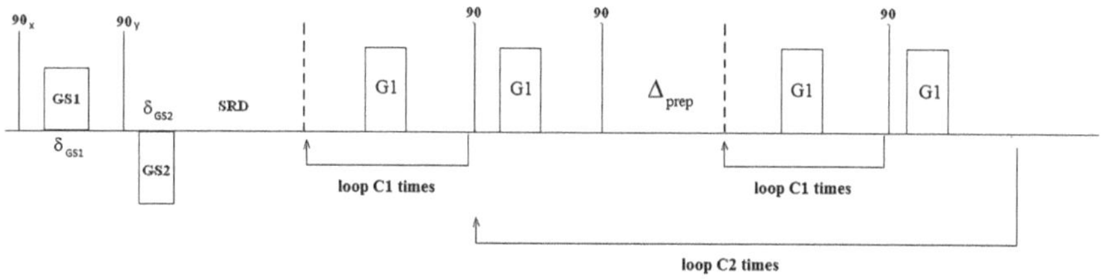

Fig. 7.36 A typical preparation sequence for suppression of the oil component and if necessary the bulk water phase

several orders of magnitude. This span means different methods are required in order to resolve the water signal from the oil signal.

In Fig. 7.36 a PFG NMR sequence is shown which illustrates the different aspects of preparation (both in terms of sample magnetisation preparation and system eddy current stabilisation) to be performed prior to the measurements that are used for characterising emulsions. A spoiler recovery starts the sequence by reducing the recycle delay. After the SRD that produces a measurable signal, a C1 loop of preparatory gradients is applied to stabilise the system eddy currents before the start of the actual experiment, which consists of a 90-degree excitation pulse and a C2 loop of PFGSTE-sequences. This part is applied for the suppression of oil signal due to differences in the root mean squared displacement between the water confined in droplets and the oil and/or water in a continuous phase. The C2-loop is optional for applying convection compensation in the sequence (i.e. C2 = 2). In Fig. 7.36, the monopolar version of the PFGSTE is shown, but a bipolar version could be used as well. The sequence is flexible as it also is applicable for separating the oil and water signal using significant difference in T_1-relaxation times. This is achieved by applying G1 values with the strengths necessary to measure diffusion of the emulsified water phase regardless of the presence of an oil signal, and using the duration of Δ_{prep} to suppress the oil signal.

After having resolved the required signal, Fig. 7.37 provides the necessary sequences for characterising the emulsion if it is not already implicit in the pulse sequence in Fig. 7.36. The first part is the PFG NMR-sequence that determines the (S/V) (the short observation time approach) or the average of the squared droplet radius (the asymptotic approach). If necessary, as for determining larger droplets (>20 μm), the bipolar PFGSE sequence should be replaced by a bipolar PFGSTE sequence. After having performed the motional encoding/decoding, the sequence is split in three; the upper part is the one for measuring the apparent diffusion coefficient of the resolved signal, in the middle a diffusion weighted profile is recorded, and in the lower part a diffusion weighted CPMG-attenuation is recorded. The z-storage delay Δ is included to ensure eddy current transients to having decayed satisfactorily.

When measuring the apparent diffusion of the water phase confined in the droplets several factors may influence the result. One of the most important is the effect from restricted diffusion during the gradient pulse. As the duration of the gradient

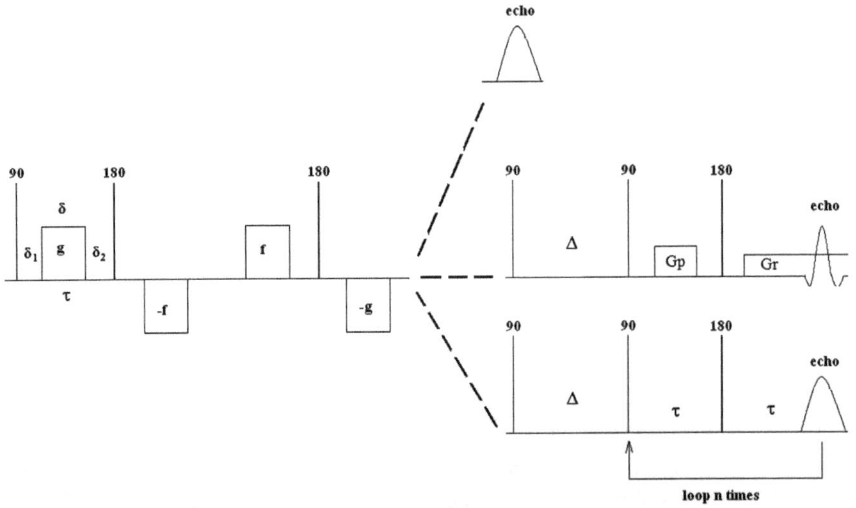

Fig. 7.37 Typical PFG-NMR sequences for measuring droplet size distributions and/or profiles of the surface to volume (S/V) ratio

pulse used to generate the PFG NMR data presented here is approximately 0.5 ms, the root of the mean squared displacement in one dimension and at room temperature is approximately 1.5 μm. As the volume based droplet size distributions for a large fraction of emulsions investigated in this chapter peaks up around 5–10 μm, the RMD is therefore less than the typical droplet size measured, but it is not necessarily negligible. From the concept of centre of mass (Chap. 3), it is evident that the diffusion measurements on the smaller droplets, due to restricted diffusion during the gradient pulses, will be biased toward determining apparently smaller droplets than the actual ones. Nordgård et al. [60] corrected for this effect by measuring the apparent diffusion at two different gradient pulse lengths, and extrapolated back to zero observation time. Here the dependency on the apparent diffusion coefficient was assumed to be of the square of the applied gradient pulse, which probably would have been more accurately represented by a series expansion

$$\frac{D(t)}{D_0} = \frac{D(t)}{D_0}_{(\delta=0)} + c_1\delta + c_2\delta^2 + \Phi(\delta^3) \tag{7.22}$$

Instead of applying a fit to (7.22), a single measurement accompanied by the corrected observation time proposed by Fordham et al. [61] could also have been employed. Their approach takes into account the restricted diffusion during the gradient pulses, and defines a corrected observation time that can be used in the short observation time approach. For the asymptotic approach, this will not apply as the observation time is far from the short observation time approach, but the series expansion shown in (7.22) still applies. Alternatively, the duration of the gradient pulse is set so long that an asymptotic attenuation is reached even during the gradient pulse as follows [62]

$$\ln\left(\frac{I}{I_0}\right) = -\frac{16\gamma^2 g^2 \delta R^4}{175 D_0} \tag{7.23}$$

Then the center of mass approaches a δ-function at the center of the cavity, and the average droplet radius may be determined by measuring the slope at different values of δ or the applied gradient strength.

When the droplet sizes measured are down to 5 μm, one approaches the limit of where the apparent diffusion at short observation times is valid. Given a typical observation time of a few milliseconds, the apparent diffusion coefficient has then attenuated so much that higher order term in the short observation expansion of the apparent diffusion coefficient may affect the measurements [35]. In order to check the validity of the two methods in this regime, both the short observation time and the asymptotic approaches were used to determine a droplet size distribution from an emulsion where the average droplet diameter was approximately 8 μm. The apparent diffusion coefficient was measured to be a fraction 0.5 of the bulk diffusion coefficient, and thus the effect from curvature and surface relaxivity may affect the determined diffusion coefficient as shown in the following equation

$$\frac{D(t)}{D_0} \approx 1 - \frac{4}{9\sqrt{\pi}}\sqrt{D_0 t}\frac{S}{V} - \frac{D_0 t}{6R_0}\frac{S}{V} + \frac{\rho t}{6}\frac{S}{V} \tag{7.24}$$

In Fig. 7.38 the number based droplet size distribution obtained from the short observation time expansion ($D(t)/D_0 = 0.5$) is shown together with the one obtained from using the asymptotic limit with an observation time of 1 s. Using an observation time Δ of 1 s, the RMD is 70 μm, and the asymptotic behaviour is valid. As can be seen from Fig. 7.38 there is hardly any discrepancy between the two approaches for determining the droplet size distributions, except for a small shift towards smaller droplets using the short observation time approach. As the value of surface relaxivity is small ($<10^{-6}$ m/s), it is the curvature term that contributes significantly to the correction term as $D(t)/D_0$ decreases. Consequently, the presence of the curvature is to increase the value of the determined (S/V) ratio found when applying the first order correction term only. In this case, the apparent measured droplet radius will be smaller than the actual one, and this corresponds to what is seen in Fig. 7.38, a small shift towards smaller droplets when applying the short observation time approach.

Even though there is a small shift in measured droplet size when comparing the two methods, there seems to be an intermediate region of droplet diameters, ~5–10 μm, where both methods apply reasonably well. Below this region the asymptotic approach is the appropriate choice of method while above this region the short observation time approach provides the best solution for obtaining the droplet size distributions.

Both the short observation time and the asymptotic approaches rely on motional encoding/decoding to measure the root of the mean squared displacement. While the short observation time has to include the knowledge on bulk diffusivity of the water phase, the asymptotic approach does not require this as an input. When aiming at having robust methods for characterising the emulsions,

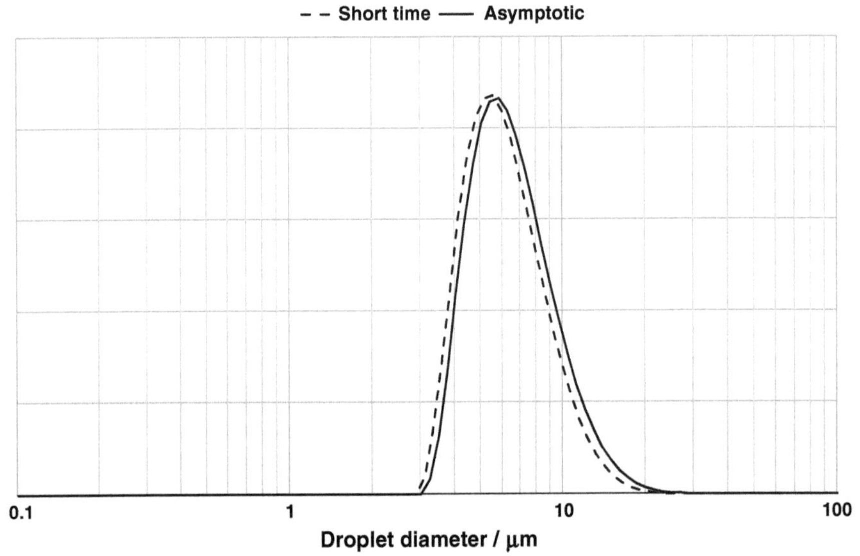

Fig. 7.38 Comparison of the droplet size distributions obtained from a water in oil emulsion using the short observation time and the asymptotic limit approximations

the asymptotic approach therefore seems to be the best choice of method. Not only can the value of the bulk diffusion coefficient be neglected, but the same settings at elevated temperatures can also be applied. For the short observation time approach, information on the bulk diffusion coefficient is needed at every temperature measured. Compared to the asymptotic approach the short observation time approach benefits when the emulsion is unstable. In these situations flow or even higher order convection terms may occur, that are not cancelled out by the double PFGS(T)E methods. The short observation time approach provides the shortest convection compensating pulse sequence, the double 11-interval PFGSE sequence, and the convection compensation is then the most effective. The longer duration of the pulse sequence, the greater will the impact from higher order convection or turbulence terms be. In most cases, as indicated in Fig. 7.38, both approaches will characterise the emulsions properly, but evidently, there are situations where one is preferred over the other.

As many of the emulsions prepared are unstable, an unwanted water bulk phase may evolve as a function of time. When choosing the method for determining the droplet size distribution one should be aware of the effect of the presence of such a phase during the PFG NMR experiments. In Fig. 7.39, the droplet size distribution is acquired with and without suppression of the water bulk phase. Obviously, both the distribution and average droplet size may change significantly towards including apparently larger droplets. This is a consequence of the measurement of a too high apparent diffusion coefficient due to the interference from a bulk fraction of water.

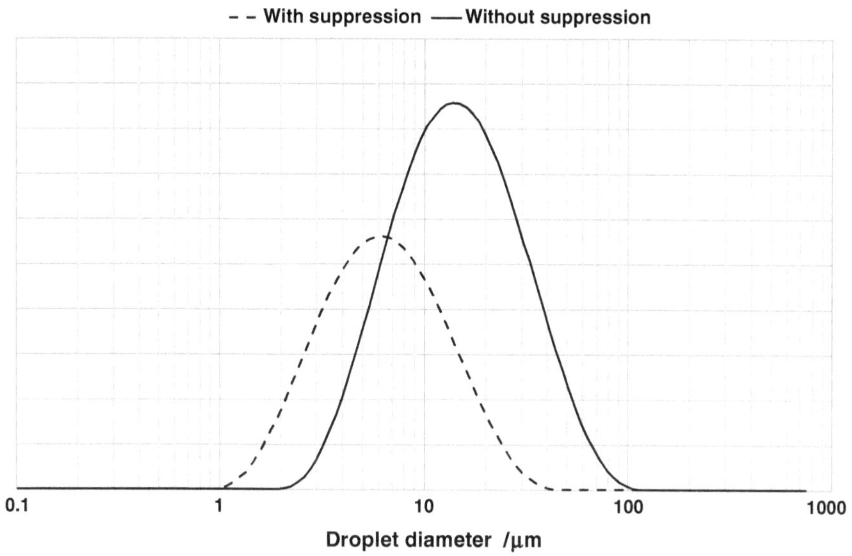

Fig. 7.39 The droplet size distributions obtained with and without the suppression of a water continuous phase

In Fig. 7.40, a water in oil emulsion with a bulk water phase at the bottom of the sample, the $(S/V)^{-1}$ profile has been measured with or without suppression of the bulk water signal. The bulk water was suppressed by using a PFGSTE diffusion sequence in the RMD-approach. For this particular experimental set-up, the emulsified water is also suppressed by a significant amount, as can be seen from the enhanced noise in the $(S/V)^{-1}$ profile when the bulk water signal is being suppressed. In this situation, the $(S/V)^{-1}$ profile also returns slightly lower values for the average droplet diameter. This confirms the existence of a distribution of droplets in the emulsion and the suppression of bulk water leads to a bias of the $(S/V)^{-1}$ profile towards the smaller droplets with the lowest apparent diffusion coefficients. An increased z-storage delay Δ in the RMD-approach would have made the $(S/V)^{-1}$ more equal, as the attenuation of the emulsified water would be less due to a reduction in the apparent diffusion coefficient.

By combining different approaches for characterising emulsions, it is possible to resolve the emulsified water regardless of the oil viscosity and the bulk fraction of the water phase. However, the use of low viscous oils when preparing water in oil emulsions, such as decane, introduces other effects that may alter the determined droplet size significantly. In Fig. 7.41, the average droplet size is determined at various observation times in the diffusion sequence. For the emulsion with high viscosity oil as the continuous phase there is a small increase towards larger droplet as the observation time is increased. This effect is due to a T_1 weighting of the diffusion experiment; at the longer observation times, the measured average droplet diameter will be biased towards larger droplets with longer

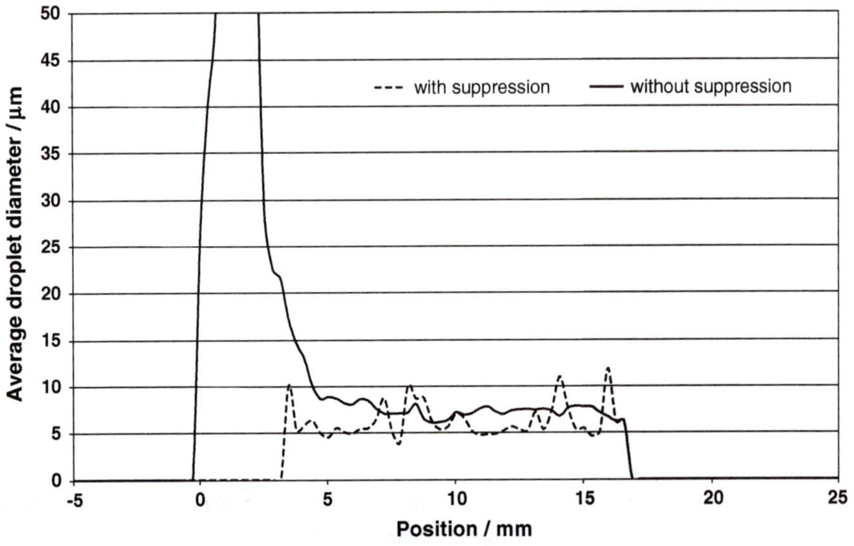

Fig. 7.40 The $(S/V)^{-1}$ profile obtained with and without suppression of a water continuous phase

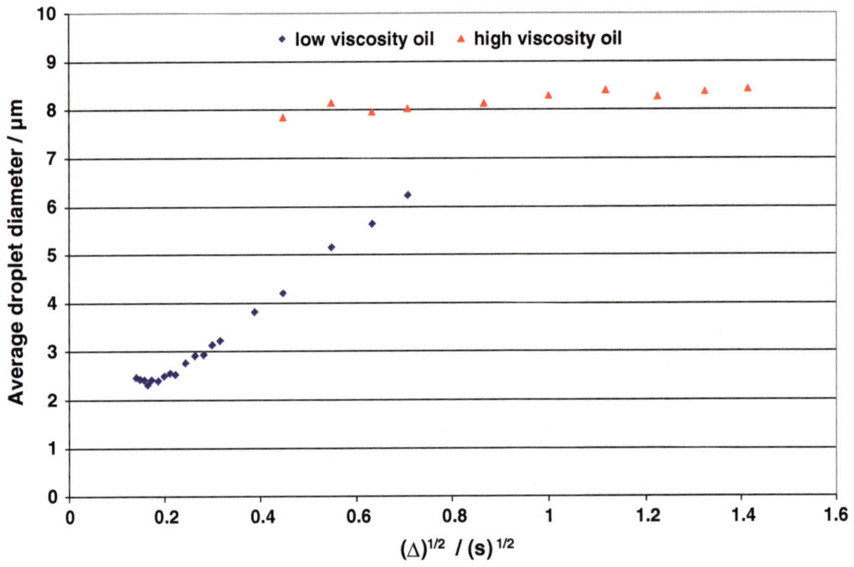

Fig. 7.41 The measured volume average droplet radius as a function of observation time Δ

T_1. However, the observation time dependency of the obtained diameter is small and is nearly within the noise value of the experiment. For the emulsion with low viscosity oil as the continuous phase, the situation is very different. The emulsion containing low viscosity oil was prepared by mixing the decane and water phases at the ambient temperature (22 ± 1 °C). As can be observed in Fig. 7.41, there is a strong Δ dependency on the measured average droplet size, but as the sequence used is a convection compensated sequence, this dependency is not likely to arise from sedimentation of the droplets.

Due to the square root of time dependency of the determined average droplet size, a more plausible case for this dependency would be a Brownian-like movement of the droplets themselves. As the droplets turned out to be quite small, the thermal movement of the droplets is more likely to appear when they are confined in a low viscosity oil as decane. The emulsion system containing the low viscosity oil exhibit a strong dependency on Δ down to approximately 50 ms, then it levels out. Measurements below 25 ms are not conducted as in this regime the asymptotic approach is no longer valid ($RMD_{DSD} \ll RMD_{bulk}$), i.e. the apparent measured droplet diameter starts to increase. Consequently, care must be taken when choosing Δ for determining the average squared droplet diameter and it is recommended to check for any possible unwanted Δ-dependency. Indeed, if brine droplets were allowed to move freely a linear dependency of measured root of the mean squared displacement should be expected, but in Fig. 7.41 it reaches a plateau value. This is probably due to the fact that the system measured on had a water cut of 60 % and it continued to sediment until it had a closed sphere packing of 70 % brine droplets. The picture of Brownian movement does not fit to the system at the shortest observation times. In this case, the importance of the packing of the spheres must somehow be taken into account, for example as a correlation time for droplets exchanging location within the densely packed layer.

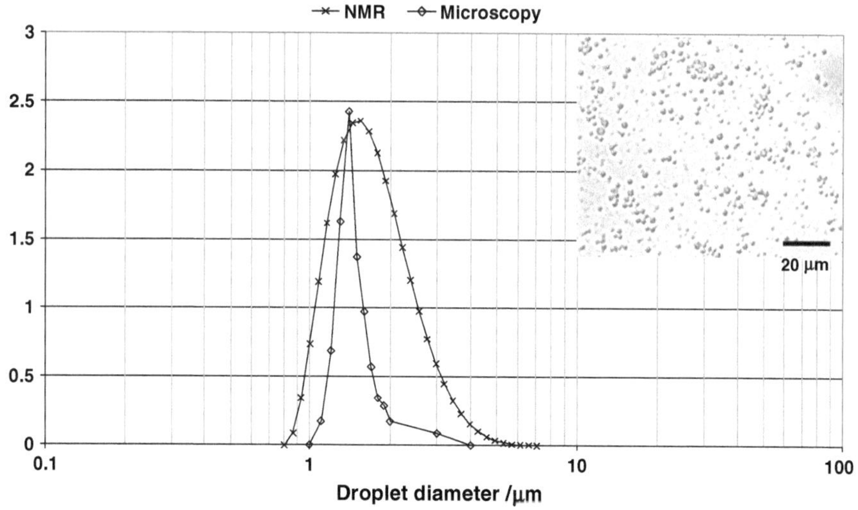

Fig. 7.42 The number based droplet size distribution from PFG-NMR and microscopy. Scale bar in the microscopy image is 20 μm. 306 droplets were measured to produce the number based droplet size distribution from a microscopy image

The z-storage delay Δ used for determining the actual droplet size distribution of the water in decane emulsion was 35 ms, but as the droplets were quite small, this observation time was sufficient to apply the asymptotic limit approach. When comparing with microscopy results reasonable agreement between the two methods is found (Fig. 7.42). The distribution from the PFG NMR method seems to be more broadened. However, as the number based droplet size distribution from microscopy data arises from the counting of a small numbers of droplets and considering the fact that the droplet diameters are close to the resolution limit of the microscope, it is more likely that the actual droplet size distribution is as broadened as obtained by the PFG NMR method. Most important is that the average droplet diameters from the two number based distributions are found to be quite similar, 1.5 μm from microscopy versus 1.8 μm from PFG NMR.

What if the droplets were so large that the asymptotic limit was not valid and the movement of the droplets prevented a plateau region to be reached as shown in Fig. 7.41? In this situation, it would more convenient to apply the short observation time approach to extract the droplet size distribution [35], i.e. application of the 11-interval PFGSE sequence in order to reduce the observation time down to just a few milliseconds. A PFGSTE sequences could have been placed in the preparation sequence for suppression of low viscosity oil and bulk water, i.e. application of the RMD-approach for preparation purpose. Conclusion is that the PFG NMR method provides a very flexible tool for characterizing emulsions regardless of the oil viscosity.

References

1. K. Holmberg et al., "*Emulsions and Emulsifiers*" *Surfactants and Polymers in Aqueous Solution* (Wiley, New York, 2003), pp. 451–471
2. F. Leal-Calderon, J. Bibette, V. Schmitt, *Double Emulsions*, in *Emulsion Science* (Springer, New York, 2007), pp. 173–199
3. A. Bhardwaj, S. Hartland, Dynamics of emulsification and demulsification of water in crude oil emulsions. Ind. Eng. Chem. Res. **33**(5), 1271–1279 (1994)
4. D.J. McClements, *Food Emulsions: Principles, Practice, and Techniques* (CRC Press, Boca Raton, 2005)
5. J. Sjoblom, *Encyclopedic Handbook of Emulsion Technology* (Taylor & Francis, UK, 2001)
6. F. Nielloud, *Pharmaceutical Emulsions and Suspensions: Second Edition, Revised and Expanded* (Taylor & Francis, UK, 2000)
7. G.H. Sorland, Characterization of emulsions by PFG-NMR. Magn. Reson. Porous Media **1330**, 27–30 (2011)
8. N.V. Opedal, G. Sorland, J. Sjoblom, Emulsion stability studied by nuclear magnetic resonance (NMR). Energy Fuels **24**, 3628–3633 (2010)
9. G.H. Sorland et al., Determination of total fat and moisture content in meat using low field NMR. Meat Sci. **66**(3), 543–550 (2004)
10. Y. Asano, K. Sotoyama, Food Chem. **66**, 327–331 (1999)
11. Y. Otsubo, R.K. Prud'homme, Rheol. Acta **33**, 303–306 (1994)
12. E.S. Basheva et al., Langmuir **15**, 6764–6769 (1999)
13. M. Chakraborty, C. Bhattacharya, S. Datta, Colloids Surf., A **224**, 65–74 (2003)
14. J.N. Coupland, D. Julian McClements, Droplet size determination in food emulsions: Comparison of ultrasonic and light scattering methods. J. Food Eng. **50**(2), 117–120 (2001)

15. J.N. Coupland, D.J. McClements, J. Food Eng. **50**, 117–120 (2001)
16. P.V. Hemmingsen et al., *Droplet size distributions of oil-in-water emulsions under high pressures by video microscopy*. 2nd edn., ed. by J. Sjöblom. Emulsion and Emulsion Stability, vol. 132 (Taylor & Francis, Boca Raton, 2006)
17. R. Bernewitz, G. Guthausen, H.P. Schuchmann, NMR on emulsions: Characterisation of liquid dispersed systems. Magn. Reson. Chem. **49**(1), S93–S104 (2011)
18. M.A. Voda, J. van Duynhoven, Characterization of food emulsions by PFG NMR. Trends Food Sci. Technol. **20**(11–12), 533–543 (2009)
19. O. Söderman, *Pulsed field gradient NMR studies of emulsions: Dsroplet sizes and concentrated emulsions*, ed. by K. Kawasaki, B. Lindman, H. Okabayashi. Formation and Dynamics of Self-organized Structures in Surfactants and Polymer Solutions (Steinkopff, Germany, 1997), pp. 34–41
20. K.J. Packer, C. Rees, Pulsed NMR studies of restricted diffusion. I. Droplet size distributions in emulsions. J. Colloid Interface Sci. **40**(2), 206–218 (1972)
21. S. Simon et al., Separation profile of model water-in-oil emulsions followed by nuclear magnetic resonance (NMR) measurements: Application range and comparison with a multiple-light scattering based apparatus. J. Colloid Interface Sci. **356**(1), 352–361 (2011)
22. F. Civan, Viscosity-temperature correlation for crude oils using an Arrhenius-type asymptotic exponential function. Pet. Sci. Technol. **24**(6), 699–706 (2006)
23. P.M. Kampmeyer, The temperature dependence of viscosity for water and mercury. J. Appl. Phys. **23**(1), 99–102 (1952)
24. E.O. Stejskal, Use of spin echoes in a pulsed magnetic-field gradient to study anisotropic, restricted diffusion and flow. J. Chem. Phys. **43**(10), 3597–3603 (1965)
25. E.O. Stejskal, J.E. Tanner, Spin diffusion measurements: Spin echoes in the presence of a time-dependent field gradient. J. Chem. Phys. **42**(1), 288–292 (1965)
26. J.E. Tanner, E.O. Stejskal, Restricted self-diffusion of protons in colloidal systems by the pulsed-gradient, spin-echo method. J. Chem. Phys. **49**(4), 1768–1777 (1968)
27. I. Fourel, J.P. Guillement, D. Le Botlan, Determination of water droplet size distributions by low resolution PFG-NMR: II. "Solid" emulsions. J. Colloid Interface Sci. **169**(1), 119–124 (1995)
28. D. Gabriele et al., Characterisation of dairy emulsions by NMR and rheological techniques. Food Hydrocolloids **23**(3), 619–628 (2009)
29. J.P.M. van Duynhoven et al., Scope of droplet size measurements in food emulsions by pulsed field gradient NMR at low field. Magn. Reson. Chem. **40**(13), S51–S59 (2002)
30. C.P. Aichele et al., Water in oil emulsion droplet size characterization using a pulsed field gradient with diffusion editing (PFG-DE) NMR technique. J. Colloid Interface Sci. **315**(2), 607–619 (2007)
31. S. Kiokias, A.A. Reszka, A. Bot, The use of static light scattering and pulsed-field gradient NMR to measure droplet sizes in heat-treated acidified protein-stabilised oil-in-water emulsion gels. Int. Dairy J. **14**(4), 287–295 (2004)
32. B. Balinov, O. Söderman, T. Wärnheim, Determination of water droplet size in margarines and low-calorie spreads by nuclear magnetic resonance self-diffusion. J. Am. Oil Chemists' Soc. **71**(5), 513–518 (1994)
33. P.T. Callaghan, K.W. Jolley, R.S. Humphrey, Diffusion of fat and water in cheese as studied by pulsed field gradient nuclear magnetic resonance. J. Colloid Interface Sci. **93**(2), 521–529 (1983)
34. A.A. Peña, G.J. Hirasaki, Enhanced characterization of oilfield emulsions via NMR diffusion and transverse relaxation experiments. Adv. Colloid Interface Sci. **105**(1–3), 103–150 (2003)
35. P.P. Mitra, P.N. Sen, L.M. Schwartz, Short-time behavior of the diffusion coefficient as a geometrical probe of porous media. Phys. Rev. B **47**(14), 8565–8574 (1993)
36. M.H. Cohen, K.S. Mendelson, Nuclear magnetic relaxation and the internal geometry of sedimentary rocks. J. Appl. Phys. **53**(2), 1127–1135 (1982)
37. K.S. Mendelson, Nuclear magnetic resonance in sedimentary rocks: Effect of proton desorption rate. J. Appl. Phys. **53**(9), 6465–6466 (1982)
38. A. Valfouskaya et al., Nuclear magnetic resonance diffusion with surface relaxation in porous media. J. Colloid Interface Sci. **295**(1), 188–201 (2006)
39. J. Uh, A.T. Watson, Nuclear magnetic resonance determination of surface relaxivity in permeable media. Ind. Eng. Chem. Res. **43**(12), 3026–3032 (2004)

40. G.H. Sorland, D. Aksnes, L. Gjerdaker, A pulsed field gradient spin-echo method for diffusion measurements in the presence of internal gradients. J. Magn. Reson. **137**(2), 397–401 (1999)
41. J.J. Van Vaals, A.H. Bergman, Optimization of eddy-current compensation. J. Magn. Reson. (1969), **90**(1), 52–70 (1990)
42. J.M. Fauth et al., Elimination of unwanted echoes and reduction of dead time in three-pulse electron spin-echo spectroscopy. J. Magn. Reson. (1969), **66**(1), 74–85 (1986)
43. N. van der Tuuk Opedal, G. Sørland, J. Sjöblom, Methods for droplet size distribution determination of water-in-oil emulsions using low-field NMR. Diffusion fundamentals, **7**(diffusion-fundamentals.org), 1–29 (2009)
44. R. Sandnes et al., Optimization and validation of low field nuclear magnetic resonance sequences to determine low water contents and water profiles in W/O emulsions. Colloids Surf., A
45. J. Sjoblom, *Emulsions and Emulsion Stability: Surfactant Science Series/61* (Taylor & Francis, UK, 2012)
46. S. Less, A. Hannisdal, J. Sjöblom, An electrorheological study on the behavior of water-in-crude oil emulsions under influence of a DC electric field and different flow conditions. J. Dispersion Sci. Technol. **29**(1), 106–114 (2008)
47. K.J. Lissant, *Demulsification* (Marcel Dekker, New York, 1983)
48. L. Schramm Laurier, *Petroleum emulsions*, in Emulsions (American Chemical Society, 1992), pp. 1–49
49. J.-L. Salager et al., Current phenomenological know-how and modeling of emulsion inversion. Ind. Eng. Chem. Res. **39**(8), 2665–2676 (2000)
50. A. Barrabino et al., Phase inversion in emulsions studied by low field NMR. Colloids Surf., A **443**, 368–376 (2014)
51. R.F. Karlicek Jr, I.J. Lowe, A modified pulsed gradient technique for measuring diffusion in the presence of large background gradients. J. Magn. Reson. (1969), **37**(1), 75–91 (1980)
52. P.T. Callaghan, *Translational Dynamics & Magnetic Resonance* (Oxford University Press, Oxford, 2011), p. 547
53. A. Almela, M.P. Elizalde, R. Benito, The aggregation of Span 80 in toluene. J. Solution Chem. **22**(3), 231–241 (1993)
54. G.H. Sørland, Short-time PFGSTE diffusion measurements. J. Magn. Reson. **126**(1), 146–148 (1997)
55. A.T. Florence, D. Whitehill, Stabilization of water/oil/water multiple emulsions by polymerization of the aqueous phases. J. Pharm. Pharmacol. **34**(11), 687–691 (1982)
56. A.T. Florence, D. Whitehill, Some features of breakdown in water-in-oil-in-water multiple emulsions. J. Colloid Interface Sci. **79**(1), 243–256 (1981)
57. B.W. Brooks, H.N. Richmond, Dynamics of liquid–liquid phase inversion using non-ionic surfactants. Colloids Surf. **58**(1–2), 131–148 (1991)
58. A.W. Pacek, A.W. Nienow, I.P.T. Moore, On the structure of turbulent liquid–liquid dispersed flows in an agitated vessel. Chem. Eng. Sci. **49**(20), 3485–3498 (1994)
59. N. Garti, Double emulsions—scope, limitations and new achievements. Colloids Surf., A **123–124**, 233–246 (1997)
60. E.L. Nordgård, G. Sørland, J. Sjöblom, behavior of asphaltene model compounds at w/o interfaces. Langmuir **26**(4), 2352–2360 (2009)
61. E.J. Fordham, P.P. Mitra, L.L. Latour, Effective diffusion times in multiple-pulse PFG diffusion measurements in porous media. J. Magn. Reson. Ser. A **121**(2), 187–192 (1996)
62. L.Z. Wang, A. Caprihan, E. Fukushima, The narrow-pulse criterion for pulsed-gradient spin-echo diffusion measurements. J. Magn. Reson. Ser. A **117**(2), 209–219 (1995)

Chapter 8
Special Core Analysis (SCAL)

Abstract Due to the challenges of performing well-controlled NMR experiments down in the borehole, equivalent laboratory techniques for the characterisation of the rock core removed from the well and brought to the surface have emerged. Along with various other core characterisation techniques, the NMR laboratory analysis of cores is referred to by the oil industry as special core analysis (SCAL), and has become a valuable tool in characterising the reservoir properties. In this chapter, applications that have been developed for better understanding of the formation from which the rock core plugs has been drilled are presented with examples. The applications used will depend on the state of the rock core plugs, defined by type of fluids and their saturation in the pore space of the plug. The sections will be devoted to native state plugs and cleaned plugs saturated with various fractions of brine and oil. One of the most important states to measure during the SCAL program is the cleaned rock core plugs saturated with brine. With PFG NMR, information on characteristic properties of the porous system will be extracted, for example the pore size distribution and the connectivity of pores. By making use of the existence of internal magnetic field gradients, surface coating is also probed. In addition, on six plugs, a series of measurements at various states are presented, in order to probe the wettability.

Since they were first produced in 1958 [1], NMR logging tool have been used to provide information about the quantities of the fluids present in oil reservoir by recording of the NMR response from the surrounding borehole. Further developments in tool technology led to the first commercial logging service in 1991 with Numar's mandrel device (MRIL) [2]. In conjunction with downhole measurements, laboratory experiments have also been conducted to understand NMR behaviour of liquids present in the rock core plugs and to develop petrophysical interpretations from these data [3]. An important motivation for this interest in NMR as an analysis tool within the oil industry was (and still is) the need for better characterisation of the oil reservoir prior to the onset of oil production. Enhanced oil recovery techniques (known as EOR) have become more important

G.H. Sørland, *Dynamic Pulsed-Field-Gradient NMR*,
Springer Series in Chemical Physics 110, DOI 10.1007/978-3-662-44500-6_8

as the remaining crude oil reservoirs are less amenable to production, contain a higher portion of heavy crude oil fractions (such as asphaltenes), and there is less amount of reserve for each well.

When the production potential of new areas is being assessed, the NMR logging tool has now become a critical instrument, which can be used both in borehole logging following the drilling and also in so called measuring while drilling (MWD) applications, where the NMR signal is acquired simultaneously while the borehole is drilled [4]. The NMR tool can be used to record the NMR relaxometry data from the vicinity of the borehole as a function of drilling depth. Thus, one may characterise the formation, and information on the fluid properties and the rock morphology can be extracted. Due to the hardware limitations and the hostile environment into which the NMR logging tool is lowered, the NMR experiments that can be conducted are limited. The most common sequence used is the ordinary CPMG sequence for the recording of transverse relaxation times [5]. Data are usually acquired while the logging tool is in motion, and a large number of scans is required to acquire an acceptable signal to noise ratio. Consequently, there will be a trade off situation between vertical resolution and signal to noise in the CPMG data that are processed. The left part of Fig. 8.1 shows T_2 distributions, i.e. CPMG data that are inverted using the one-dimensional inverse Laplace technique, from a NMR log as a function of the drilling depth.

Even though the CPMG is the most common method, there are other and more sophisticated methods employed to extract more information on the fluids confined in the wall of the drilling hole, as the so called DDIF developed by Schlumberger [6, 7]. This application is devoted to fluid typing, i.e. separating the response from different fluid types due to their differences in molecular mobility. The result of the application of DDIF is shown in the right part of Fig. 8.1, where the different fluids or gases are defined to be within regions. The molecular mobility at a given temperature and pressure sets the boundary of these regions.

Due to the challenges of performing well-controlled NMR experiments down in the borehole, equivalent laboratory techniques for the characterisation of the rock core removed from the well and brought to the surface have emerged. Along with various other core characterisation techniques, the NMR laboratory analysis of cores is referred to by the oil industry as special core analysis (SCAL), and has become a valuable tool in characterising the reservoir properties. Core material brought to the surface is transported to the laboratories, where it is tested for various physical parameters, such as porosity and permeability. Mineralogical content is also assessed. Amongst the SCAL instrumentation, a low field NMR rock core analyser can be used for calibration and verification of the interpretation of the data produced from the NMR logging tools. In addition to the experiments that emulate those performed by downhole tools, the access to pulsed field magnetic gradients makes it possible to perform special core analysis (SCAL) in the laboratory that is not feasible using a logging tool, as the conditions for setting up PFG NMR instrumentation are better in the laboratory than downhole. The RF-field homogeneity can be improved substantially compared to the ones present in the

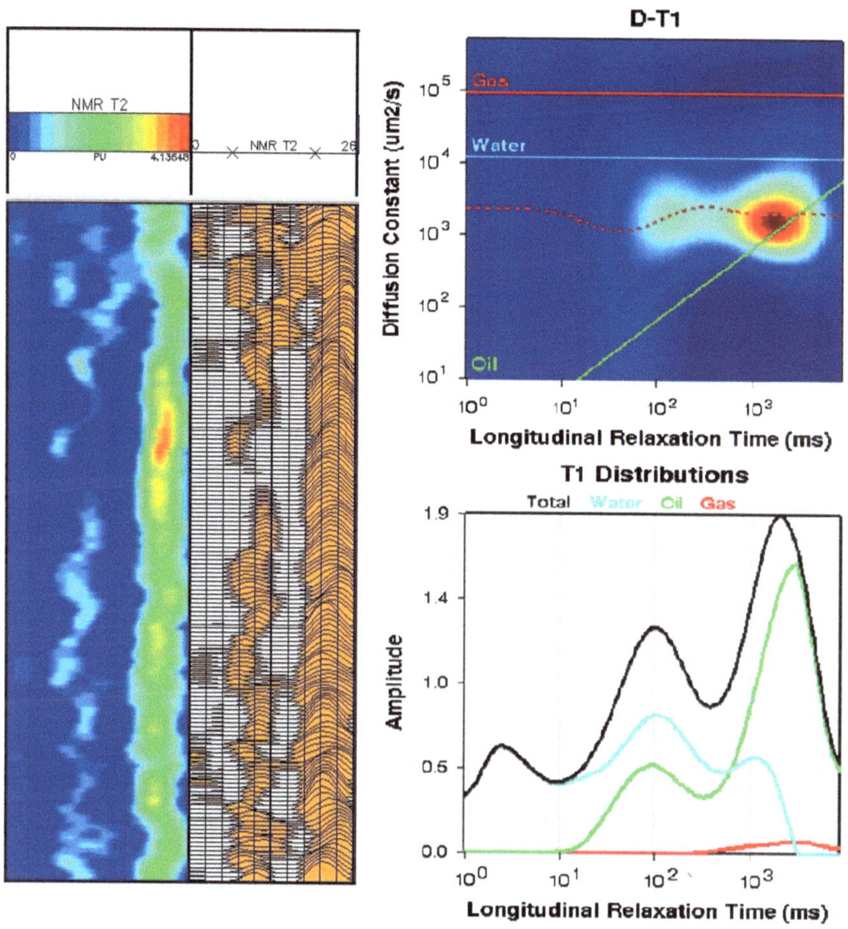

Fig. 8.1 Processed results from NMR logging tools: The *left* part is a set of processed T_2 distributions from CPMG data acquired at different depths in the well. The *right* part is a processed D-T_1 distribution from DDIF-data acquired at a given depth

logging tools, the gradient accessories are able to produce much stronger gradients than the static gradients available in the logging tool, and can also be made dynamic. In addition, the operating magnetic field can be increased to improve signal to noise ratio, provided susceptibility effects are taken into account. Also important is the time available to conduct the NMR measurements in a laboratory, which is less restricted compared to the available time when performing a real time logging in the well.

In the following section, some applications that have been developed for better understanding of the formation from which the rock core plugs has been drilled will be presented with examples. The applications used will depend on the state of the rock core plugs, defined by type of fluids and their saturation in the pore space of the plug. The sections will be devoted to native state plugs and cleaned plugs saturated with various fractions of brine and oil. On six plugs, a series of measurements at various states will also be presented, in order to probe the wettability.

8.1 Native State

Once the rock cores are retrieved from the well, they are still saturated with their original fluids, and are said to be in their native state. Depending on the mud used in the production of the well, the core may be invaded by the water or components from oil based mud used for drilling [8]. As the drilling mud can be characterized independently, the effect of the mud on the results can be determined and isolated, leaving only the reservoir fluids of interest. In the following section, methods for fluid typing and quantification of crude oil and brine simultaneously present in the rock core plug are presented.

8.1.1 Fluid Typing

As shown in Fig. 8.1 there is an application provided by some logging tools that provide information of the type of fluids present in the well. The method combines the inherent magnetic field gradient available in the tool with a CPMG measurement, to produce a two-dimensional Diffusion-T_2 distribution [6]. Knowing the experimental conditions (such as pressure and temperature in the well) the diffusion coefficient measured will indicate whether a fluid consists of hydrocarbons or brine. Often in the application, a cut-off is defined where all signal with mobility higher than this value is defined as the brine or gas signal, while the remaining signal is the crude oil.

In the laboratory, there exist numerous possibilities for identifying the fluid within the rock core plug. In the following experiment, a synthetic sample containing quartz sand, crude oil and brine was measured at 35° under atmospheric pressure by a 20 MHz PFG NMR spectrometer. The first method used was the so-called oneshot method (upper left corner in Fig. 8.2), which is a multi PFG Spin Echo method (m-PFGSE) [9]. The attenuation of the echo signal is written

$$\mathrm{I} = \sum_i \sum_n \mathrm{I}_i \mathrm{e}^{-\frac{2n\tau}{T_2^i}} \mathrm{e}^{-n\gamma^2 \mathrm{g}^2 \delta^2 \mathrm{D}^i (\tau - \delta/3)} = \sum_i \sum_n \mathrm{I}_i \mathrm{e}^{-\frac{2n\tau}{\mathrm{T}_2^{*i}}} \tag{8.1}$$

n refers to the echo number and i to the component within the sample. When collecting the diffusion terms into an apparent relaxation time, T_2^*, the attenuation is equivalent to that achieved from a T_2-experiment, and a 1D-ILT can be performed on the acquired data. However, with the ability to adjust the applied gradient strength, the apparent T_2^* may be altered. As the applied gradient is increased, the

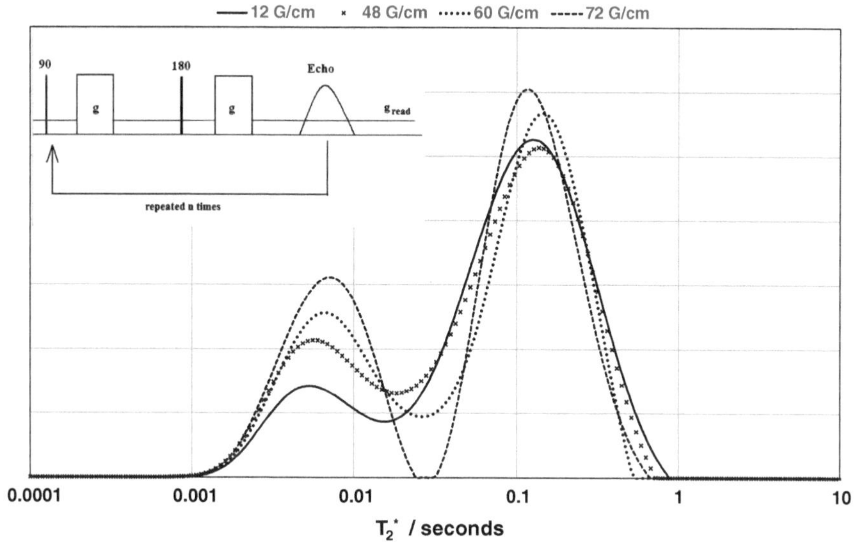

Fig. 8.2 The inverted (1D-ILT) data from the oneshot method at various gradient strengths. In the *upper left* corner the oneshot pulse sequence is displayed

variation in T_2^* is dependent on both the true T_2 and the diffusion coefficient. The movement towards shorter values is stronger for the components with higher diffusion coefficients, provided their real T_2^i satisfies the following relation

$$T_2^i > \frac{2\tau}{\gamma^2 g^2 \delta^2 D^i (\tau - \delta/3)} \tag{8.2}$$

Equation 8.2 is valid when the diffusion term in (8.1) provides a significant attenuation compared to the term containing the true T_2^i relaxation. In Fig. 8.2 the apparent T_2^* distribution is shown for different values of the applied gradient strength. The inter echo spacing is 6.7 ms and the number of echoes acquired is 128. At the lowest value of the gradient strength, the signals from the crude oil and brine overlap in T_2^* as there are no resolved peaks. At stronger applied magnetic field gradients, there is a relatively larger fraction of a short T_2^* component present, and finally at the strongest applied gradient, there are two resolved peaks. As a crude oil is believed to contain a continuum of molecular sizes, it is likely to produce a continuous T_2 or diffusion coefficient distributions, and since the oneshot method is able to resolve a peak with short T_2^* values, it is likely that this peak originates from brine only. Because of the resolved peaks, crude oil and brine can be quantified.

Two-dimensional NMR measurements, such as the DDIF, can also identify the different fluids within the sample. This is done by running the Diffusion-T_2/T_1 correlated experiments described in Chap. 1. In Figs. 8.3 and 8.4, the plots are shown from the 2D-ILT algorithm applied to the PFG NMR data acquired from

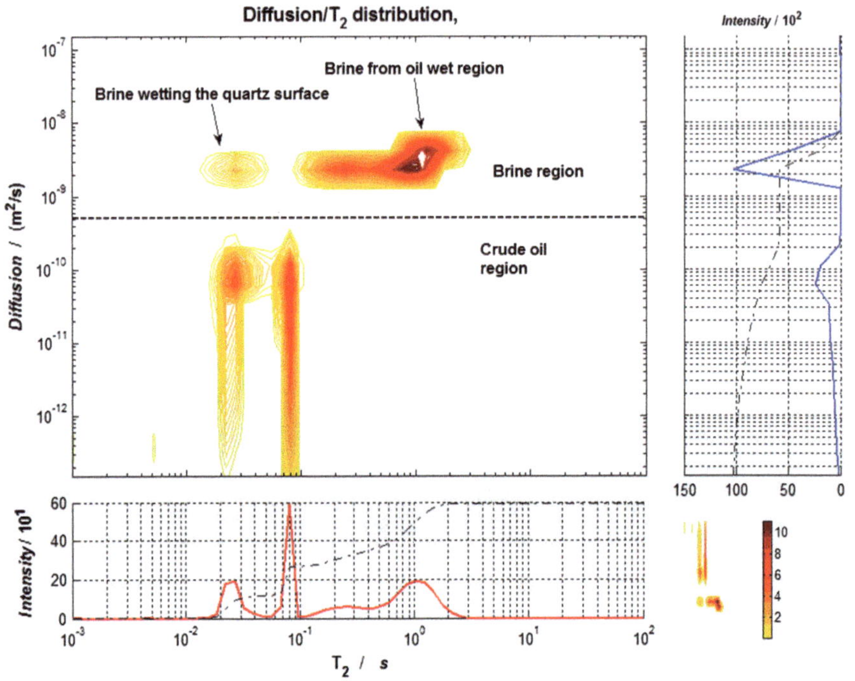

Fig. 8.3 The processed data from a Diffusion-T_2 correlated experiment acquired on a synthetic sample containing crude oil, brine and partly oil and water wetting quartz sand

the synthetic sample of quartz sand containing crude oil and brine. As the measurement temperature was 35 °C, the apparent diffusion coefficient of the brine should be of the order of 10^{-9} m^2/s. When the observation time is as short as 4 ms, the apparent brine diffusion is less affected by the confinements of the solid matrix. Consequently, the measurement is conducted in the short observation time regime (D(4 ms) $\approx D_0$). In the Diffusion-T_2 correlated experiment, there are components at higher mobility that are well resolved from the fraction of much slower mobility, and therefore these signals may be assigned to the brine fraction. In addition, within the brine fraction there is one resolved component at short T_2 values and a larger fraction at longer T_2 values. Also in the Diffusion-T_1 correlated experiment there is a well resolved fraction at an apparent diffusion of ~10^{-9} m^2/s (again the observation time in the diffusion part is 4 ms), and as for the Diffusion-T_2 this fraction can be assigned to brine. Within this fraction, there are again two components of different relaxation time, one of them a small component with shorter T_1. The relaxation times of the two components are much closer in value, which indicates that the small component with the shorter T_1 and T_2 values has a T_1/T_2 ratio that is larger than the component with the longer values. Thus, the smaller component is in a different environment than the larger

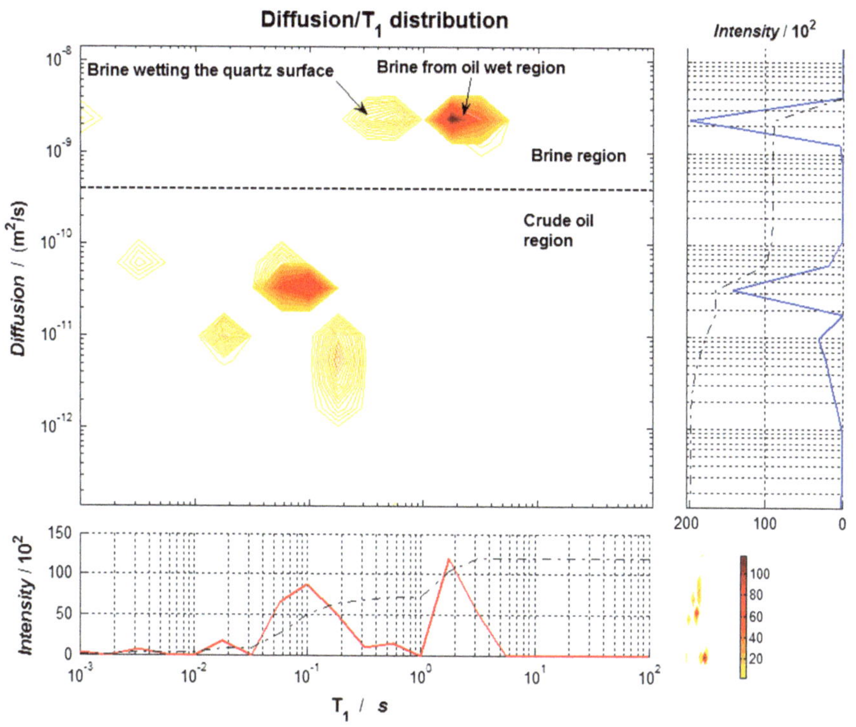

Fig. 8.4 The processed data from a Diffusion-T_1 correlated experiment acquired on a synthetic sample containing crude oil, brine and partly oil and water wetting quartz sand

component, and is the brine component that is wetting the surface of the quartz. When the brine is in contact with the surface of the quartz grains, i.e. wetting the surface, it experiences a stronger surface relaxivity and internal gradient effects. Consequently, its apparent T_2 relaxation is reduced compared to the brine in a region where crude oil is wetting the surface of the quartz. The conclusion is then that the sample contains both crude oil and brine, where the brine is partly wetting the surface of the solid matrix. Due to the slow mobility of the crude oil (~10^{-11} m^2/s), it is less sensitive to the existence of a porous material, and its Diffusion, T_1, and T_2 characteristics are more or less the same as if they had been measured on a sample containing only the crude oil.

8.1.1.1 Fluid Typing on a Rock Core Plug in Its Native State

In Figs. 8.5, 8.6 and 8.7, the 2D-ILT inversion has been applied on data from Diffusion-T_2/T_1 and T_2-T_1 correlated experiments on a native plug where the type of fluids is unknown.

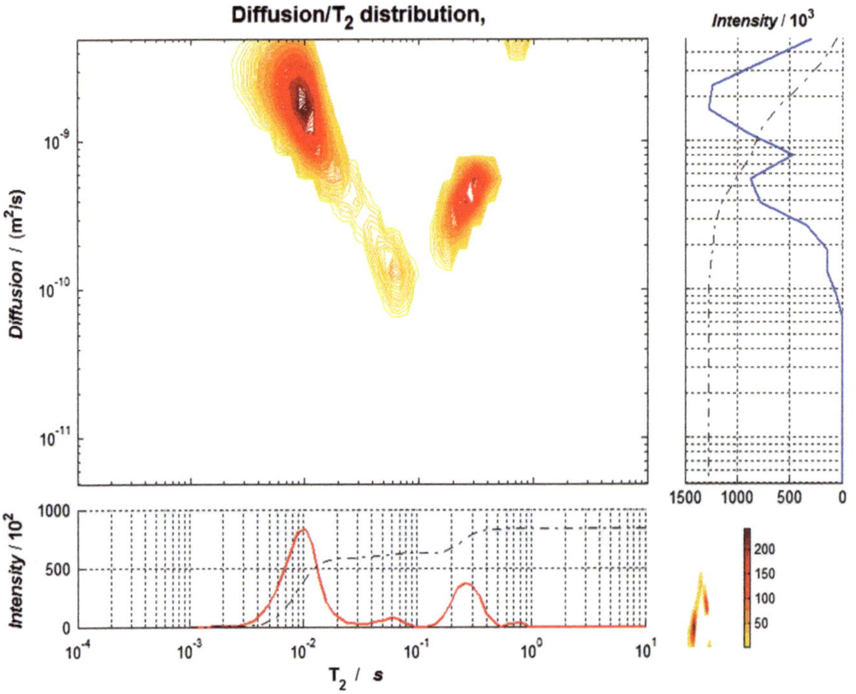

Fig. 8.5 The processed data from a Diffusion-T_2 correlated experiment acquired on a 1.5-inch rock core plug at native state

The data were acquired on a 12 MHz rock core analyzer [10], where the observation time in the diffusion experiments was 4 ms and the applied gradient strength was varied from 1 up to 125 G/cm. The sample was tempered to 35 °C. In all 2D-plots a component of short T_2 ($\leq$10 ms) of relatively high mobility appears (apparent diffusion of ~10^{-9} m^2/s), and in the T_1-T_2 correlated plot it appears with T_1/T_2 close to unity. The origin of this component is most likely clay bound brine for the following reasons: By definition, in colloid science, clay includes particles with effective diameters of less than 1 μm [11]. With a τ-value of 0.15 ms, the root of the mean squared displacement of the clay bound brine is ~1 μm, and it is thus experiencing a fluctuating internal magnetic field gradients during the τ interval, T_i^{clay}, of both polarities. Consequently, the term including the internal gradient in the equation describing apparent T_2 relaxation is much smaller for clay bound brine than brine surrounded by grains (grain diameter ≫ 1 μm => internal magnetic field gradient G^{grain})

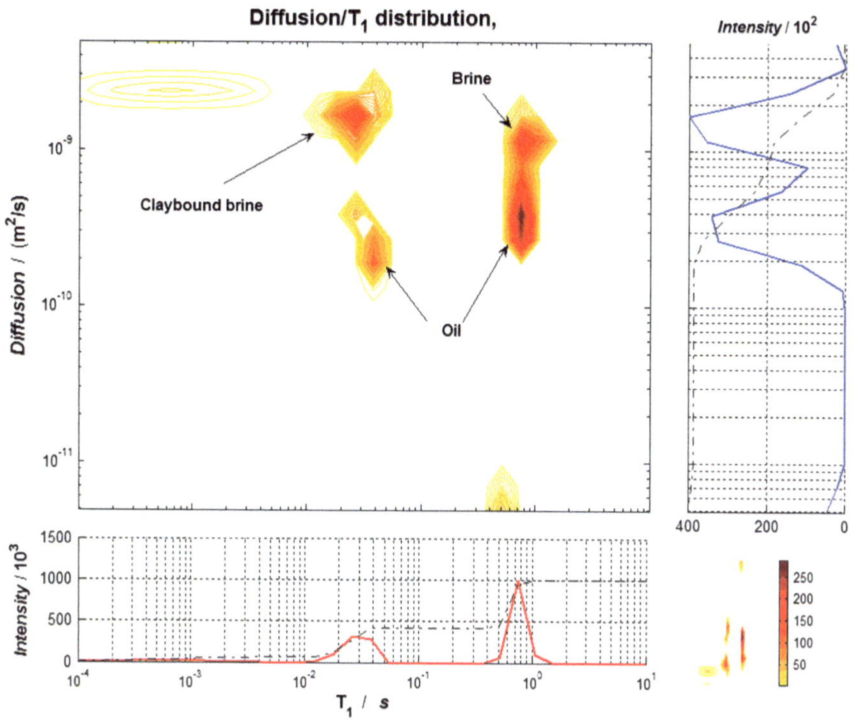

Fig. 8.6 The processed data from a Diffusion-T_1 correlated experiment acquired on a 1.5-inch rock core plug at native state

$$\gamma \int_0^{\tau} G_i^{clay}(t)\, z(t)\, dt \ll \gamma \int_0^{\tau} G_i^{grains}(t)\, z(t)\, dt \qquad (8.3)$$

From the consideration in the D-T_1/T_2 measurements, it is therefore very likely that the component with shortest T_2 and highest mobility arises from clay. Regarding the rest of the signals, there is one fraction at relatively short T_1 and T_2 (50–100 ms) and with an apparent diffusion coefficient of ~1×10^{-10} m^2/s. Due to T_1/T_2 ratio close to 1 and low mobility the most probable origin of this signal is from oil. The remaining signal from 100 ms and above divides into two fractions both in T_1/T_2 and mobility. It is therefore likely that the origin of this signal is from a combination of oil and brine. The origin of the different components is indicated in Fig. 8.6. By combining the diffusion and relaxation time correlated experiments it is possible to type the different fluids with a certain degree of accuracy.

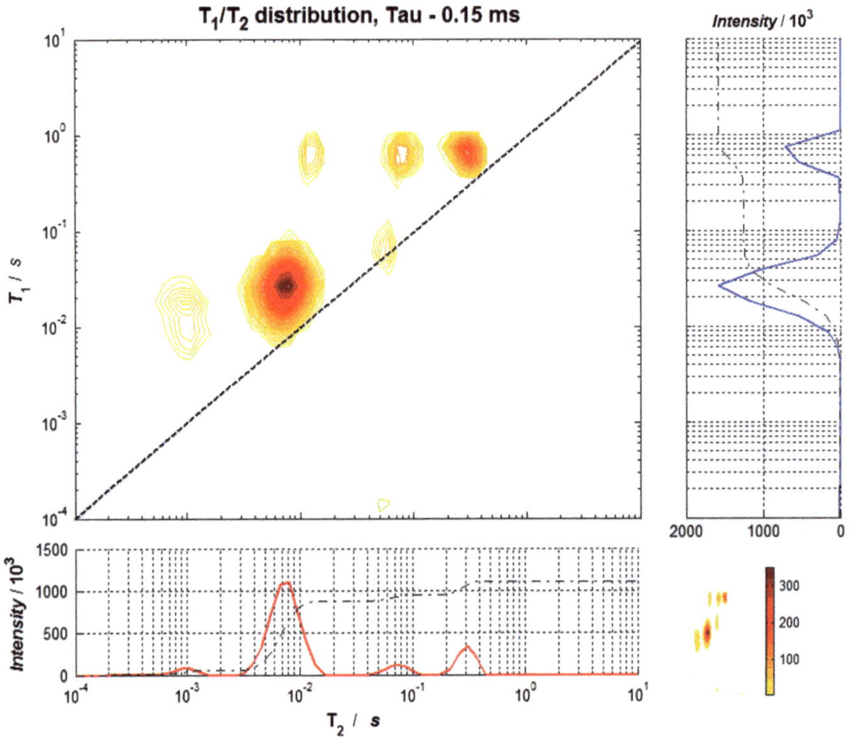

Fig. 8.7 The processed data from a T_1-T_2 correlated experiment acquired on a 1.5 inch rock core plug at native state. The *dotted line* indicates the diagonal where $T_1 = T_2$

8.2 Brine Saturated State

One of the most important states to measure during the SCAL program is the cleaned rock core plugs saturated with brine. When applying the PFG NMR technique, it is possible to extract information on the physical properties of the porous system. This information is not obtainable when the rock core plugs are in their native state. As will be shown in the following section it is then possible achieve a measure for the pore size distributions and the connectivity of pores. In addition, surface coatings will be probed.

8.2.1 Pore Size Distributions

Ever since Mitra et al. [12] introduced the short observation time expansion of the observed diffusion coefficient, resolving the surface to volume ratio (*S*/*V*) from the surface relaxivity ρ, efforts have been made to measure the true surface to volume ratio of core samples. However, a major obstacle has been the lack of hardware properly dedicated to this purpose, leading to loss of contribution from the smallest pores due to long values of the motional encoding/decoding intervals in the pulsed field gradient sequences. Also, it has been difficult to measure points at short enough observation times for the square root of time expansion to be valid [12, 13]. In order to reduce the observation/diffusion time a 12 MHz bench top rock core analyser was developed aiming at improving the characterisation of 1.5″ rock core plugs at laboratory conditions [10]. The unshielded gradient coils were mounted on the magnetic pole piece to prevent vibration coupling to the sample when pulsing the gradient coils. The maximum gradient strength is approximately 125 G/cm, and the 11 interval Pulsed Field Gradient Spin Echo sequence [14] and the 13 interval Pulsed Field Gradient Stimulated Echo sequence [15] were optimized with preparatory gradient pulse pairs such that eddy current dead times down to 400 μs were achieved without active shielding or pre-emphasis adjustment. Hence, reliable diffusion measurements on brine saturated plugs with motional encoding/decoding intervals as short as 1 ms and with a shortest observation time of approximately 1.5 ms, is achieved. With as short encoding/decoding intervals the CPMG sequence can be run with the same inter echo spacing (2τ) as used in the diffusion experiment. As a result, the T_2 distribution derived from a CPMG experiment can be directly converted to an absolute pore size distribution. By optimizing the experimental set up to generate a pore size distribution, pore sizes *(V/S)* down to just a few μm may be resolved. The use of the T_2 distribution derived from a multi exponential fit of the CPMG-echo attenuation is a significant approximation itself. However, when comparing the PFG NMR approach with conventional methods, such as mercury porosimetry techniques, it can be argued that despite its implicit approximations the method reveals new and vital information not extractable by other means.

In systems with significant contributions from T_2 components of 1 ms or less, the CPMG sequence with $\tau = 1$ ms does not represent the complete T_2 distribution of the system. To be able to relate the measured surface to volume ratio from the diffusion experiment to a pore size distribution using an ordinary CPMG experiment with an arbitrary τ value, an important assumption must be made: The surface relaxivityρ does not change significantly with pore size. With this approach, it is possible to resolve pore sizes down to less than 1 μm.

8.2.1.1 The Short Observation Time Expansion of the Diffusion Coefficient

As shown by Mitra et al. [12], there is a situation where the surface relaxation term is absent in the solution of the diffusion propagator, i.e. the short time expansion. By assuming piecewise smooth and flat surfaces and that only a small fraction of the diffusing particles are sensing the restricting geometries, the restricted diffusion coefficient D(t) can be written as shown in Chap. 2

$$\frac{D(t)}{D_0} \approx 1 - \frac{4}{9\sqrt{\pi}}\sqrt{D_0 t}\frac{S}{V} + \varphi(\rho, R, t) \tag{8.4}$$

$D(t)$ is the time dependent diffusion coefficient, D_0 is the unrestricted diffusion coefficient in the bulk fluid, and t is the observation time. The higher order terms in t, $\varphi(\rho, R, t)$, hold the deviation due to finite surface relaxivity and curvature (R) of the surfaces. At the shortest observation times, these terms may be neglected such that the deviation from bulk diffusion depends on the surface to volume ratio alone. In a porous system, a large span in pore sizes must be assumed, but (8.4) is expected to be valid also for a heterogeneous system. If ξ_i is the volume fraction of the pores with surface to volume ratio $(S/V)_i$, (8.4) can be expressed as

$$\sum_i \xi_i \frac{D_i}{D_0} \approx \sum_i \xi_i \left[1 - \frac{4}{9\sqrt{\pi}}\sqrt{D_0 t}\left(\frac{S}{V}\right)_i\right] = \left(1 - \frac{4}{9\sqrt{\pi}}\sqrt{D_0 t}\overline{\left(\frac{S}{V}\right)}\right) \quad (8.5)$$

Measurements of the early departure from bulk diffusion combined with a linear fit of the experimental data to the square root of time will thus result in a value for the average surface to volume ratio $\overline{(S/V)}$.

8.2.1.2 Transforming a T_2 Distribution to a Pore Size Distribution

Assuming that the brine is probing the pores within a porous rock as shown in Fig. 8.8, and neglecting bulk relaxation, there is a simple relation between T_2 values and the pore sizes [16]

$$T_2 \approx \frac{V}{S\rho} \quad (8.6)$$

Fig. 8.8 A two-dimensional slice of a rock core plug acquired with scanning electron microscopy (*SEM*). The *black* regions are to the voids while the *grey* regions correspond to the grains. Regions of very small grains (<1 μm) are identified as clay

This shows that the surface to volume ratio is coupled to the surface relaxivity ρ, and the assumption that (8.6) holds for any pore size is made. With ξ_i being the volume fraction of pores with surface to volume ratio $(S/V)_i$ and corresponding relaxation time T_{2i}, the average of the inverse of the transverse relaxation time can be written following Uh and Watson [17]

$$\sum_{i=1}^{n} \xi_i \frac{1}{T_{2i}} = \sum_{i=1}^{n} \xi_i \rho_i \left(\frac{S}{V}\right)_i \approx \rho \sum_{i=1}^{n} \xi_i \left(\frac{S}{V}\right)_i = \rho \overline{\left(\frac{S}{V}\right)} \tag{8.7}$$

Here the basic assumption is made that the surface relaxivity ρ is independent of pore size. The left hand side of (8.7) is the harmonic mean $\overline{(1/T_2)}$ of the T_2-distribution weighted by the fraction ξ_i of nuclei with relaxation time T_{2i} and n is the number of subdivisions of pore sizes. This average can be calculated from the T_2-distribution obtained in a CPMG measurement where the magnetization attenuation $M^{obs}(t)$ is converted to a T_2 distribution by in a standard way solving an inverse problem using e.g. an Inverse Laplace Transform (ILT) routine. Then the surface relaxivity ρ can be calculated from (8.7) if the average surface to volume ratio $\overline{(S/V)}$ is already found from the diffusion experiment as described in Chap. 1. Finally, the measured T-distribution can be transformed into an absolute pore size distribution (V/S) by means of the relationship inherent in (8.6).

One should note that in order to equate the surface to volume ratio from the diffusion experiment in (8.4) to the surface to volume ratio from the relaxation time experiment (8.6), the inter echo spacing of the two NMR experiments should be equal. The presence of position-dependent strong internal magnetic field gradients will reduce the number of molecules probing the surface. As seen from the NMR experiment there will not be a constant distribution of nuclear spins within the cavities. To make use of (8.5–8.7) the initial distribution of nuclear spins must therefore be equal, i.e. the inter echo spacing in the PFGSE and CPMG must be equal. Otherwise, the experiments will probe apparently different pore structures. However, the presence of internal magnetic field gradients will not have an influence on the leading order term in the short observation time expansion. To summarise, the procedure for deriving absolute distributions is as follows:

(1) The average surface to volume ratio $\overline{(S/V)}$ is found from fitting (8.5) to a set of diffusion measurements at short observation times.
(2) The average (S/V) can be correlated to the average ($1/T_2$) found from a CPMG experiment. From (8.7), (8.6) can be written as

$$\frac{\overline{1}}{T_2} \approx \rho \overline{\left(\frac{S}{V}\right)} \Rightarrow \rho = \overline{\left(\frac{1}{T_2}\right)} \cdot \overline{\left(\frac{S}{V}\right)}^{-1} \tag{8.8}$$

hence the surface relaxivity ρ which is assumed to be pore size independent is found.

(3) Under the assumption of pore size independency of the relaxivity the value of ρ can then be used in (8.6) and produce a linear relation between T_2 and the volume to surface ratio which is a measure of the pore size. By multiplying the T_2 distribution by the calculated surface relaxivity, the distribution is normalized to a pore size distribution in absolute length units.

In a situation where there is a significant fraction of components with T_2 of 1 ms or shorter, the CPMG with τ of 1 ms will underestimate the contribution from this fraction in an ordinary T_2 distribution. This is also the situation when the coupling to an internal magnetic field gradientG_i adds an extra term to equation as follows

$$\frac{1}{T_2} \approx \rho\frac{S}{V} + \frac{2}{3}\gamma^2\tau^2 DG_i^2 \tag{8.9}$$

The T_2 distribution will be weighted towards a relatively larger contribution from apparently smaller T_2's, i.e. changing the shape of the distribution. One method to include signal and information that is initially lost or weighted differently with a τ-value of 1 ms is to make the assumption that the total distribution integral with a τ-value 1 ms corresponds to the integral from right to left from a CPMG at $\tau = 0.1$ ms, starting at point j in the distribution (see Fig. 8.9). When the distribution is constructed from n points and the entire distribution integral for $\tau = 1$ ms equals the distribution integral at 0.1 ms when summated from $i = j$ to n, then one may replace the $\overline{(1/T_2)}$ value at $\tau = 1$ ms with the one at $\tau = 0.1$ ms as follows

$$\overline{\left(\frac{1}{T_2}\right)}_{\tau=1\text{ ms}} \approx \rho_{\tau=1\text{ ms}}\overline{\left(\frac{S}{V}\right)} \text{ and } \sum_{i=j}^{n}\xi_i\frac{1}{T_{2\,\tau=0.1\text{ ms}}^{i}} \approx \rho_{\tau=0.1\text{ ms}}\overline{\left(\frac{S}{V}\right)} \tag{8.10}$$

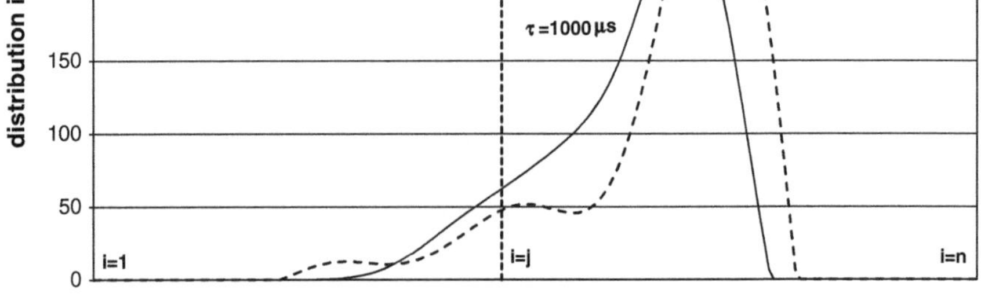

Fig. 8.9 T_2 distributions of a brine saturated rock core sample at two τ values

$$\Rightarrow \quad \rho_{\tau=0.1\ \mathrm{ms}} = \frac{\sum_{i=j}^{n} \xi_i \frac{1}{T_2^i}{}_{\tau=0.1\ \mathrm{ms}}}{\left(\frac{1}{T_2}\right)_{\tau=1\ \mathrm{ms}}} \rho_{\tau=1\ \mathrm{ms}} \tag{8.11}$$

Substituting this surface relaxivity for the T_2 distribution at τ value of 0.1 ms yields a pore size distribution that takes into account the component with T_2 values of ~1 ms or shorter.

This method for determining pore size distributions was applied on two brine saturated samples, an outcrop sandstone sample and a reservoir carbonate sample using the 12 MHz rock core analyser described earlier [10]. The operating temperature used was 35 °C and the bulk diffusion coefficient of the brine was $D_0 = 2.9 \times 10^{-9}$ m^2/s. Due to finite rise time on the gradient amplifier the shortest duration of the applied gradient pulse was set to $\delta = 0.6$ ms. The eddy current settling time was set to $\delta_1 + \delta_2 = 0.4$ ms yielding an inter echo spacing of $\tau = 1$ ms. G_i is the internal magnetic field gradient while g is the applied magnetic field gradient. For comparison, the pore size distribution was also determined by the means of mercury porosity measurements [18].

Two NMR experiments must be conducted to obtain a pore size distribution, the diffusion and the relaxation time experiment. The diffusion experiment was performed using the 11-interval bipolar PFGSE sequence [14] and the relaxation time experiment was conducted using the ordinary CPMG-sequence [5]. Figure 8.10 shows the results from the short observation time diffusion experiment on the two brine saturated samples where the observation time t is varied.

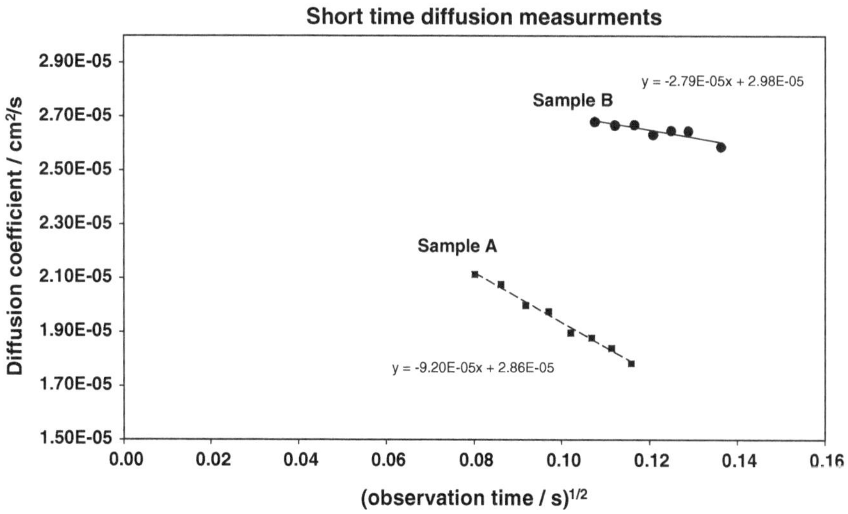

Fig. 8.10 Diffusion measurements of two brine saturated core samples at short observation times

The carbonate core is denoted sample A while the sandstone core is denoted sample B. According to (8.4) the measured diffusion coefficients show a linear attenuation when plotted against the square root of the observation time, and the slope of the attenuation yields the surface to volume ratio, (S/V). From the two attenuations, it can be observed that the deviation from bulk diffusion is higher for sample A at equal observation times. As the first order deviation term is dependent on the bulk diffusion and the surface to volume ratio only, it may be concluded that sample A on average contains smaller pores than sample B. Even though their pore sizes differ significantly, the extrapolation back to zero observation time yields the bulk diffusion coefficient of water $D_0 = (2.9 \pm 0.1) \times 10^{-9}$ m^2/s. A successful determination of the surface to volume ratio relies on the determination of the apparent diffusion coefficient at observation times where the deviation from bulk diffusion is small. Otherwise, second order correction terms to the short observation time expansion may interfere. A check for measured values for $D(t)$ being within the limits of the short time approximation as given in (8.4), is to extrapolate the diffusion coefficient $D(t)$ back to zero observation time. If the fitted diffusion coefficient $D(t)$ at zero observation time deviates significantly from the true bulk diffusion coefficient of the brine, the short time expansion is not valid.

In order to produce a pore size distribution similar to the one obtained by mercury porosity measurements, an ordinary T_2 relaxation time experiment followed by a one dimensional Inverse Laplace Transform (ILT) on the real time attenuation was performed. From the ILT data the average of $(1/T_2)$ is found and may then be used to calculate the surface relaxivity in accordance with (8.6). Substitution of the surface relaxivity then transforms the ILT into a pore size distribution shown in Fig. 8.11 for sample A, where V/S has been multiplied with

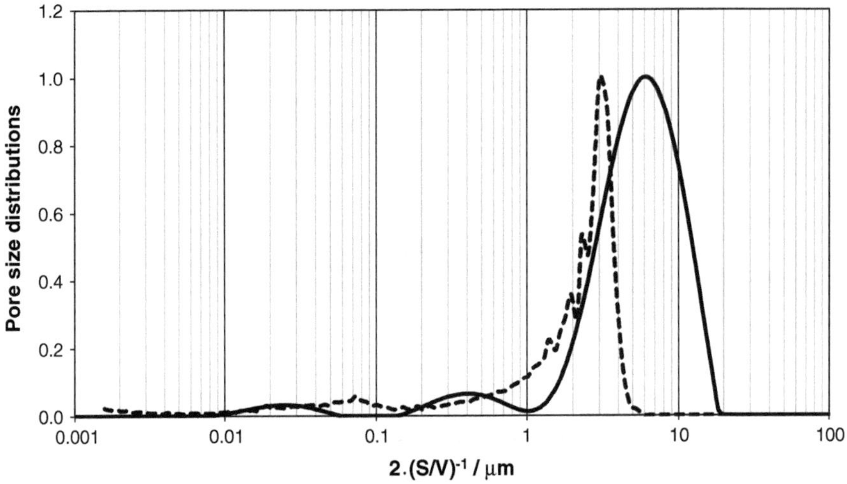

Fig. 8.11 Pore size distributions of sample A from the NMR method (*solid line*) and from mercury porosimetry (*dashed line*)

a geometry factor of 2 in order to correspond to cylindrical pores and thus be comparable to the result from mercury porosimetry. When comparing the pore size distribution derived from the two methods the determined mean pore size is found to be larger from the NMR measurements than from mercury porosimetry. It should be noted that the smoothed distribution achieved from PFG NMR data is due to the regularization parameter that is applied when performing the ILT-routine on the raw data.

While the distribution curves from mercury porosimetry are relatively uniform, i.e. a narrow peak with a left shoulder, the pore size distributions from the NMR measurements varies more in shape from sample to sample. The average pore radius is 6.0 μm from the NMR measurements and 2.0 μm from mercury porosimetry when assuming cylindrical pores. The somewhat higher average radius and broader distribution from NMR can be explained as a result of fundamental differences in the measuring techniques. Mercury porosimetry is based on injecting mercury into the sample at successively higher pressures while recording the volume of fluid injected into the sample at each step. Pore sizes are then calculated through the relation between capillary pressure and pore radius [18], and the volume of each pore size is calculated from the volumes injected. When the connectivity of the medium deviates from ideality, the technique tends to underestimate the larger cavities within the rock as the volumes of larger pores will not be probed at the correct corresponding pressure due to e.g. pore throat effects, and will be assigned to smaller pores than actually present. The result is often a narrow distribution of pore sizes shifted towards smaller pore sizes. The NMR experiments monitor molecules that are probing different cavities throughout the NMR measurement, resulting in a response depending on the size of the individual cavities. In theory, the NMR technique should then be more sensitive to a variation in pore size distributions within the rock core as compared to the mercury porosimetry technique.

Figures 8.8 and 8.12 show a rock core plug sample examined with mercury porosimetry (Fig. 8.12), scanning electron microscopy (SEM) (Fig. 8.8) and PFG NMR (Fig. 8.12). The distribution extracted from the PFG NMR experiment represents 2 times the $(S/V)^{-1}$ of the pores. Again, the number 2 is included as it then can be directly compared against mercury porosimetry, which assumes cylindrical pores. The highest peak from the PFG NMR data then appears at approximately 8 μm, but if spherical pores are assumed the peak would appear at a radius of approximately 12 μm and correspondingly at a diameter of approximately 24 μm. The cut off diameter, where the largest pore size is observed in the distribution and assuming a spherical diameter, is found to be at 90 μm. The picture in Fig. 8.8 clearly shows a large fraction of pores with diameter larger than this cut off value, average diameter in two dimensions for the largest pores seems to be approximately 200 μm. This difference can be explained by the fact that the pore space is not made of spherical cavities with smooth surfaces. The model of spherical pores with smooth surfaces yields the smallest sphere radius as seen by the NMR method. The rough surfaces

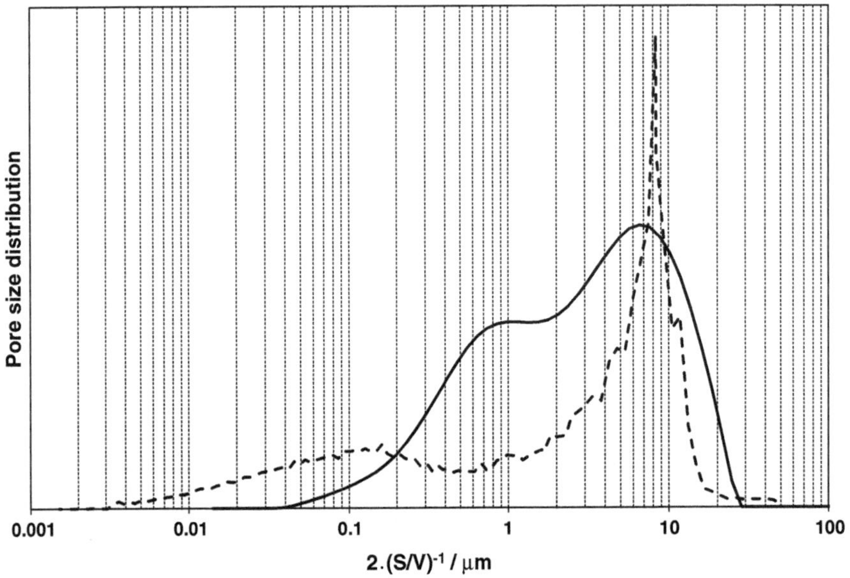

Fig. 8.12 Pore size distributions of a rock core plug from the PFG-NMR method (*solid line*) and from mercury porosimetry (*dashed line*)

sensed by the NMR measurements are much larger in area than the surface in the smooth surface model

$$\left(\frac{S}{V}\right)_{NMR} > \left(\frac{S}{V}\right)_{SPHERE-MODEL} = \frac{3}{r} \tag{8.12}$$

When imposing the model, the larger surface areas therefore yield smaller radii than the actual dimension of the pores in the real system. This could explain why the NMR cut off diameter with the spherical model is less than the diameter of the largest pores seen in the SEM image in Fig. 8.8.

Another feature is found in the T_1-(V/S) correlated distribution as shown in Fig. 8.13. This is the T_1-(V/S) correlated distribution extracted from the rock core plug also described in Figs. 8.8 and 8.12. The plot shows that there is additional information at smaller values of V/S. One fraction is found at smaller T_1, which is equivalent to T_1/T_2 ratio close to 1, while one fraction appears at larger T_1. This is equivalent to T_1/T_2 ratio larger than 1. In clay material small pores with a T_1/T_2 ratio close to 1 is expected [19]. In surface layers, channels and areas where internal gradients do not change polarity over the distance the molecules are diffusing during the inter echo spacing in the NMR experiment, a T_1/T_2 ratio is expected that is larger than 1. Consequently, the brine found at small V/S values can be identified to different locations within the porous medium.

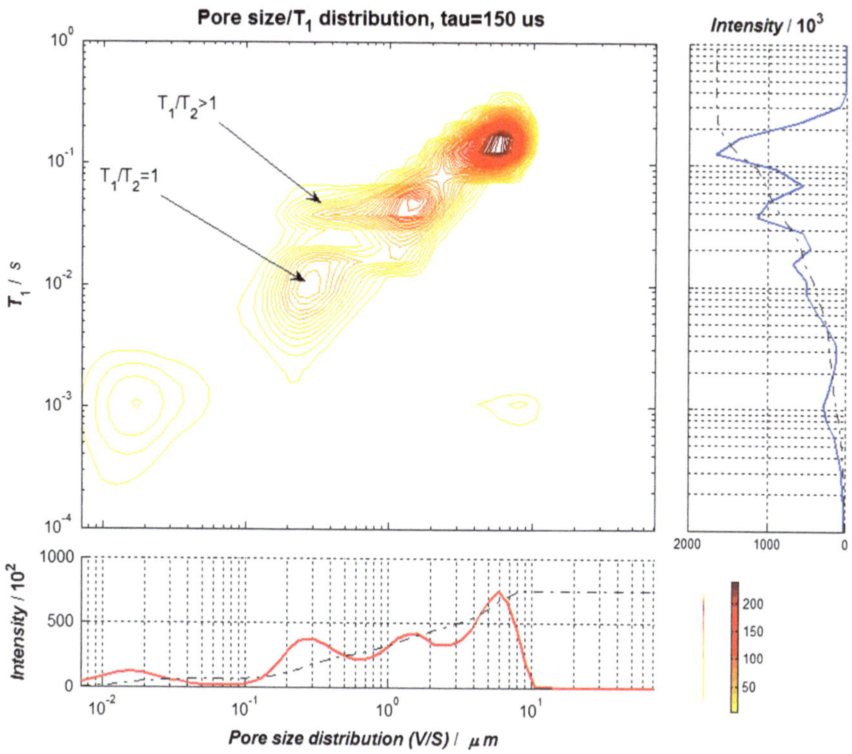

Fig. 8.13 T1-Pore size correlated distributions of a rock core plug extracted from the PFG-NMR method

8.2.1.3 On the Approximations in the Pore Size (*V*/*S*) Distribution Model

To be able to process the NMR measurements one must make some approximations prior to transforming the NMR relaxomtery measurements to a length scale. The literature makes use of three length scales, the diffusion length, the dephasing length and the pore length [7]. In the short time expansion developed by Mitra et al. [12] it was assumed piecewise smooth surfaces existed that only a small fraction of the probing molecules experienced, i.e. the diffusion length was much less than the pore length. Therefore, the square root of time attenuation of the diffusion coefficient could be applied to obtain the surface to volume ratio. However, in systems mainly consisting of small pores, as in clays or zeolites, this model fails. If it is apparent that the measured diffusion coefficient is attenuated significantly at the shortest observation time possible (around 1 ms), the *S*/*V* ratio cannot be found accurately. To measure the *S*/*V* for such a system by NMR another larger probing molecule with lower mobility as compared to the brine should be used.

To make use of the expression that relates the T_2 distribution to the surface relaxivity and the surface to volume ratio, it is assumed that the system is probed in the fast diffusion limit, i.e. has a diffusion length larger than the pore length. In the early part of the CPMG-attenuation, the diffusion length is clearly not larger than the pore length. This may lead to an overestimated $(S/V)^{-1}$, as a significant portion of the probing molecules have not sensed the restricting geometries yet and therefore contribute as a bulk component while at larger echo numbers, the assumption is fulfilled. If the life time of the magnetisation is long there will be an exchange between different pore classes at higher echo numbers, as the probing molecule diffuses from pore to pore without relaxation taking place. Under this effect, the pore size distribution will be smeared out and leads to broader peaks in the pore size distribution. Latour et al. [20] performed 1D-ILT on CPMG attenuations of water-saturated sandstones at different temperatures, and found little effect on the distribution due to more exchange between regions because of higher mobility at higher temperature, and consequently they were in the fast diffusion limit. In general, (8.6) should be applied with care as it is indeed not fulfilled for a CPMG attenuation initially. Still, this equation is used extensively in studies because it does probe the heterogeneity of the porous samples. When comparing the NMR results with conventional techniques, the NMR approach probes the whole pore space while methods such as mercury porosimetry probe the pore throats more or less independently of the pore cavities. Thus, it will not be as sensitive to bi- or tri-modal structures as the NMR approach, as shown in Fig. 8.13.

From an experimental point of view, it is also important that the diffusion coefficient extracted from NMR pulse sequence truly reflects the mean squared displacement. As the logarithm of attenuation from a pulsed field gradient experiment within a heterogeneous system is not necessarily linear with respect to the square of the gradient strength, the diffusion coefficient must be fitted within the regime where the second cumulant approximation is valid [21]. If the gradient pulses are long, one should also consider taking into account restricted diffusion during the duration of the gradient pulses [22, 23]. The results presented here have not taken into account restricted diffusion during the gradient pulses. This is evidently not correct for the clay bound brine, but the contribution from this fraction when determining the average S/V is small due to it short relaxation times [24]. It is when the CPMG attenuation with $\tau = 100\ \mu s$ is applied the clay bound brine is included in the pore size distribution. Inclusion of restricted diffusion during the gradient pulses would lead to a smaller and more correct S/V ratio, i.e. a small shift of the $(S/V)^{-1}$ distribution towards higher values.

8.2.2 Micro-connectivity

For the characterisation of brine saturated rock core plugs it is shown that PFG NMR may measure the surface to volume ratio and pore size distribution. The surrounding pore walls restrict the diffusional displacement of brine. When there is a degree of connectivity between the cavities in the porous system, it is possible for the molecules to diffuse between pores. As the observation time is increased, the molecules may probe larger distances and experience the restrictions of the porous medium. Having found the surface to volume ratio (S/V), an observation time may be chosen for the measurement of the diffusion coefficient such that the mean square displacement at bulk diffusion is, for example, $3 \times (S/V)^{-1}$. With a connected pore space one would then expect to measure a diffusion coefficient placed between the two extreme cases; bulk diffusion and diffusion within closed spheres of radius (3 times $(S/V)^{-1}$), in the case of spherical pores.

Previous attempts to measure connectivity, which in turn may be related to permeability using NMR, have failed because the life time of the NMR signal is not long enough for the diffusing molecules to probe the heterogeneity of the porous rock [25]. Unfortunately the NMR signal is more likely to having decayed substantially before the tortuosity limit is reached, i.e. *D*(observation time)/*D*(bulk) = constant. The approach presented here is different as the aim is to measure the diffusion coefficient at an observation time that is predefined or scaled against the measured $(S/V)^{-1}$ for the specific rock core sample. Thus, a diffusion coefficient is measured that can be compared against the extreme cases, zero connectivity and no restrictions. The apparent diffusivity, or time-dependent diffusion coefficient, in terms of the mean-square displacement of the molecules during a time t is

$$D(t) = \frac{\left\langle [r(t) - r(0)]^2 \right\rangle}{6t} \tag{8.13}$$

$\mathbf{r}(t)$ is the position of the spin polarized molecule at time t and the average is over the entire ensemble of such molecules in the pore. As shown in Chap. 2 this can be approximated as

$$\frac{D(t)}{D_0} \approx 1 - \frac{4}{9\sqrt{\pi}} \sqrt{D_0 t} \frac{S}{V} \tag{8.14}$$

D_0 is the unrestricted diffusion coefficient, in bulk fluid, and t is the observation time. Equation (8.14) is the first term of a perturbation expansion of the time-dependent diffusion coefficient. The implied approximation is valid at short observation times, i.e. when only a small amount of the fluid is probing the restrictions (pore walls). Then the deviation from bulk diffusion is dependent on the surface to volume ratio alone and not on the surface relaxivity. Thus by conducting diffusion experiments at short observation times, a linear fit to the attenuation will provide the surface-to-volume ratio *S*/*V*. This can be combined with relaxation time measurements to produce a pore size distribution of the rock core sample as shown in the previous section. However, knowing the surface to volume ratio, the time dependency in (8.14) may be rescaled by using the root of the mean squared displacement

$$\sqrt{D_0 t} = n \left(\frac{S}{V}\right)^{-1} \tag{8.15}$$

where n is an arbitrary number. Substitution of (8.15) into (8.14) yields

$$\frac{D(t)}{D_0} = 1 - \frac{4}{9\sqrt{\pi}} n \tag{8.16}$$

By plotting the observation time dependent diffusion coefficient as a function of n, i.e. in units of $(S/V)^{-1}$, a slope of the initial attenuation is achieved that is independent of the sample being investigated, i.e. a universal slope equal to $-4/(9\sqrt{\pi})$. However, using the same scaling at longer observation times, the attenuations may deviate from the universal slope to a degree depending on the porous structure of the sample. Assuming that the observation time dependent diffusion coefficient is determined by the pore space geometry only, the given scaling is expected to provide overlapping curves only for identical structures. With this scaling, the molecules in two identical structures of different dimensions will probe identical pore spaces at the same value of n. Thus, their curves should overlap at all numbers of n. This situation is equivalent to measuring diffusion in the same system but at different temperatures [13, 26]. Then a plot of $D(t)/D_0$ as a function of the mean squared displacement is independent of temperature. In Fig. 8.14 the results on compact polystyrene spheres of diameter 45 and 98 μm are plotted as a function of the root of the mean squared displacement and normalized against $(S/V)^{-1}$. The initial decay coincides with the universal slope $-4/(9\sqrt{\pi})$, while at higher values it approaches a tortuosity limit of approximately 0.7. When scaling the x-axis with the inverse of the surface to volume ratio, the two curves tend to overlap. This is to be expected when the packing of the spheres is similar. Then the pore space should have the same characteristic features, such as connectivity between pores, and the two identical systems of different scale overlap when the scale is normalized out

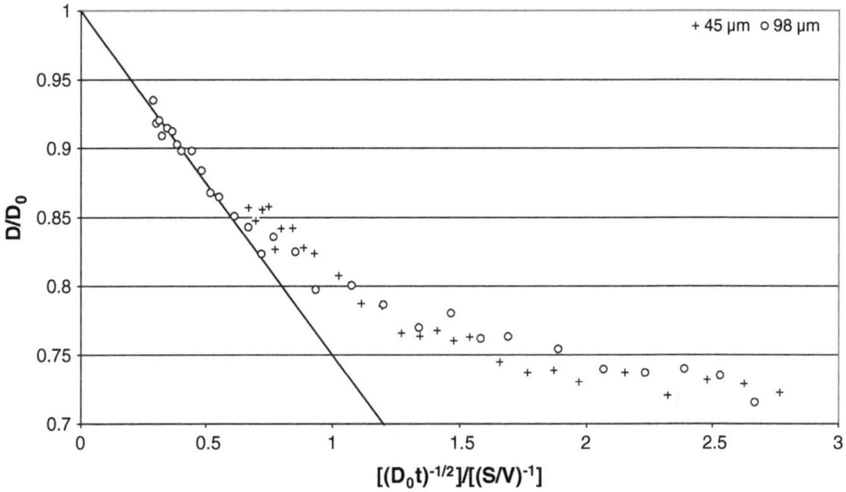

Fig. 8.14 Observation time dependent diffusion coefficients of compact mono-sized spheres of diameter 48 and 98 μm, plotted in units of the inverse of the surface-to-volume ratio. The *solid line* indicates the universal slope given by (8.16)

Table 8.1 Data summary from on a set of rock core samples

Sample ID	Porosity (%)	Permeability (mD)	V/S (μm)	D/D_0 (n = 10)
1	4.1	0.03	0.99	0.23
2	5.7	0.04	1.02	0.17
3	12.9	5.60	1.06	0.30
4	10.6	0.91	1.16	0.32
5	10.3	0.47	1.37	0.33
6	14.3	1.85	1.19	0.30
7	6.9	0.06	0.95	0.20
8	15.9	9.70	1.68	0.30
9	13.5	15.53	1.54	0.39
10	17.3	15.18	1.83	0.44
11	23.7	50.44	1.57	0.40
12	17.1	81.55	2.00	0.53
13	24.3	17.20	2.57	0.53
14	19.8	1713.92	3.01	0.59
15	31.1	3010.47	2.80	0.60
16	18.9	61.17	1.60	0.42

along the abscissa. A deviation from the universal slope will be a measure of the so-called micro-connectivity between the pores. The larger the deviation, the better the pores are connected. The upper limit is unrestricted or bulk diffusion, while the lower limit could be diffusion within closed cavities.

The method for determining pore size distributions and plotting the time dependent diffusion coefficient in units of the inverse of the surface-to-volume ratio were applied to 16 brine saturated carbonate core samples using the aforementioned 12 MHz rock analyser [10]. The permeability of the 16 rock core samples were significantly different, as seen in Table 8.1, and span 5 orders of magnitude in mD (millidarcy). The operating temperature was 35 °C, and the bulk diffusion coefficient of the brine was $D_0 = 2.9 \times 10^{-9}$ m^2/s. Due to finite rise time on the gradient amplifier, the shortest duration of the applied gradient pulse was set to $\delta = 0.5$ ms. The eddy current settling time was set to $\delta_1 + \delta_2 = 0.4$ ms, thus yielding an inter echo spacing of $\tau = 0.9$ ms. The pulse sequences applied are the CPMG [5], the 11 interval PFGSE [27], and the 13 interval PFGSTE [28].

Figure 8.15 shows the pore size distributions for two of the rock core plugs, #5 and #7. The pore size distribution in #5 is significantly larger, but also contains a left shoulder, which indicates a large span in pore sizes. For #7 there is one distinct peak and some minor signal at smaller pore sizes. To obtain a clearer picture of the pore size distribution, the T_1-pore size correlated distributions are visualized in Figs. 8.16 and 8.17. In the T_1-pore size correlated distributions the pore size distribution for plug 5 and 7 are more similar, with 3 (#5) and 2 (#7) main peaks. The two largest peaks in #5 are of similar volume while a smaller peak appears at shorter pore size. For #7 there is one major peak and a small one at smaller pore

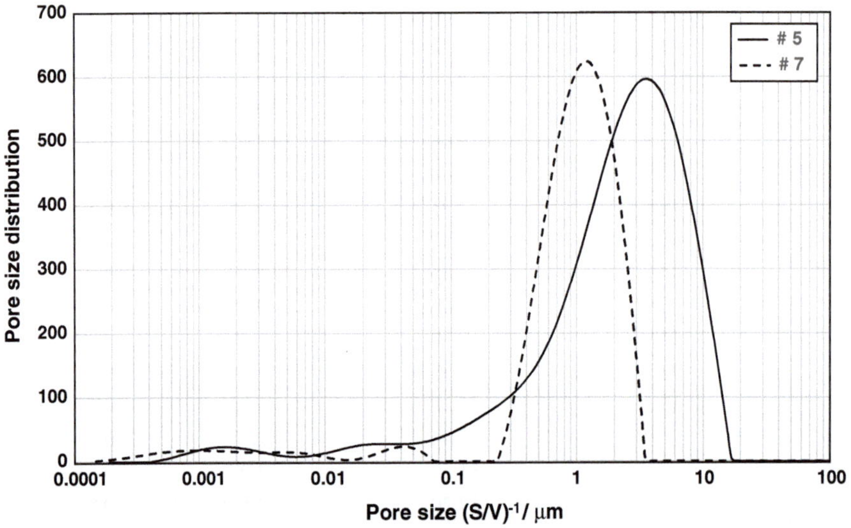

Fig. 8.15 Pore size distributions of two of the rock core plugs using the PFG NMR method

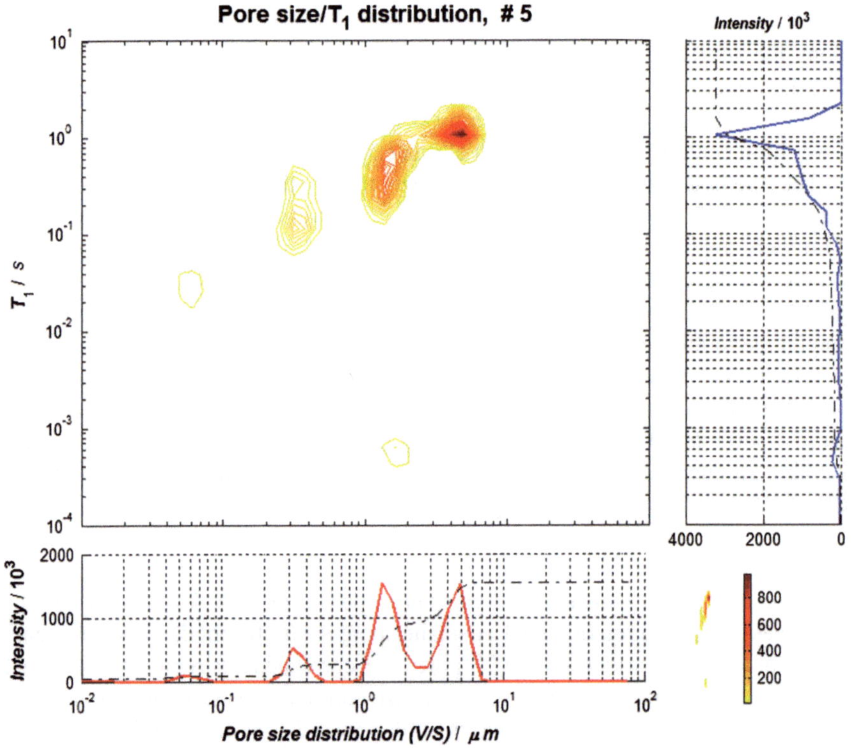

Fig. 8.16 T_1-Pore size correlated distributions of a rock core plug #5

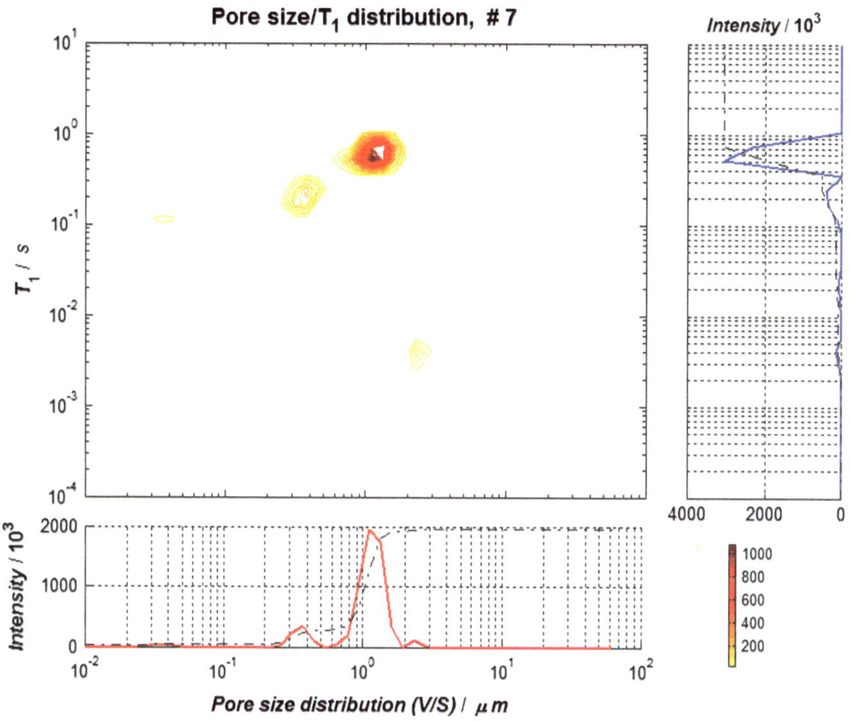

Fig. 8.17 T_1-Pore size correlated distributions of rock core plug #7

sizes. Thus, #7 seems to be more well defined than #5, but there is no information on the connectivity between the cavities. To probe the connectivity diffusion measurements were conducted covering observation times from the shortest possible (approximately 1.5 ms) to the order of 1 s. The surface to volume ratio *S/V* was fitted for each plug, and the measured apparent diffusion coefficients were plotted against the observation time normalized against $(S/V)^{-1}$ [as shown in (8.15)–(8.16)].

Figure 8.18 shows the time dependent diffusion coefficients for 7 of the brine saturated rock core plugs, where the apparent diffusion coefficients are plotted as a function of the root of the mean squared displacement and divided by the inverse of the surface to volume ratio, $(S/V)^{-1}$. The solid line indicates the universal slope $-4/(9\sqrt{\pi})$, which all brine saturated rock core samples should follow at the shortest observation times. All plugs seem to depart from the universal slope after the first measurement point, but with very different slopes. The question is whether the different slopes of the plugs can provide information on the connectivity or even better the permeability of the porous system. For all the 16 plugs the apparent diffusion coefficients at 10 times $(S/V)^{-1}$ were evaluated, and are listed in Table 8.1 together with corresponding porosity and permeability data. If there was no measuring point exactly at $n = 10$, the nearby measuring points were used to achieve an approximate value. At this individual observation time for each plug, it is likely that the diffusing

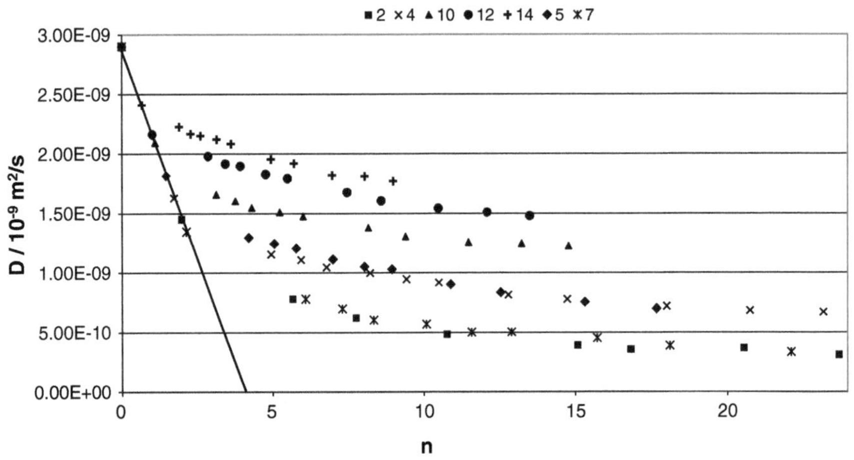

Fig. 8.18 Observation time dependent diffusion coefficients of 7 of the rock core plugs, plotted in units of n

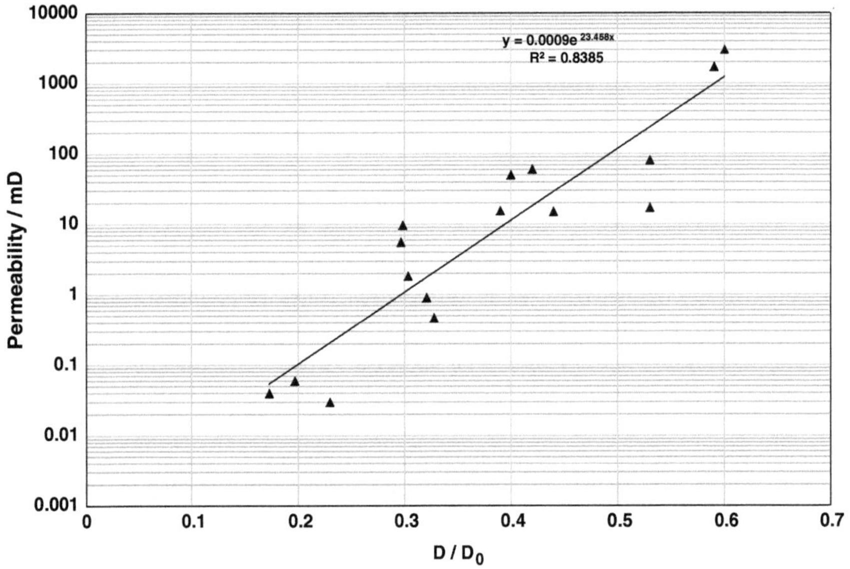

Fig. 8.19 The apparent diffusion coefficient at 5 times $(S/V)^{-1}$, normalized against the bulk diffusivity and plotted against the permeability of the corresponding plugs

molecules to a large degree have probed the heterogeneity of the porous plug, as the root of the mean squared displacement then exceeds the typical $(S/V)^{-1}$ within the plug. In Fig. 8.19 the apparent diffusion coefficients normalized to the bulk diffusivity at n = 10 are plotted against the permeability data found in Table 8.1.

The plot in Fig. 8.19 shows that it is possible, for this type of rock, a carbonate, to achieve a correlation between the micro-connectivity as measured by PFG NMR and the permeability found by other means [29]. Even though the 16 plugs differ significantly both in average pore sizes and shape of the pore size distributions, from mono- to bi- or even trimodal distributions, there is a reasonably good correlation between the normalized apparent diffusion coefficient (at $n = 10$) and the permeability. In the presence of stronger and fluctuating internal magnetic field gradients, the measured apparent diffusion coefficient may be influenced by factors not directly related to connectivity or permeability. Then the method for correcting for the impact of fluctuating internal magnetic field gradients should be applied [26]. Other models could probably improve the correlation between permeability and PFG NMR measurements, for example plotting of the apparent diffusion coefficient against the $(S/V)^{-1}$ from the most significant peak in the pore size distribution instead of the average value of $(S/V)^{-1}$.

8.2.3 Coupling to the Internal Magnetic Field Gradients

One of the major challenges in applying NMR measuring techniques is to choose the right applications for a given task. The result from a NMR experiment is particularly sensitive to the viscosity of the components confined in the cavities of the porous rock, the electromagnetic properties of the rock and the dimension and type of cavities within the porous rock. NMR diffusion experiments can be used to probe the viscosity of the fluid components, NMR relaxometry experiments (CPMG) with a variable inter echo spacing can be used to probe the electromagnetic properties, and in the combination of these experiments pore size distribution can be produced. In the following section focus will be on an application that is optimised for indicating the nature of the surface coating within a rock core plug, i.e. develop a surface coating indicator that is sensitive to the mineral composition on the surfaces of the porous rock.

The typical T_2 measurement of a water saturated core plug results in a distribution containing a main peak with a significant left shoulder. Occasionally this shoulder may appear as a peak, which may indicate a separate region at shorter T_2 values. The ordinary T_2 experiment usually correlates to the product of surface relaxivity and the pore size distribution [30]

$$\frac{1}{T_2} = \frac{1}{T_2^{bulk}} + \rho\frac{S}{V} \tag{8.17}$$

ρ is the surface relaxation strength and S/V is the surface to volume ratio (which correlates to the pore sizes). An internal magnetic field gradient$\mathbf{G_i}$ is generated due to the magnetic susceptibility differences throughout the sample when it is placed in an external homogeneous magnetic field [31]

$$G_i \approx \frac{B_0 \Delta\chi}{R} \tag{8.18}$$

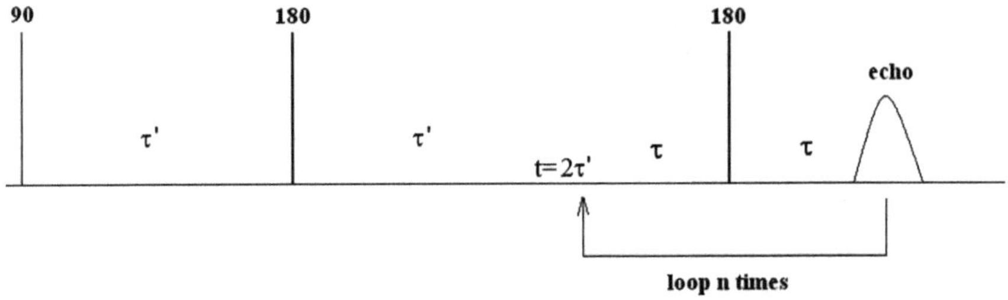

Fig. 8.20 The Single Spin Echo CPMG (SE-CPMG) experiment

R is the distance over which the magnetic field varies, B_0 is the applied external magnetic induction, and $\Delta\chi$ is the difference in magnetic susceptibility between the surface of the solid matrix and the fluid. Thus the presence of internal magnetic field gradients requires another relaxation term to be added to (8.17); dephasing of the nuclear magnetisation due to diffusion in the presence of an internal magnetic field gradient $\mathbf{G_i}$:

$$\frac{1}{T_2} = \frac{1}{T_2^{bulk}} + \rho\frac{S}{V} + \frac{\tau^2(\gamma G_i)^2 D}{3} \tag{8.19}$$

γ is the gyromagnetic ratio, D is the diffusion coefficient, and τ is the inter echo spacing illustrated in Fig. 8.20. In principle the internal magnetic field gradient is measured by varying the inter echo spacing for the first spin echo in the CPMG echo train, while keeping the other inter echo spacing at a constant value. Applying a 2 Dimensional Inverse Laplace Transform (2D-ILT) on the data sets would then result in the following values for the T_2 and ψ dimensions (assuming $T_2^{bulk} \gg T_2$)

$$T_2 = \frac{r_{pore}}{\rho} \text{ and } \Psi = \frac{3}{(\gamma G_i)^2 D} \tag{8.20}$$

By assuming that the diffusion coefficient is close to the bulk value throughout the sample and inter echo spacing used, a value for $\mathbf{G_i}$ can be found from ψ in (8.20). This will correlate to the inverse of the typical distance defined in (8.18). As the pore radii characterise the typical distance over which the magnetic field varies ($\mathbf{R} \approx \mathbf{r_{pore}}$), a distribution in $\mathbf{G_i}$-values yields direct information on the porous structure. The product $\mathbf{G_i} \times \mathbf{r_{pore}}$ is thus mainly dependent on the magnetic susceptibility differences between the solid matrix and the brine (see 8.18)

$$\begin{aligned} &G_i \cdot R \approx G_i \cdot r_{pore} \approx B_0 \cdot \Delta\chi \\ &\Downarrow \\ &G_i \cdot \frac{r_{pore}}{\rho} \approx B_0 \cdot \left(\frac{\Delta\chi}{\rho}\right) \\ &\Downarrow \\ &G_i \cdot T_2 \approx B_0 \cdot \left(\frac{\Delta\chi}{\rho}\right) = K_{Gi} \end{aligned} \tag{8.21}$$

Equation (8.21) defines the surface coating indicator K_{Gi}. It should be noted that K_{Gi} is dependent only on the interface between the solid matrix and the liquid medium, $\Delta\chi$. Consequently for a given pore structure, a variation in the product between $\mathbf{G_i}$ and T_2 must be caused by a difference in properties at the surface of the porous rock and/or the interface between the solid and the liquid in the pores within the solid matrix. In the following $\mathbf{G_i}$ is assumed to be given by (8.18). This is truly an approximation as there evidently will be a variation of $\mathbf{G_i}$ dependent on the complexity of the porous structure and on the distribution of solid compounds (as for example chlorite) within the porous rock [32, 33]. In addition, there will be an averaging effect on $\mathbf{G_i}$ due to diffusion during the inter echo spacing in the CPMG sequence. However, as will be shown in the following, new and important information is extractable from NMR data acquired on the rock core plugs using the approximations leading to (8.21).

Figure 8.20 displays a NMR pulse sequence that will extract information regarding the porous structure and paramagnetic impurities on the surface of the solid matrix. The optimisation of this experiment is based on varying the first inter echo spacing in the sequence (τ'), and to fit all experiments as a function of observation time to produce the initial NMR signal undistorted by relaxation at the shortest spacing (τ). All information obtainable is thus preserved, which is especially important in systems exhibiting T_2 distribution shifted towards short T_2 values. For a single component system the attenuation is written

$$I = I_0 \, \mathrm{e}^{-(2\tau')\left(\frac{1}{T_2^{bulk}} + \rho\frac{S}{V} + \frac{\tau'^2(\gamma G_i)^2 D}{3}\right)} \mathrm{e}^{-(2n\tau)\left(\frac{1}{T_2^{bulk}} + \rho\frac{S}{V} + \frac{\tau^2(\gamma G_i)^2 D}{3}\right)} \tag{8.22}$$

where the first exponential represents the attenuation due to the first inter echo spacing while the second exponential arises from the ordinary CPMG echo train of the sequence. The experiment will return a data matrix where the attenuation is due to T_2 along the first dimension (as a function of observation time ($2n\tau$)), while the attenuation is due T_2 and internal magnetic field gradients along the second dimension. When assuming that the T_2 is not significantly dependent on the inter echo spacing, the attenuation along the second dimension will be a function of τ'^3.

Figures 8.21 and 8.22 show the processed T_2-distributions acquired from the pulse sequence shown in Fig. 8.20 and by varying τ' from 0.1 up to 8 ms. The distributions for plug #1 in Fig. 8.21 are most affected at the shortest T_2-values, but this loss of signal is also partly due to the loss of measuring points at the shortest observation times. As τ' is increased the first measuring point in the CPMG train is shifted to larger times. Interpreting T_2-peaks below 1 ms as experiencing strong internal gradients only may therefore be wrong. In Fig. 8.22, the attenuation of the T_2-distributions for plug #4 is significantly different. Here the distributions loose large signal fractions at all T_2 values. The apparent drift in the T_2 peak at the largest T_2 values is a consequence of coupling to the internal gradient in the $2\tau'$-interval, as the CPMG-trains are identical at all acquisitions, except for a change in τ'. The conclusion is that the CPMG echo attenuations from plug #4 is more affected by the presence of internal gradients, than the CPMG echo attenuations from plug #1. The loss of the component at the shortest T_2-values is partly a consequence of loss of signal information at the shortest observation times.

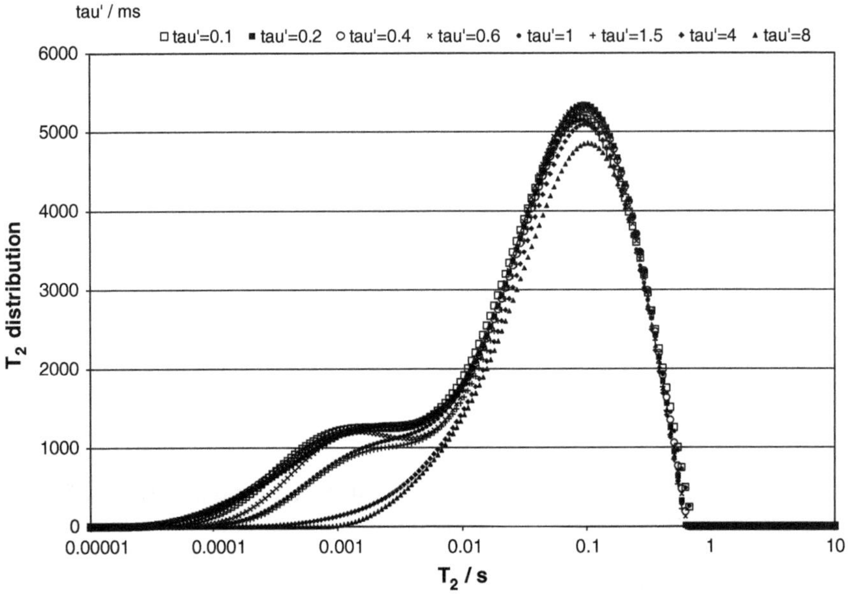

Fig. 8.21 The processed T_2 distribution of the acquired CPMG-data at various values of τ' from a brine saturated rock core plug #1

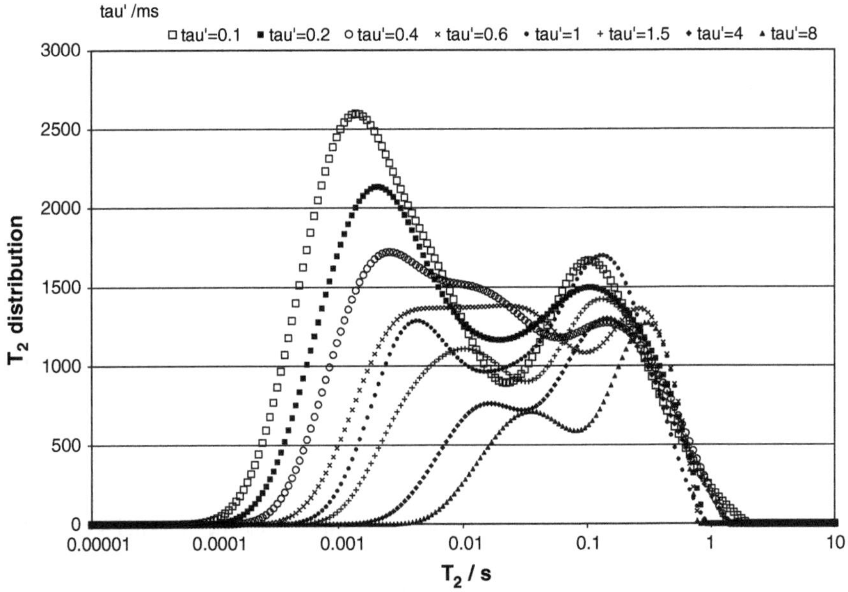

Fig. 8.22 The processed T_2 distribution of the acquired CPMG-data at various values of τ' from a brine saturated rock core plug #4

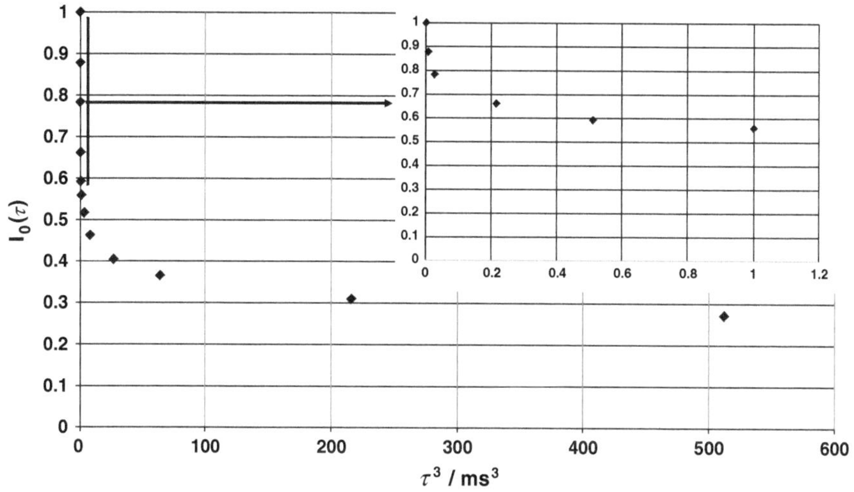

Fig. 8.23 The attenuation of the fitted initial signal at t = 0 from the acquired CPMG-train as a function of τ'^3

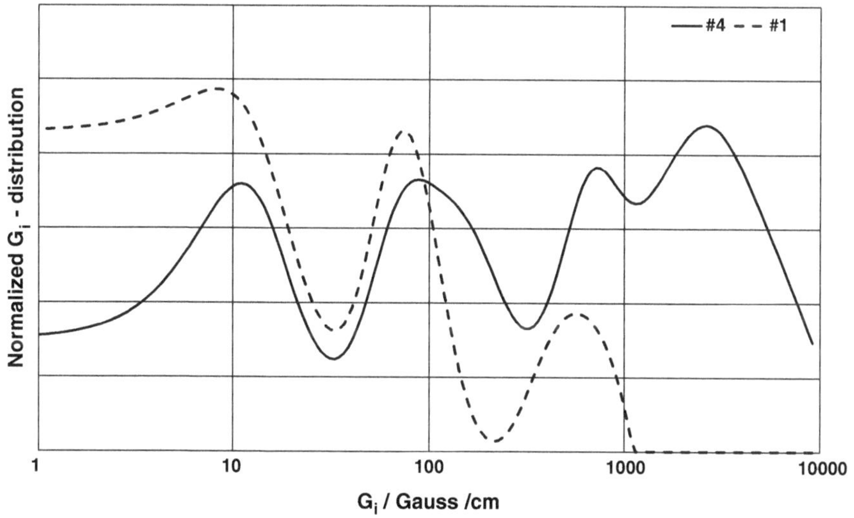

Fig. 8.24 The distribution of Gi-values for two rock core plugs based on the attenuation initial signal at t = 0 from the acquired CPMG-train as a function of τ'^3

In Fig. 8.23, the total integral of the T_2-distributions from plug #4 is plotted as a function of τ'^3, and it indicates that the signal attenuates with time constants that span a range of G_i-values that cover several orders of magnitude. Performing an inverse Laplace transform on this attenuation leads to a distributions of G_i-values, which is shown in Fig. 8.24 for plugs #1 and #4.

The Gi-distributions for plugs #1 and #4 in Fig. 8.24 shows that plug #4 span internal gradient values from practically 0 to several 1,000 G/cm, while for plug #1 most of the G_i-values are found to be less than 100 G/cm. The small component between 200 and 1,000 G/cm for plug #1 may just as well originate from the before mentioned loss of signal information as τ' is increased. Due to the finite time of 0.2 ms before the first measuring point, there is a limit to how strong G_i-values one may measure using this technique. If the real G_i value is more than 10,000 G/cm, the apparent T_2^* will be less than 50 μs. Then the signal will be almost completely attenuated at the first observation time (i.e. 200 μs when $\tau = 100$ μs), and information on the strongest Gi-values are lost in the CPMG-experiment. In such cases, the shape of the free induction decay (FID) may be used for gathering information on G_i-values not detectable from the CPMG-experiment.

In Table 8.2, the major results from a set of brine saturated plugs are displayed. Due to different time constants in the attenuation shown in Fig. 8.23, 4 subclasses have been defined depending on the value of the fitted internal magnetic field gradient G_i. This can be seen for plug #4 in Fig. 8.24, where four regions within the G_i-distribution are partly resolved. For plug #1 there are only 3 regions present, and thus the region with the strongest internal gradient is not present in Table 8.2. The region of T_2 values have been chosen in accordance with the fraction of signal from the different regions of G_i. It is then assumed that the values with shorter T_2 correlates to the region of stronger G_i while the longest T_2 fraction corresponds to the region of the weakest G_i. Consequently, when for example 30 % of the G_i-distribution is classified in region A, 30 % of the T_2-distribution, starting with the shortest component and upwards, is defined to be in region A. The average T_2 is then calculated from this region. From the G_i and T_2 data, the average value for K_{Gi} is calculated in accordance with (8.21).

A striking effect from the definition of K_{Gi} is that the plug where the strongest internal magnetic field gradient is measured does not have a high value of K_{gi}. Thus, it is not expected to have the largest difference in magnetic susceptibility between the surface of the solid matrix and the brine, indicating that a measure of

Table 8.2 Mean values of internal magnetic field gradients (Gauss/cm), T_2 (ms) and the K_{Gi} averaged over all regions. (No sig. => No detectable signal)

Sample ID	G_i Region A	G_i Region B	G_i Region C	G_i Region D	T_2 Region A	T_2 Region B	T_2 Region C	T_2 Region D	K_{Gi} Average
1	No sig.	473.4	63.5	0.90	No sig.	1.5	11.1	233.5	0.33
2	1913	690	101.7	1.10	1.1	2.8	29.2	223.0	1.31
3	1033	1033	155.6	1.12	2.1	2.1	22.2	267.9	1.10
4	2310	600	77.8	1.03	0.4	1.2	5.3	230.3	0.53
5	1934	153.4	153.4	1.08	0.7	6.2	6.2	245.0	0.48
6	547.4	547.4	75.6	1.43	4.3	4.3	52.2	199.2	1.96
7	2,160	381	50.7	1.18	2.5	12.2	63.9	283.7	3.07

G_i is given in Gauss/cm while T_2 is given in ms

G_i only is not a good surface coating indicator. When determining K_{Gi} the measured internal magnetic field gradient is scaled against the expected pore size, but K_{Gi} is made independent on pore size (see 8.21). For this reason, K_{Gi} is mainly sensitive to a varying magnetic susceptibility at the interface between the pore cavity and the solid matrix of the porous rock. When combining a single spin echo having variable inter-echo spacing with a train of spin echoes with fixed inter-echo spacing, it is then possible to distinguish between brine saturated rock core plugs having different mineral surface coating. The reason why the experiment displayed in Fig. 8.20 returns more information than an ordinary CPMG experiment is found in the attenuation of the spin echo NMR signal as a function of the square of the internal magnetic field gradient and the τ'^3 dependency. As there is a correlation between the internal magnetic field gradient and the pore structure, the square of this dependency will resolve different porous structures better than the CPMG experiment, where there is a linear dependency between T_2 and the porous structure. Due to a τ'^3 dependency on the time parameter in the spin echo attenuation, it is possible to span most of the distribution in the internal gradients, even though the time domain in the CPMG varies from 0.2 ms up to seconds. When τ' varies from 0.2 to 10 ms, it produces a variation in τ'^3 from 0.008 to 1,000, which is two order of magnitude broader than the span from an ordinary CPMG experiment.

It has been shown previously by Rueslåtten et al. [34] that the average T_1/T_2 correlates to the chlorite content at the surface of the grains. To verify that the K_{Gi} actually does work as a surface coating indicator, the T_1/T_2 ratio was measured using the T_1/T_2—correlated experiment provided in Chap. 1. In Fig. 8.25 K_{Gi} is plotted against T_1/T_2, and there is evidently a correlation between T_1/T_2 and K_{Gi}. As the combined spin echo-CPMG can be made significantly less time consuming as compared to a T_1/T_2—correlated experiment, the determination of K_{Gi} is

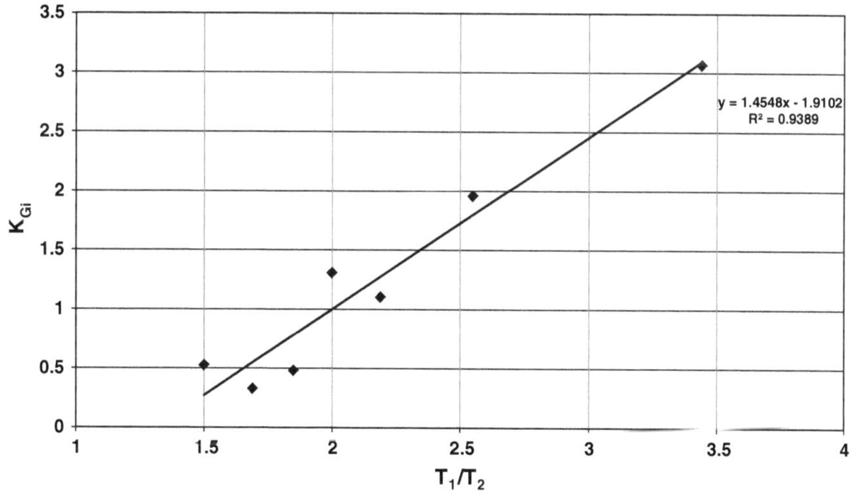

Fig. 8.25 The mean values of (T_1/T_2) versus K_{Gi}. A linear correlation yields $R^2 = 0.94$

a highly competitive alternative to the determination of T_1/T_2. In addition, the G_i distribution resolves the different sub regions, and K_{Gi} provides more information from these regions than a single (T_1/T_2) map.

8.3 Exploring the Separate Responses from Crude Oil and Brine

In the analysis of transverse relaxation (T_2) time curves from a CPMG-experiment on a multi-component system originating from measurements of oil and brine in rock cores, where internal magnetic field gradients broaden the line widths significantly, there may be little direct information to be extracted of the different components contributing to the total T_2 relaxation time curve. When basing the analysis of the T_2 data on the Inverse Laplace Transformations (ILT), the obtained CPMG attenuation is transformed to a distribution of T_2 values. The procedures of inverting relaxation data involve several uncertainties, and it may be difficult to give a satisfactory and reliable interpretation of the obtained results, especially when two components are overlapping in relaxation times. It will be shown that an optimized experimental set-up enables the extraction of resolved fluid signals from the NMR response, which would not be possible with the ordinary CPMG experiments. This set-up combines Pulsed Field Gradient (PFG) methods with the CPMG experiment utilizing data from both rock cores and bulk oil and brine. Then it becomes feasible to separate the signals from oil and brine in systems where both one and two-dimensional Inverse Laplace Transforms appear to fail in resolving the individual components.

8.3.1 The Separation of the Oil and Brine Signals Before Numerical Treatment

The application of the ILT-algorithm is ill-posed, meaning that several and very different solutions may fit the experimental data equally well [35]. An experimental approach where the distribution from oil and brine could be separated before the numerical inversion should make the inversion less prone to error. This can be achieved using the combined PFGSE CPMG method as proposed in [36, 37], in which one may obtain the separate T_2 contributions for oil and brine in a porous system. The basic requirement leading to a method that resolves the relaxation information from different components such as oil and brine, is that the components exhibit a sufficient difference in molecular diffusivity. In a system consisting of n components of different mobility, the echo attenuation for the bipolar PFGSE CPMG sequence, as provided in Chap. 1, is written

$$I = \sum_{i=1}^{m} \sum_{k} \rho_{ik} e^{-\frac{2(n\tau+\tau')}{T_2^{ik}}} e^{-\gamma^2 g^2 \delta^2 D_{ik} (\tau' - \delta/3)} \tag{8.23}$$

In a heterogeneous system there will in general be a distribution of values of both D and T_2 within each component i. This can be expressed as a sum over k exponentials, and the set of coefficients ρ_{ik} are the corresponding intensities in this distribution of decay times. The first exponential term represents effects from relaxation, as given in an ordinary CPMG experiment, while the second exponential term represents the diffusion weighting. By varying the gradient strength, g, from zero and upwards, there will be a weighting of the components within the system in the combined PFGSE CPMG experiment. As the gradient strength is increased, the faster diffusing component within the system contributes less to the echo signals as compared to the slower diffusing one. The attenuation from the PFGSE CPMG pulse sequences assuming two components, oil and brine, is written

$$I = \sum_{i=1}^{2}\sum_{k} \rho_{ik} e^{-\frac{2(n\tau+\tau')}{T_2^{ik}}} e^{-\gamma^2 g^2 \delta^2 D_{ik}(\tau'-\delta/3)} \quad (8.24)$$

If the diffusivities, D_{1k} and D_{2k} differ significantly for all k, for example $D_2 = 50D_1$, it is possible to choose the applied magnetic field gradient, g, and the time constant τ, such that signal from component 2 is suppressed to below 1 % of the original signal. Then there will still be a fraction of 0.9 left of component 1, for which the relaxation curve is recorded. When the mobility of all components within the oil phase is significantly lower than the mobility of brine, it is possible to subtract a scaled version of the CPMG decay of the oil only, from the combined PFG CPMG sequence with the pulsed field gradients set to zero. The resulting T_2 attenuation will then originate from the brine phase only. Following such a procedure for separation of multi component attenuation, it is possible to measure the T_2 contributions from oil and brine separately in oil and brine saturated systems.

In the following analysis, the slow diffusing substance is chosen as component number 1, and the fast diffusing substance as component number 2 in a two-component system. Using the combined PFG CPMG experiment the T_2 attenuation curve of the slow diffusing component can be analysed separately from the fast diffusing one. By applying a sufficiently high gradient strength, the signal from the fast diffusing component is suppressed to an insignificant level, and the echo attenuation for the slow diffusing component can be written

$$I = \sum_{k} \rho_{1k} e^{-\frac{2(n\tau+\tau')}{T_2^{1k}}} e^{-\gamma^2 g^2 \delta^2 D_{1k}(\tau'-\delta/3)} \quad (8.25)$$

Denoting $I_0^{1k*} = \rho_{1k} e^{-\gamma^2 g^2 \delta^2 D_{1k}(\tau'-\delta/3)}$, the diffusion weighted T_2 relaxation-curve for component 1 can be fitted to

$$I = \sum_{k} I_0^{1k*} e^{-\frac{2(n\tau+\tau')}{T_2^{1k}}} \quad (8.26)$$

To determine the relaxation curve for the fast diffusing component (component 2), a series of PFGSE CPMG experiments at varying gradient strengths were performed.

Using the part where the signal from component 2 is satisfactorily suppressed, the obtained data sets can be fitted to (8.25). From this, the curve for component 1 is extrapolated to zero gradient strength. However, it is important to take into consideration that component 1 may contain molecules of different sizes and different intrinsic correlation times. A distribution of both diffusion coefficients and T_2 values is therefore expected. As the echo becomes more T_2 attenuated, there will be a bias toward smaller, faster diffusing, molecules, and each echo in the set of T_2 attenuations will be diffusion weighted to different degrees. This has to be taken into consideration in the analysis of the experimental data. The intensity of each echo in the T_2 attenuations has to be extrapolated separately with respect to the diffusion weighting. Using this procedure, the obtained curve, which represents the T_2 attenuation for component 1 at zero gradient strength, is given by

$$I = \sum_k \rho_{1k} e^{-\frac{2(n\tau+\tau')}{T_2^{1k}}} \tag{8.27}$$

The attenuation curve for component 2 is determined by using PFGSE CPMG data at zero strength of the applied gradient pulses, i.e. an ordinary CPMG except for a different duration of the first τ-intervals, which equals the τ' interval in the combined PFGSE CPMG. The extrapolated curve (8.27) is then subtracted from the attenuation obtained in the “zero-gradient” experiment. The resulting curve represents the T_2 attenuation for component 2, and is written as

$$I = \sum_k \rho_{2k} e^{-\frac{2(n\tau+\tau')}{T_2^{2k}}} \tag{8.28}$$

In Figs. 8.26 and 8.27 the results from separation of the two components' contribution to the decaying curve following the prescribed procedure is shown together with the inverse Laplace transform of all the decays. The system used for visualisation is comprised of crude oil and brine in quartz sand of mixed wettability. While the crude oil is rather insensitive to the solid matrix and its wetting state, the T_2-distribution of the brine is bimodal. The fraction of short T_2-values originates from the water wet part of the quartz sand of relatively strong surface relaxivity and internal magnetic field gradients, while the fraction of long T_2's originates from an oil wet region where the brine is not in contact with the surface.

Provided a successful extrapolation of the crude oil signal as a function of the applied gradient strength, the two components may be separated for all values of the gradient strength. Consequently, a two-dimensional analysis of the separated data may be performed and compared against a two-dimensional analysis of the unresolved data set.

When using a combined PFGSE-sequence with a CPMG-sequence one should be aware that phase and magnitude of the recorded spin echo signals are very sensitive to eddy current field transients following the gradient pulses. One must therefore make sure that these transients have no significant effect on the CPMG echo attenuation. When τ is short (~150 μs) and the eddy current transients are

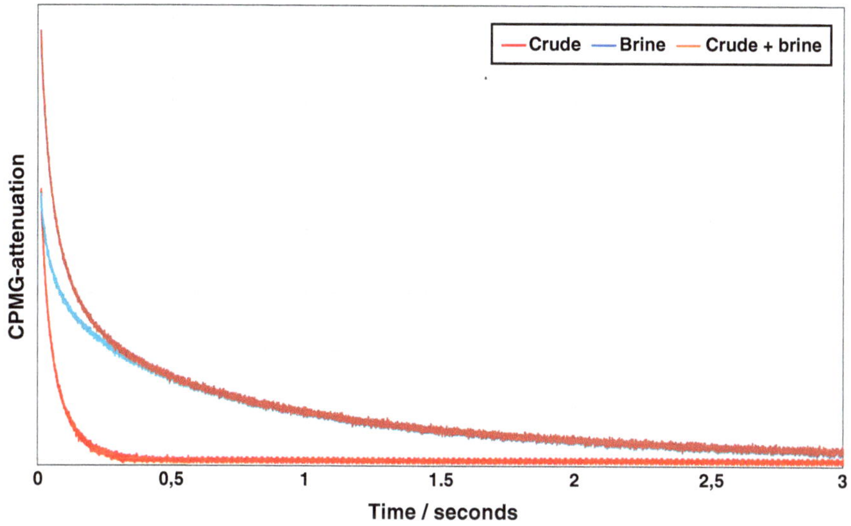

Fig. 8.26 Combined PFGSE-CPMG measurements on a sample containing crude oil, brine and quartz sand of mixed wetting, with (*orange*) or without (*brown*) brine suppression. The attenuation in *blue* colour corresponds to the resolved PFGSE-CPMG-attenuation from brine only

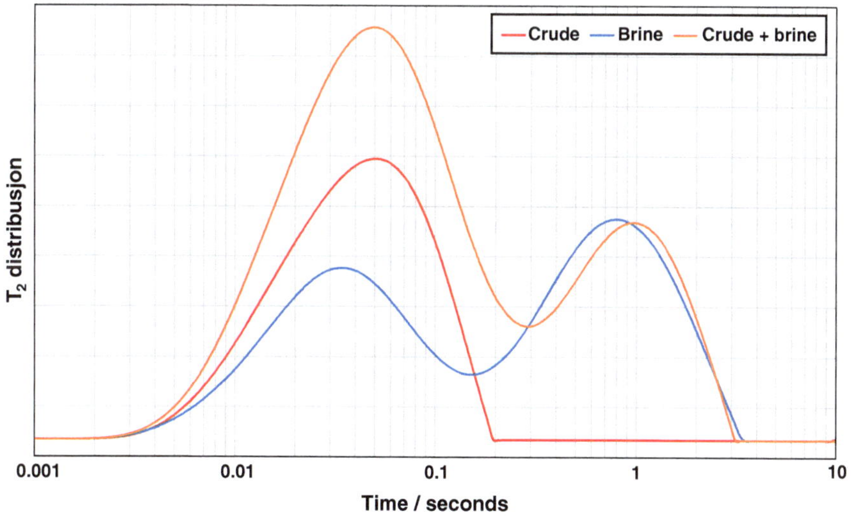

Fig. 8.27 The T_2-distributions arising from the attenuations in Fig. 8.26

significantly reduced by the eddy current delay following the last gradient pulse in the PFGSE-part of the experiment, the measured attenuation should be unaffected by the eddy current transients.

8.3.2 Application Examples

The combined 11 interval PFGSE CPMG experiments were implemented on a 0.047 T system from Resonance Instruments tempered at 25 °C. The 1.5″ probe contained actively shielded gradient coils with which it was possible to achieve stable pulsed field gradient strength up to approximately 100 G/cm. In all experiments, a τ value of 150 μs in the CPMG was employed. In the PFGSE part of the experiment the gradient pulse length δ was set to 2 ms together with an eddy current dead time of 1.0 ms. By using the given settings in both the PFG part and the CPMG echo train the transient magnetic fields following the gradient pulses had no significant effect on the echo attenuation. A time interval of 0.1 ms was used between the initial $\pi/2$ pulse and the following gradient pulse, giving a τ' value of 3.1 ms, and a total time of 12.4 ms before the start of the CPMG measurement.

To improve the signal to noise (S/N) a smoothed version of the rescaled T_2 attenuation of oil is used in the procedure for obtaining the resolved T_2 attenuation of brine. Then the S/N ratio of the T_2 attenuation of brine is minimized without any loss of information. This is obtained by fitting the data using the 1-dimensional version of the Anahess routine (Chap. 5) and subtracting the fitted signal extrapolated to zero gradient strength from the signals of interest. Such a procedure did not add any significant noise to the subtracted signal and the only assumption employed was that the sum of a set of decaying exponentials could fit the signal properly.

The combined PFGSE CPMG sequence was applied on the rock core plugs either saturated with brine only or with both crude oil and brine. In Fig. 8.28 the

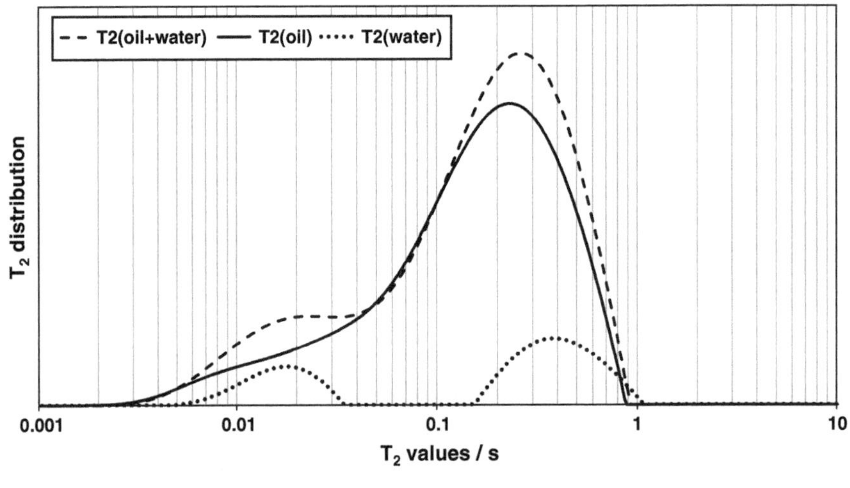

Fig. 8.28 T_2-distributions obtained from combined PFGSE-CPMG measurements on a porous rock sample containing oil and brine, oil + brine unresolved, oil resolved, and the resolved NMR response from brine only

procedure is applied on a rock core plug containing crude oil and brine. While the T_2 distribution of the oil shows one broad distribution with a left shoulder, the T_2 distribution from brine contains two well-resolved peaks centred at 20 and 400 ms, respectively. The existence of a discontinuity and large difference in transverse relaxation times of the two brine peaks is interpreted to be due to brine found in two different regions. The left peak is believed to originate from a water-wet region where the surface relaxivity is high and internal magnetic field gradients are the strongest, i.e. the apparent T_2 is short (surface region), while the right peak originates from a region in which the brine is not in contact with the surface of the porous rock. In this bulk region, the T_2 is long compared to the surface region. The existence of such a clearly separated bulk region is only possible when there is oil present to wet the surface and thus prevent the bulk brine from interacting with the surfaces within the rock core. Hürlimann et al. [38] has also assigned such a bulk region of brine to a non-wetting phase within the porous rock.

Two sandstone rock core plugs of equal porosity (0.30) were also saturated with crude oil and brine and subjected to the combined PFGSE-CPMG NMR measurements. A normal Amott test displacement cycle was performed on fresh samples using synthetic formation brine and crude oil (8.14). The Amott wettability index was found to be −0.26 for the sample denoted PL11 and +0.11 for the sample denoted PL39. Thus, PL11 is slightly oil-wet while PL39 is slightly water-wet.

In Fig. 8.29 the resolved T_2 distribution from brine in rock core plug (PL11) is shown either as the only component or in the presence of oil. When brine is present as the only component a broad distribution of T_2 values is found, and there are no resolved peaks that could be assigned to a surface region and a bulk region, respectively. When oil is injected into the system, two regions appear, and there is a shift towards larger T_2 values compared to those in the plug with brine only.

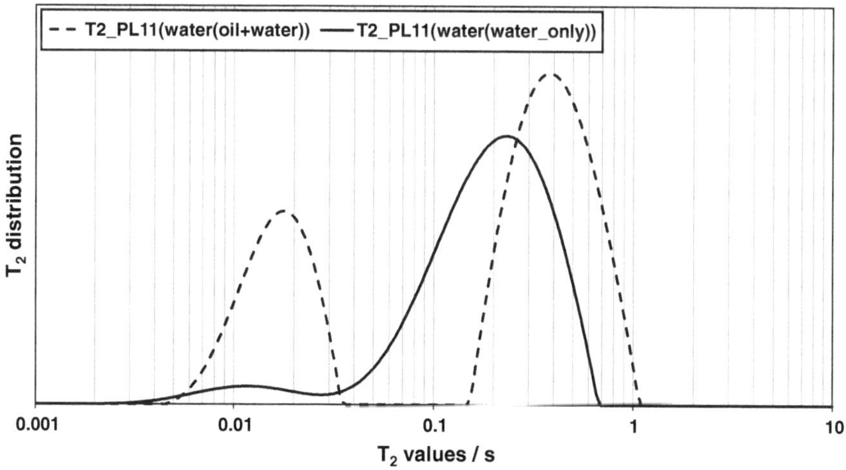

Fig. 8.29 T_2 distribution of brine in PL11 containing brine only and in the presence of crude oil

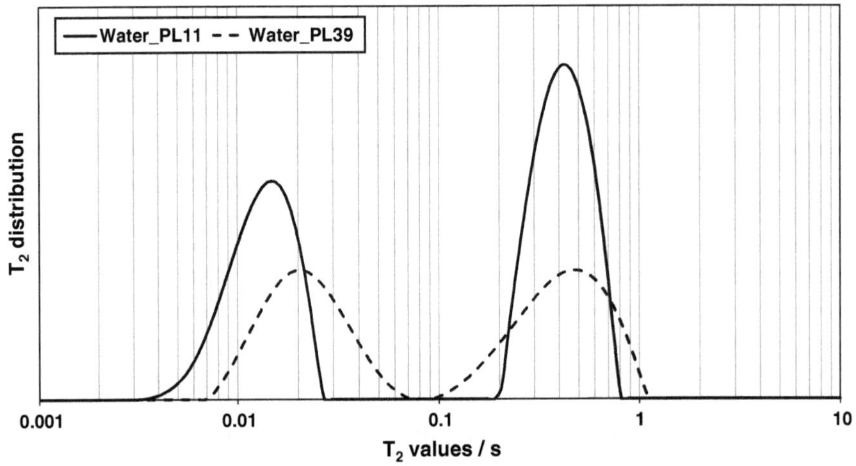

Fig. 8.30 T_2 distribution of brine in PL11 and PL39 in the presence of crude oil

Again, this trend supports the picture of the existence of a bulk region of brine separated from the solid surfaces within the rock core plug.

As long as the surface and bulk regions of brine within a rock core plug are distinguishable by T_2's, as in Fig. 8.28, the amount of bulk brine may be quantified and related to the total amount of brine present in the sample. Consequently, it should be possible to probe the wettablity. Figures 8.29 and 8.30 show the brine distribution arising from two rock core plugs with different wettabilities, but where the total amount of brine in the plugs is the same within a fraction of ±0.01. Due to the PFG part of the PFGSE CPMG sequence, the first measurement point in the echo decay is at 12.4 ms. During this time a significant amount of the brine signal originating from the surface region has decayed and thus makes it difficult to quantify the amount of brine associated with the surface. However, NMR signal from the bulk region where the T_2 distribution is found at approximately 800 ms, will not be significantly affected by the finite duration of the PFG part of the experiment. As the total amount of brine is known through the controlled injection of brine into the plug, a quantification of the bulk region yields an indirect measure of the wettability in the system. Taking into account the total amount of brine in the two plugs it can be concluded that the amount of bulk brine is less per unit pore volume in PL39 than in PL11. Due to the larger amount of brine originating from the surface region, one may conclude that PL39 is more water-wet than PL11. This is in accordance with the Amott wettability index results obtained which revealed that PL39 is in fact more water-wet than PL11.

In the study of the two rock cores saturated with different amounts of crude oil and brine, it has been shown that with an optimized experimental set-up where PFGSE is combined with the CPMG experiment, it is possible to extract information from the NMR response that is difficult to obtain via other methods.

8.4 Wettability

Identifying the wettability of a formation is crucial for the decision on how to extract as much crude oil as possible from the reservoir. In the laboratory a normal Amott test displacement cycle is regarded as the state of the art [39]. The method combines two spontaneous imbibition measurements, spontaneous displacement of one fluid by another immiscible fluid, and two forced displacement measurements. These tests probe the tendency of a porous rock to exchange the oil by brine or vice versa, and provides a number between -1 and $+1$ which describes the surface wetting. The number varies from -1 to -0.3, completely oil wet (-1), mixed wet from -0.3 to 0.3 (i.e. gradually less oil wet) until it enters the region $+0.3$ to $+1$, which is defined as water wet (8.14).

In this section, six rock core plugs have been examined by PFG NMR through different saturation and ageing states, aiming at probing the wettability of the plugs and compare it against the Amott index. The plugs are classified into two subgroups, calcites and dolomites. As seen from the microscopy pictures in Fig. 8.31, the calcites contain large pores and narrow throats (if any) while the dolomites have a pore network of which it is difficult to distinguish between the pores and the channels connecting the pores. A small but detectable NMR signal from plugs that have been cleaned and dried, where all fluids should have been removed, indicated that the plugs contain isolated pores from which it is not possible to extract the confined fluid.

Through the measurement program, the plugs went through the following saturation states; 100 % brine saturated, then flooded with synthetic oil, crude oil, centrifuged, and subjected to high pressure and temperature in the simulation of an ageing process. At all states PFG NMR measurements were conducted in order to resolve the oil signal from the brine. To achieve good resolution of the brine signal, the crude oil was exchanged with the synthetic oil after the ageing process.

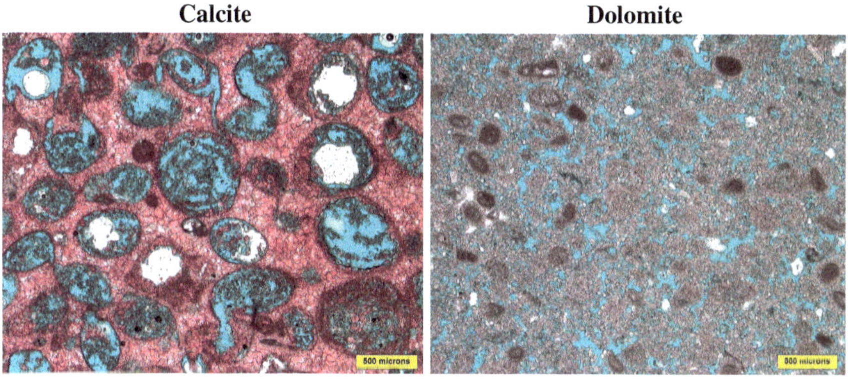

Fig. 8.31 Microscopy pictures of thin sections representing the two rock types, calcites (*left*) and dolomites (*right*). The rocks have been impregnated with *blue* epoxy to visualize the pore space

Table 8.3 Saturation and ageing states of the calcite and dolomite rock core plugs

Stage	Saturation state	Procedure	NMR sequences
1	Brine 100 %	Dried, cleaned and saturated	*CPMG, STE_CPMG*
2	Synthetic oil + brine (80–20 %)	Flooded with synthetic oil	*CPMG, PFGSE-CPMG STE_CPMG, ONESHOT*
3	Synthetic oil + brine (80–20 %)	Aged with crude oil at 100 °C and 10 bar for 2 weeks, then re-saturated as in stage 2	*CPMG, PFGSE-CPMG STE-CPMG, ONESHOT*

This also made the comparison before and after ageing easier, as the plugs were flooded with synthetic oil before ageing. In Table 8.3, saturations and ageing states of the plugs are summarized, together with the type of NMR experiments that were conducted.

Explanations to the NMR sequences found in Table 8.3:

CPMG	Ordinary T_2 acquisition
STE_CPMG	Combined T_1 and T_2 acquisition
PFGSE-CPMG	Combined Diffusion and T_2 acquisition
ONESHOT	Method for quantification of oil and brine

The Marcol82 oil was chosen as the synthetic oil, because its mobility (diffusion coefficient ~1×10^{-10} m^2/s at 25 °C) is significantly slower than that of the brine that simulates the formation water, making it feasible to separate the T_2 distributions of oil and brine when they are simultaneously present in the rock core plugs. In the following section, each state as described in Table 8.3 will be characterised using PFG NMR.

8.4.1 Stage 1, Brine Saturated

The first saturation state was established by drying, cleaning and saturating the six rock core plugs with synthetic formation water. Some characteristics of the plugs are shown in Table 8.4, where it is seen that the permeability is significantly different for the calcites and dolomites. When knowing the volume of the plugs, the fitted initial NMR signal as follows from the oneshot method without applying magnetic field gradients for water suppression, can be transformed to a value for the porosity

Table 8.4 Characteristics of the plugs

Plug ID	D1	D2	D3	C1	C2	C3
Length (cm)	4.88	4.56	4.66	4.69	4.62	4.69
Diameter (cm)	3.72	3.72	3.72	3.73	3.73	3.73
Permeability (mD)	1,831	2,756	1,194	23.7	29	68.7
Helium porosity (frac.)	0.297	0.271	0.301	0.257	0.226	0.246
Pore volume (ccm)	15.6	13.1	14.95	13.1	11.25	12.45

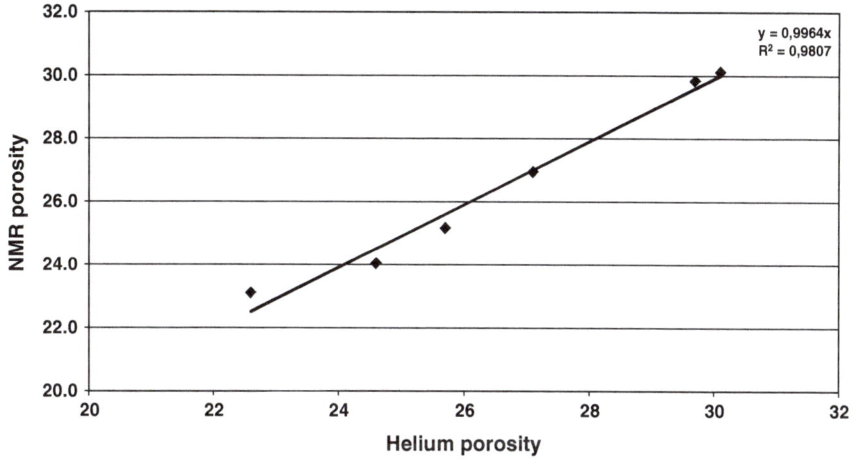

Fig. 8.32 NMR porosity found from the oneshot method versus helium porosity data

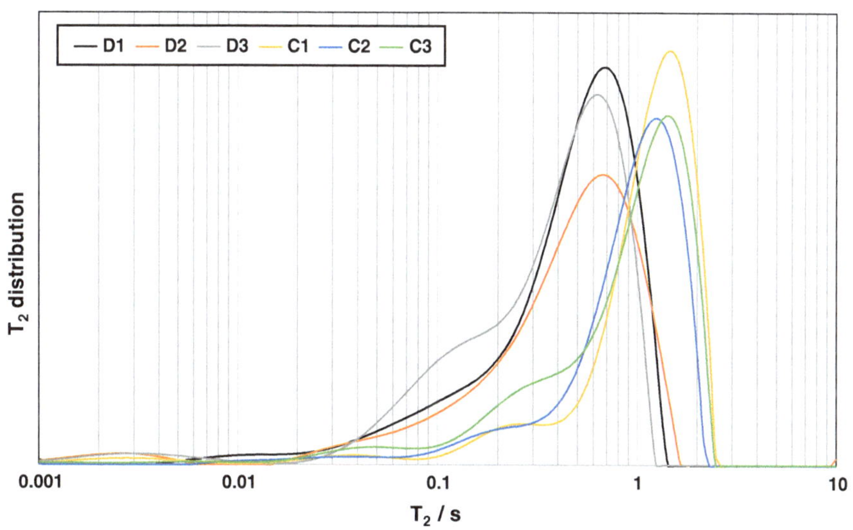

Fig. 8.33 The T_2-distributions arising from the ordinary CPMG experiment

of the sample. This is the NMR porosity, and in Fig. 8.32, it is compared against the determined Helium porosity, where they correlate well.

There is no systematic difference in NMR porosity versus helium porosity when comparing the dolomite porosities against the calcite porosities. Thus, the NMR method applies regardless of the porous structure of the rock core plugs investigated.

In Fig. 8.33, the T_2 distributions arising from the six plugs are shown, and evidently they can be grouped into two subclasses, the dolomites with the shortest

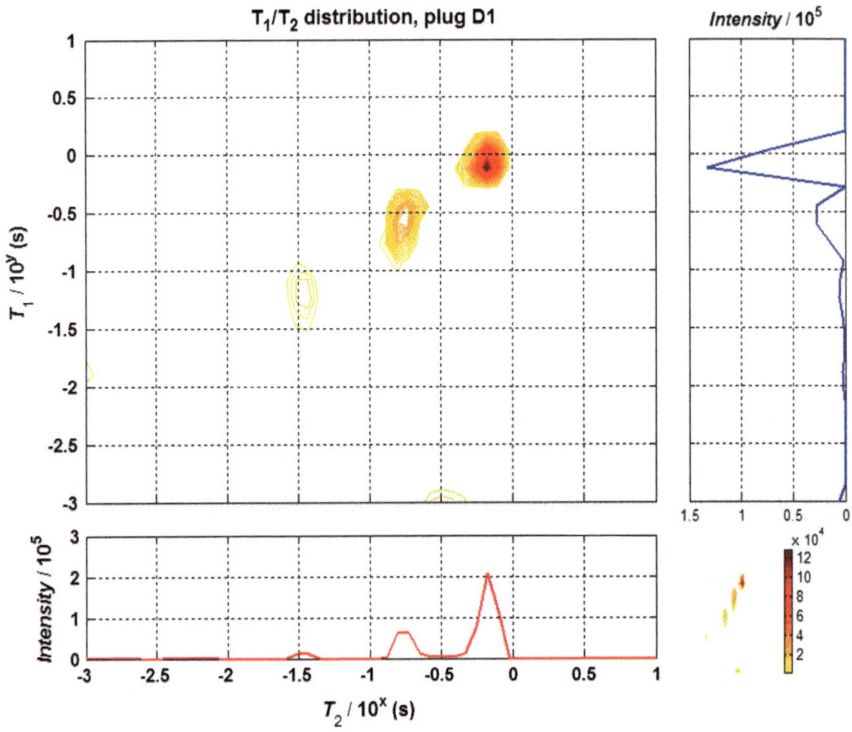

Fig. 8.34 T_1-T_2 correlated distribution for dolomite D1, brine saturated

T_2 values (smallest pores) and calcites with the longest T_2 values (largest pores). In addition, the left shoulder of the dolomite curves seems to be broader than the shoulder found from the calcite curves. This should indicate that the distribution of pore sizes is broader for the dolomites. Interestingly the permeability is lowest for the system with highest T_2-values, thus the T_2 is not a good probe for comparing the permeability of the two subclasses.

Figures 8.34 and 8.35 show the T_1-T_2 correlated plots for two representative plugs of the two major subgroups. The two T_1-T_2 plots confirm what is seen in the microscopy pictures in Fig. 8.31; that the sorting in the dolomite plug is poorer than in the calcite plug, since the fraction of the major peak is the highest for the calcite plug.

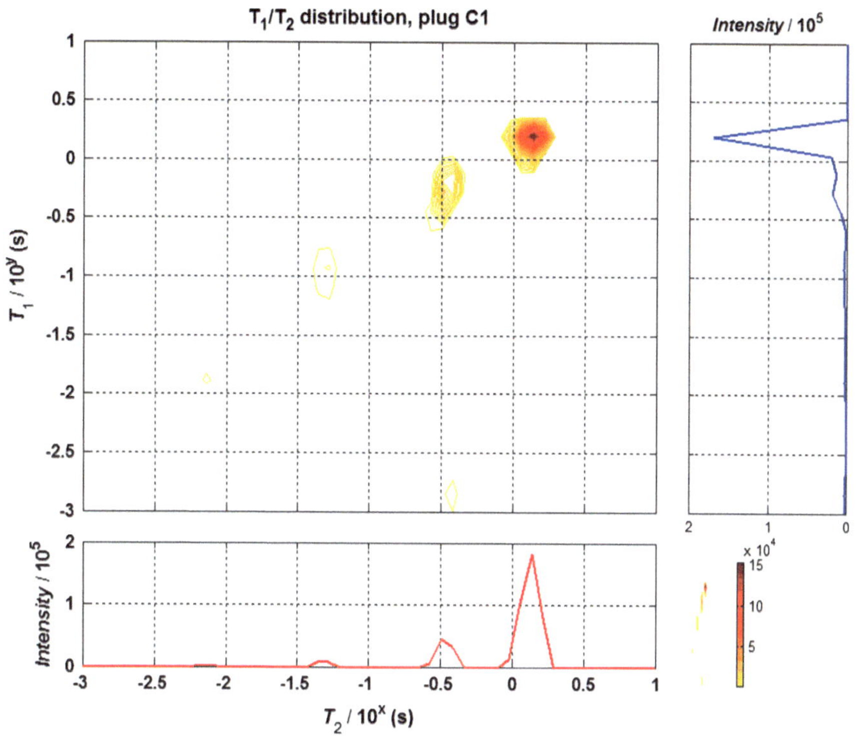

Fig. 8.35 T_1-T_2 correlated distribution for calcite C1, brine saturated

8.4.2 *Stage 2: Oil-Brine (80–20 %)*

After measuring the brine saturated plugs, they were subjected to an Amott cycle using a synthetic oil. During this process, the cycle was stopped at a time in which the saturation state was supposed to be approximately 70 % oil and 30 % brine. However, the oneshot method [9] which is implemented for the determination of oil and brine content together with the porosity, indicated a saturation state closer to 80 % oil and 20 % brine, as indicated in Table 8.5.

Figure 8.36 shows the T_2-distributions for all plugs achieved from the PFGSE CPMG sequence without applied gradients. The plugs are, according to Table 8.5, measured to contain approximately 80 % oil and 20 % brine. While there was a significant difference in the position of the T_2 peaks for the brine saturated calcite and dolomite plugs, the position of the main T_2 peak, which must originate mainly from oil, is fairly constant and found to be at a T_2 ~ 130 ms for all plugs. The small shifts in the position of the T_2-peak are probably due to the presence of a small brine component of different intensity for each plug. The oil provides an insignificant shift in T_2 as it has a much lower mobility than the brine. Thus, the mean squared displacement is less for the oil, making it a poorer probe for the

Table 8.5 Fractions of oil and brine found in the rock core plugs using the oneshot method

	NMR fraction of oil	NMR fraction of brine
D1	0.82	0.18
D2	0.86	0.14
D3	0.78	0.22
C1	0.78	0.22
C2	0.81	0.19
C3	0.80	0.20

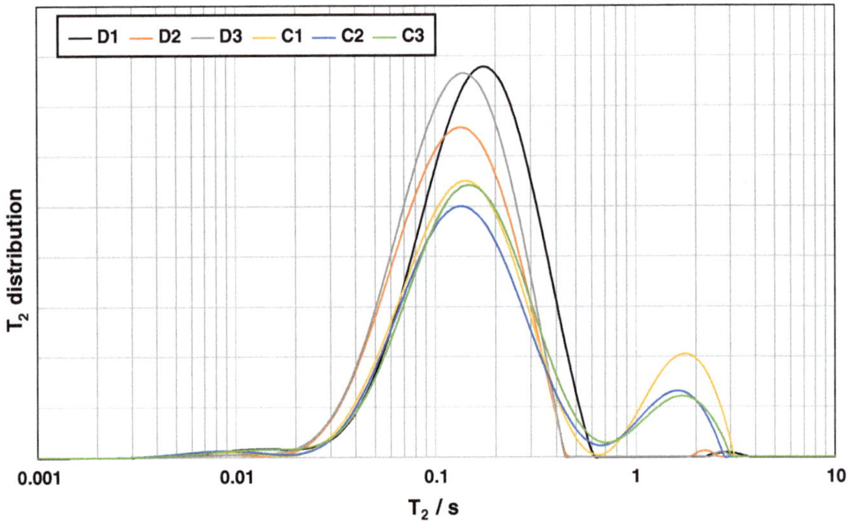

Fig. 8.36 The T_2-distributions arising from the PFGSE-CPMG experiment without gradients applied

heterogeneity of the porous sample. In other words; while the formation water may probe the heterogeneity of the porous rock, the oil NMR signal from the rock core plug is almost insensitive to the presence of the porous rock. During the lifetime of the NMR signal from the oil (~0.5 s), the root of the mean squared displacement of the oil is ~15 μm while it is 85 μm for the formation water. However, the NMR signal from the formation water lasts longer than 0.5 s, so it will probe displacements larger than 100 μm before the signal becomes insignificant.

With respect to the remaining 20 % of the brine signal, a significant difference between the calcites and dolomites is seen. While there is an extra considerable T_2 peak at approximately 1.8 s for the calcite plugs, this peak is hardly present for the dolomites. In addition, the 1.8 s peak is at slightly higher values than for the 100 % brine saturated plugs, indicating that the 1.8 s signal arises from brine now being more isolated from the solid surface of the porous rock. It should also be noted that the dolomites tend towards a larger contribution at the shortest T_2 values than the calcites. Thus, one is tempted to already conclude that the dolomites

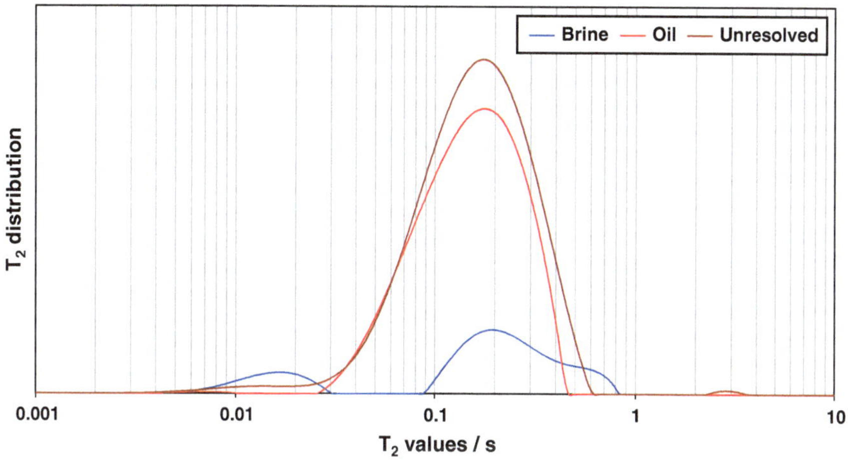

Fig. 8.37 Resolved and unresolved T_2 distributions from the combined PFGSE-CPMG experiment on dolomite D1

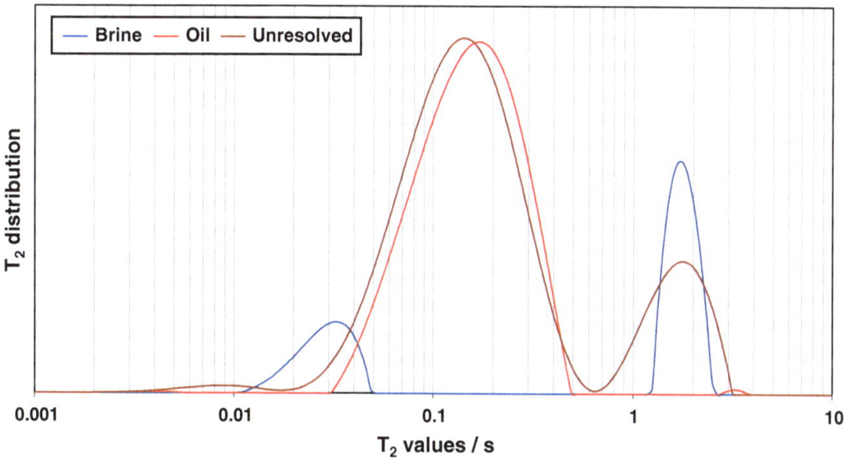

Fig. 8.38 Resolved and unresolved T_2 distributions from the combined PFGSE-CPMG experiment on calcite C1

are more water wet than the calcites. In order to verify this scenario the combined PFGSE CPMG experiment is applied to resolve the contribution of oil and brine in the T_2 distribution [36, 37].

When applying the method for separation of the oil and brine contributions to the T_2 attenuation as outlined in the previous section, the separate T distributions for oil and brine can be produced for the calcites and dolomites (Figs. 8.37 and 8.38). For the dolomites a significant part of the NMR signal from the brine connected to the porous surface is lost, due to the fact that the PFGSE CPMG starts the recording of the T_2 attenuation at 12.4 ms. This is the time necessary to suppress the brine signal during the PFGSE part and to record the oil signal for subtraction. Thus, a smaller component of T_2 values around 15 ms is seen for plug D1 in Fig. 8.37, while the major brine signal is found at T_2 values around 250 ms. This feature is less pronounced for the calcites. Here the peak for the shortest T_2-values is found between 30 and 40 ms. Thus, in the resolved T_2 distribution from brine the contribution from the brine connected to the rock surface is relatively larger for these samples than in the dolomites.

As the saturation states for all plugs are approximately the same, containing 80 % oil and 20 % brine, the fraction of brine decoupled from the surface of the porous rock should be dependent on the wettability of the sample. When quantifying the fraction of bulk brine, the dolomites and the calcites can be classified into two groups with respect to the fraction of bulk brine (fraction of brine with the longest T_2 values which is unaffected by the loss of signal due to the duration of the PFGSE-part of 12.4 ms). The fraction of bulk brine is significantly smaller for the dolomites than for the calcites, which indicates that the fraction of surface brine is the largest for the dolomites. This is a strong indication that the brine in the dolomites is wetting larger areas than in the calcites.

Instead of presenting the result as one-dimensional distributions, one may apply a two-dimensional Inverse Laplace Transform (2D-ILT) on the experimental data as a function of observation time (proportional to the number of echoes) and the applied gradient strength. This is obtained by fitting the data at an applied gradient strength where the brine signal is suppressed using the 1-dimensional version of the Anahess routine and subtracting rescaled versions of the fitted signal from the total signal. The values for the rescaling are governed by the diffusion weighting of initial oil signal in the PFGSE CPMG sequence [36]. The 2D-ILT was then employed for analysing the two dimensional data sets, either resolved or unresolved. In Figs. 8.39, 8.40, 8.41 and 8.42, the Diffusion-T correlated distributions are shown for plugs C1 and D1. In the unresolved versions (Figs. 8.39 and 8.41), the 2D-ILT is applied on the raw data while in the resolved versions the oil and brine signals were separated before subjected to the 2D-ILT routine.

The unresolved Diffusion-T_2 correlated distribution for plug D1 in Fig. 8.39 reveals one major peak where the diffusion coefficient varies by almost one order of magnitude, from 1×10^{-10} m^2/s to 1×10^{-9} m^2/s while the T_2 varies two orders of magnitude, from 10 to 1,000 ms. In the resolved Diffusion-T_2 correlated distribution for plug D1, using the approach described in Sect. 8.3, the situation is quite different. Here a major peak is found with an apparent diffusion around 1×10^{-10} m^2/s and two smaller ones at 1×10^{-9} m^2/s (shortest T_2) and 2×10^{-9} m^2/s (longest T_2). In the T_2 direction, the peaks are much narrower, only one order of magnitude each. Thus, the 2D-ILT applied on the unresolved data does not resolve the contribution of brine from the oil properly, even though the diffusion differs by one order of magnitude. Instead, the 2D-ILT returns a broader distribution that does not represent the actual components present in the plug. Still, the

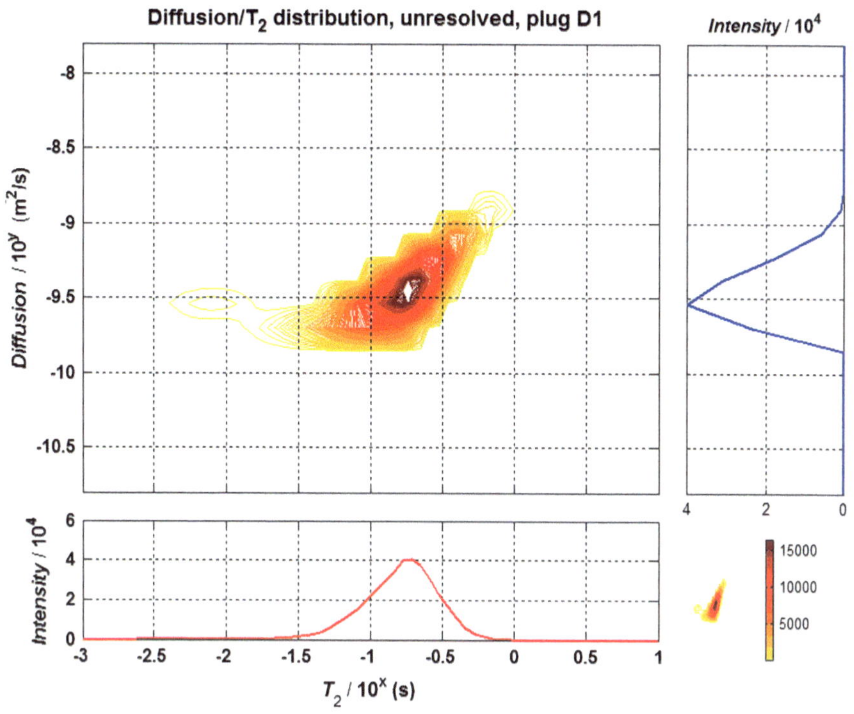

Fig. 8.39 Unresolved D-T_2 correlated distribution for dolomite D1, oil-brine (80–20)

solution for the 2D-ILT shown in Fig. 8.39 is just as good a solution as the one in Fig. 8.40, as the fit of the data set is as good as in the resolved case. As the 2D-ILT provides an infinite number of solutions to the data set, it truly reflects the lack of robustness of the 2D-ILT routine.

For plug C1 the situation is slightly different, as the 2D-ILT now is able to resolve the brine component with $T_2 > 1$ s, but the brine component of shorter T_2 is only found when the brine and oil signals are separated before being subjected to the 2D-ILT routine. This may be wrongly interpreted as a missing brine component at shorter T_2. However, the resolved Diffusion-T_2 correlated distribution indicates the presence of the brine fraction of shorter T_2's, which is confirmed in the resolved and unresolved T_2-distributions arising from the CPMG (Fig. 8.38). The unresolved T_2 peak is found at a smaller value (130 ms) than the peak arising from the oil only, which is shown to be at approximately 180 ms. This confirms the presence of a brine fraction of shorter T_2's. When it is possible to separate the contribution from brine and oil prior to a 2D-ILT, due to the significant difference in molecular mobility, the resolved Diffusion-T_2 correlated distribution may provide better and more correct information on the components found within the system than in the unresolved case.

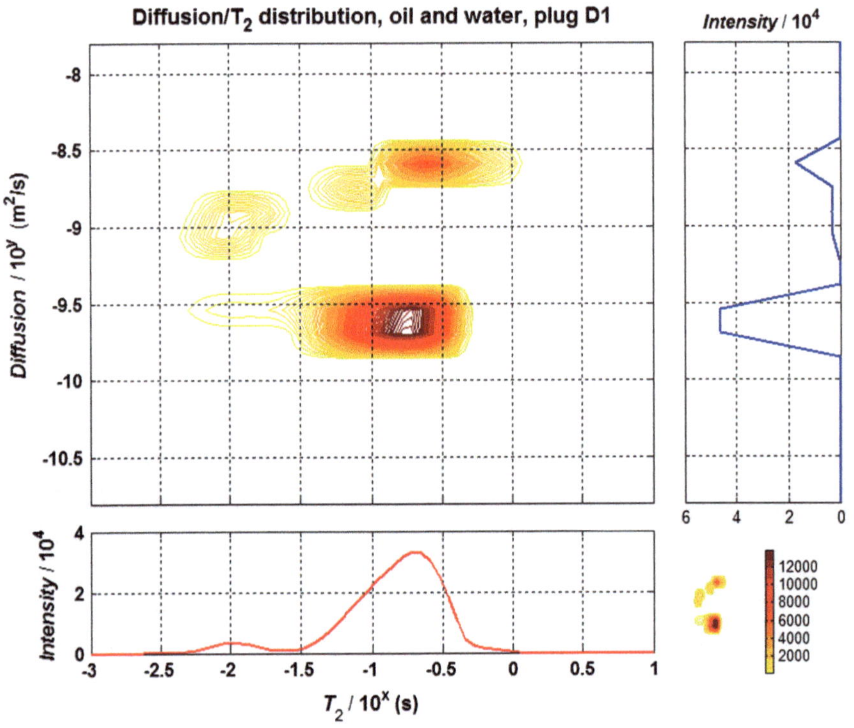

Fig. 8.40 Resolved D-T_2 correlated distribution for dolomite D1, oil-brine (80–20)

As there is evidently a relation between the fraction of the surface brine and the wettability, a measurable NMR value that reflects the fraction of surface brine found in the rock core plug could be used as an indicator for wettability. An example of such value could be the determined fraction of brine in contact with the surface normalised against the total porosity of the sample;

$$\mathbf{NMR_{wet}} = \frac{\textbf{Fraction of brine connected to the surface}}{\textbf{NMR porosity}}$$

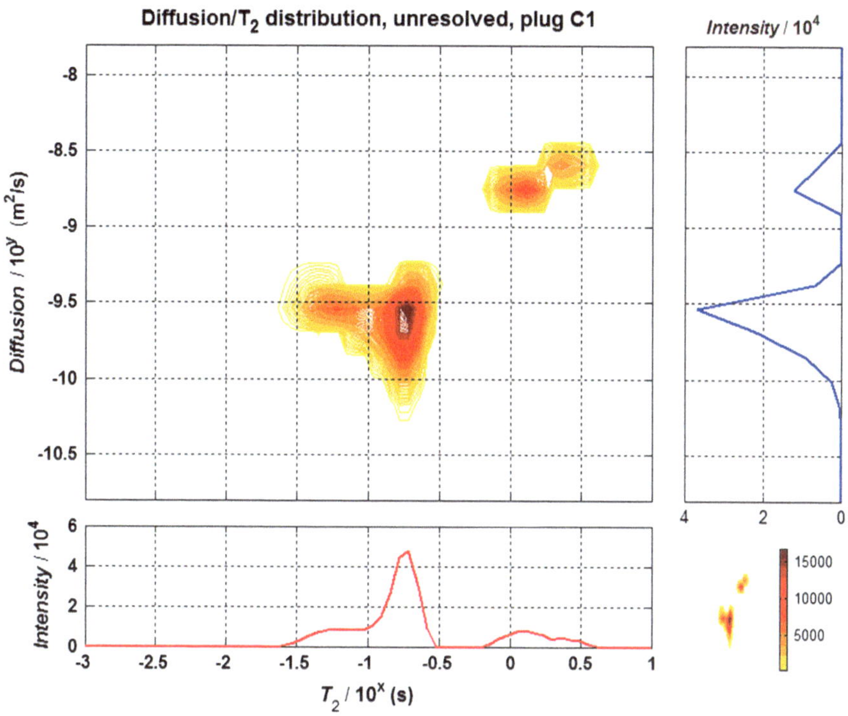

Fig. 8.41 Unresolved D-T_2 correlated distribution for calcite C1, oil-brine (80–20)

An even better indicator should normalise the fraction of brine in contact with the surface against the surface to volume (S/V). Then a variation of available surface is taken into account and should make the indicator insensitive to plugs with different surface areas, which also may produce different fractions of brine in contact with the surface. However, in the present study, the (S/V) was not measured, so the normalization against porosity is calculated instead. The NMR index for each plug is shown in Table 8.6 together with the Amott index.

As is seen from Table 8.6 the Amott indices states that the calcites are of mixed wettability while the dolomites are strongly water wet. This difference in wettability can also be extracted from the NMR_{wet}; the higher NMR_{wet} index the more water wet is the sample.

Table 8.6 Amott index versus NMR_{wet} index

Sample ID	Amott	NMR_{wet}
D1	0.65	0.60
D2	0.47	0.54
D3	0.54	0.74
C1	0.02	0.20
C2	0.12	0.27
C3	0.00	0.24

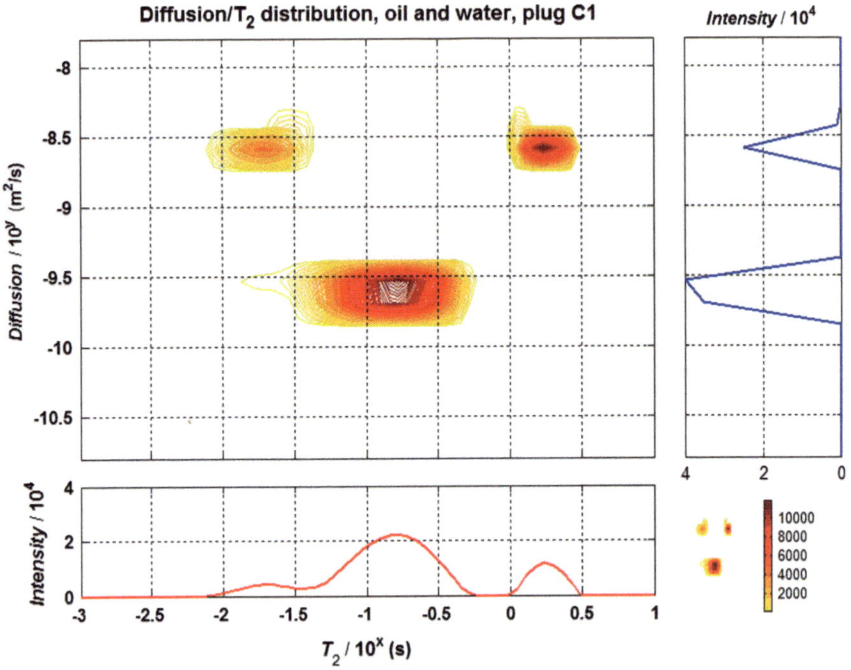

Fig. 8.42 Resolved D-T_2 correlated distribution for calcite C1, oil-brine (80–20)

8.4.3 Stage 3: Oil-Brine (80–20) After Ageing

After having aged the rock core plugs with crude oil at 100 °C and 10 bar for two weeks, the synthetic oil replaced the crude oil, and a re-saturating to 80 % synthetic oil and 20 % brine was established. In Figs. 8.43 and 8.44, the T_2 distributions extracted from the ordinary CPMG are shown for plug C1 and D1.

There are two major changes after ageing of the plugs, the appearance of a T_2-component in D1 at much larger values (>1 s) than found before ageing, and the increase of the same component in C1. Separation of the contribution from oil and brine before producing the T_2-distribution (only shown for D1 in Fig. 8.45), still produces a brine component of shorter T_2. However, the signal amplitude of this component is smaller and appears at even shorter T_2. Most likely, it originates from a region in the plug where is difficult to change the wettability during the ageing process.

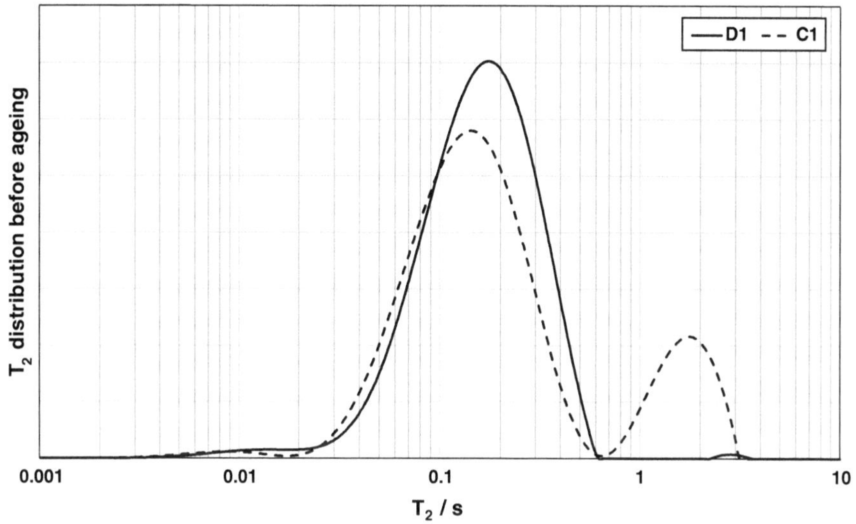

Fig. 8.43 The T_2-distributions before ageing for plugs D1 and C1

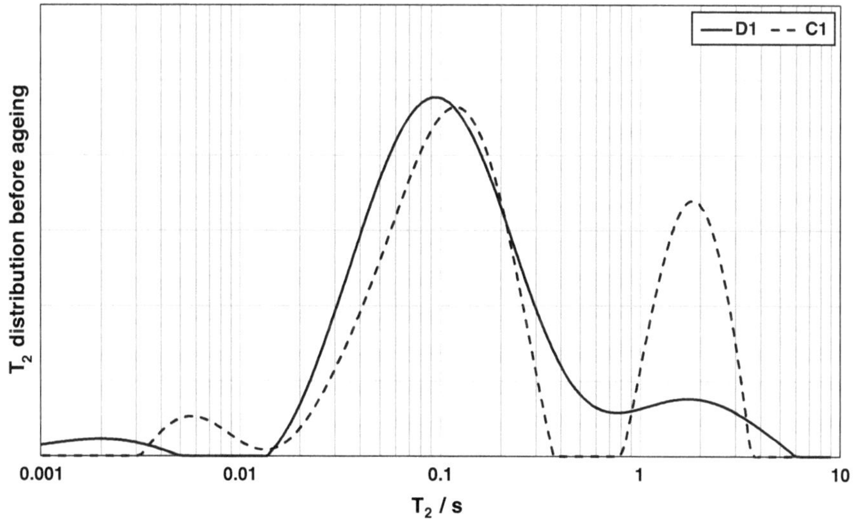

Fig. 8.44 The T_2-distributions after ageing for plugs D1 and C1

While the ageing process has significant impact on the T_2 distribution of brine, there is a small shift towards shorter T_2's for the oil signal due to the ageing process. If a crude oil component during the ageing process had deposited on the rock core surface, its T_2-value would be shorter than from the synthetic oil, and may explain the small shift of the T_2 towards shorter values.

One major draw back using the PFGSE CPMG NMR technique is that the given experimental set-up requires 12.4 ms for motional encoding/decoding (the

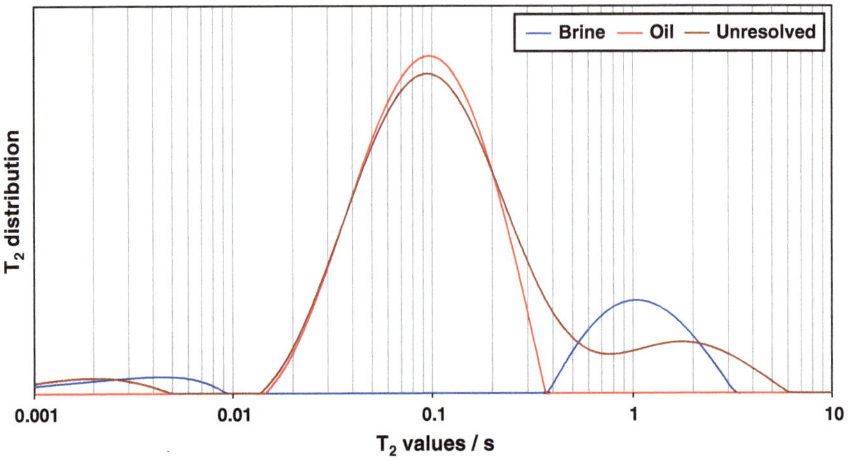

Fig. 8.45 Resolved and unresolved T_2 distributions from the combined PFGSE-CPMG experiment on dolomite D1 after ageing

PFGSE-part) before the NMR signal in the CPMG echo train is acquired. During this period, some of the information in the T_2 response from brine in contact with the surface may be lost. Still, the quantification of this component can be performed quite accurately as long as the T_2 values of the oil component are larger than 10 ms. The fraction of the surface connected brine may then be quantified accurately by for example the oneshot method. In addition, the T_2 distribution from the ordinary CPMG experiment will reveal some features of the shape of the distribution at T_2 values shorter than 10 ms. Using an inert synthetic oil it should also be possible to reconstruct the T_2 distribution of the complete brine fraction, starting at an observation time of 2 times τ. This is done by subtracting the rescaled CPMG attenuation of the synthetic oil from the CPMG instead of the PFGSE CPMG.

Using the PFGSE CPMG experimental setup it is not only possible to quantify the oil and brine content within the rock core plugs, but also to separate the contribution from the oil and brine to the T_2 distribution. Together with the resolved Diffusion-T_2 correlated distribution, this provides information on the rock core plugs that cannot be extracted using the unresolved approach.

References

1. R.J.S. Brown, The Earth's-field NML development at Chevron. Concepts Magn. Reson. **13**(6), 344–366 (2001)
2. S. Luthi, *Nuclear Magnetic Resonance Logging, in Geological Well Logs* (Springer, Berlin, 2001), pp. 159–182
3. K.J. Dunn, D.J. Bergman, G.A. LaTorraca, *Nuclear Magnetic Resonance: Petrophysical and Logging Applications* (Elsevier Science, Oxford, 2002)
4. M.G. Prammer, NMR logging-while-drilling (1995–2000). Concepts Magn. Reson. **13**(6), 409–411 (2001)

5. S. Meiboom, D. Gill, Modified Spin-Echo method for measuring nuclear relaxation times. Rev. Sci. Instrum. **29**(8), 688–691 (1958)
6. M.D. Hürlimann, L. Venkataramanan, Quantitative measurement of two-dimensional distribution functions of diffusion and relaxation in grossly inhomogeneous fields. J. Magn. Reson. **157**(1), 31–42 (2002)
7. M.D. Hürlimann, Effective gradients in porous media due to susceptibility differences. J. Magn. Reson. **131**(2), 232–240 (1998)
8. F. Civan, F. Civan, *Reservoir Formation Damage* (Elsevier Science, Oxford, 2011)
9. G.H. Sorland et al., Determination of total fat and moisture content in meat using low field NMR. Meat Sci. **66**(3), 543–550 (2004)
10. G.H. Sørland, H.C. Widerøe, Improved characterization of 1.5 inch, brine-saturated rock cores using a newly designed 12.8-MHz rock core analyzer. Magn. Reson. Imag. **25**(4), 584–585 (2007)
11. F. Bergaya, B.K.G. Theng, G. Lagaly, *Handbook of Clay Science* (Elsevier Science, Oxford, 2011)
12. P.P. Mitra, P.N. Sen, L.M. Schwartz, Short-time behavior of the diffusion coefficient as a geometrical probe of porous media. Phys. Rev. B **47**(14), 8565–8574 (1993)
13. M.D. Hurlimann et al., Restricted diffusion in sedimentary rocks. Determination of surface-area-to-volume ratio and surface relaxivity. J. Magn. Reson. Ser. A **111**(2), 169–178 (1994)
14. G.H. Sørland, D. Aksnes, L. Gjerdåker, A pulsed field gradient spin-echo method for diffusion measurements in the presence of internal gradients. J. Magn. Reson. **137**(2), 397–401 (1999)
15. G.H. Sørland, B. Hafskjold, O. Herstad, A stimulated-echo method for diffusion measurements in heterogeneous media using pulsed field gradients. J. Magn. Reson. **124**(1), 172–176 (1997)
16. K.S. Mendelson, Nuclear magnetic resonance in sedimentary rocks: effect of proton desorption rate. J. Appl. Phys. **53**(9), 6465–6466 (1982)
17. J. Uh, A.T. Watson, Nuclear magnetic resonance determination of surface relaxivity in permeable media. Ind. Eng. Chem. Res. **43**(12), 3026–3032 (2004)
18. H.L. Ritter, L.C. Drake, Pressure porosimeter and determination of complete macropore-size distributions. Ind. Eng. Chem. Anal. Ed. **17**(12), 782–786 (1945)
19. R.L. Kleinberg, S.A. Farooqui, M.A. Horsfield, T1/T2 ratio and frequency dependence of NMR relaxation in porous sedimentary rocks. J. Colloid Interface Sci. **158**(1), 195–198 (1993)
20. L.L. Latour, R.L. Kleinberg, A. Sezginer, Nuclear magnetic resonance properties of rocks at elevated temperatures. J. Colloid Interface Sci. **150**(2), 535–548 (1992)
21. J. Stepišnik, Validity limits of Gaussian approximation in cumulant expansion for diffusion attenuation of spin echo. Physica B **270**(1–2), 110–117 (1999)
22. P.P. Mitra, B.I. Halperin, Effects of finite gradient-pulse widths in pulsed-field-gradient diffusion measurements. J. Magn. Reson. Ser. A **113**(1), 94–101 (1995)
23. E.J. Fordham, P.P. Mitra, L.L. Latour, Effective diffusion times in multiple-pulse pfg diffusion measurements in porous media. J. Magn. Reson. Ser. A **121**(2), 187–192 (1996)
24. Y. Nakashima, Pulsed field gradient proton NMR study of the self-diffusion of H_2O in montmorillonite gel: effects of temperature and water fraction. Am. Mineral. **86**(1–2), 132–138 (2001)
25. L.L. Latour et al., Pore-size distributions and tortuosity in heterogeneous porous media. J. Magn. Reson. Ser. A **112**(1), 83–91 (1995)
26. J.G. Seland et al., Diffusion measurements at long observation times in the presence of spatially variable internal magnetic field gradients. J. Magn. Reson. **146**(1), 14–19 (2000)
27. G.H. Sorland, D. Aksnes, L. Gjerdaker, A pulsed field gradient spin-echo method for diffusion measurements in the presence of internal gradients. J. Magn. Reson. **137**(2), 397–401 (1999)
28. R.M. Cotts et al., Pulsed field gradient stimulated echo methods for improved NMR diffusion measurements in heterogeneous systems. J. Magn. Reson. (1969) **83**(2), 252–266 (1989)

29. Klinkenberg LJ (1941) The permeability of porous media to liquids and gases. Drilling Prod Pract. Am. Petrol. Inst. 200–213
30. K.R. Brownstein, C.E. Tarr, Importance of classical diffusion in NMR studies of water in biological cells. Phys. Rev. A **19**(6), 2446–2453 (1979)
31. L.E. Drain, The broadening of magnetic resonance lines due to field inhomogeneities in powdered samples. Proc. Phys. Soc. **80**(6), 1380 (1962)
32. G.Q. Zhang, G.J. Hirasaki, Internal field gradients in porous media. Petrophys. Houston **44**, 422–434 (2003)
33. L.J. Zielinski, P.N. Sen, Combined effects of diffusion, nonuniform-gradient magnetic fields, and restriction on an arbitrary coherence pathway. J. Chem. Phys. **119**(2), 1093–1104 (2003)
34. H. Rueslåtten, NMR studies of an iron-rich sanstone oil reservoir. The Society of Core Analysts **SCA 9821** (1998)
35. W.H. Press, *Numerical Recipes in FORTRAN Example Book: The Art of Scientific Computing* (Cambridge University Press, Cambridge, 1992)
36. G.H. Sørland et al., Exploring the separate NMR responses from crude oil and water in rock cores. Appl. Magn. Reson. **26**(3), 417–425 (2004)
37. J.G. Seland et al., Combining PFG and CPMG NMR measurements for separate characterization of oil and water simultaneously present in a heterogeneous system. Appl. Magn. Reson. **24**(1), 41–53 (2003)
38. M.D. Hürlimann, L. Venkataramanan, C. Flaum, The diffusion–spin relaxation time distribution function as an experimental probe to characterize fluid mixtures in porous media. J. Chem. Phys. **117**(22), 10223–10232 (2002)
39. E. Amott, Observations relating to the wettability of porous rock. Trans AIME **216**, 156–162 (1959)

Chapter 9
PFG NMR Applied to Biological Systems

Abstract The application of PFG NMR in characterizing biological systems may be used for the quality control of fat content in different product as meat, fish, fish and animal feed and its ingredients. In the first two sections of this chapter, focus will be on describing the quality control methods for determining total lipid content, and compare the results achieved against chemical extraction methods. The rest of the chapter will be devoted to other methods/applications for the characterization of various biological systems such as salmon, cheese, fish feed, margarine, and the tendon in a shoulder.

The application of PFG NMR in characterizing biological systems is one of the few areas where NMR equipment is located in an environment without the need for key personnel with specific knowledge on PFG NMR. For more than 15 years, instrumentation has been used for the quality control of fat content in different product as meat, fish, fish and animal feed and its ingredients. The control of the NMR instrumentation is fully automated, and the calibration is done on a 100 % oil sample. In the first two sections of this chapter, focus will be on describing the quality control methods for determining total lipid/fat content. The rest of the chapter will be devoted to other methods/applications for the characterization of various biological materials such as salmon, cheese, fish feed, margarine, and the tendon in a shoulder.

9.1 Quantification of Fat in Biological Systems

In spite of the innovative use of Near Infrared Reflectance (NIR) in food analysis, the food industry will still benefit from the development of fast methods that do not require time consuming, expensive and frequent calibration. Low field NMR (LF-NMR) is a fast and accurate alternative to NIR when the aim is to determine the content of water and fat simultaneously. LF-NMR may be used to analyse most food (i.e. fish, meat, dairy) products along the production process from the raw material until the finished product.

G.H. Sørland, *Dynamic Pulsed-Field-Gradient NMR*,
Springer Series in Chemical Physics 110, DOI 10.1007/978-3-662-44500-6_9

The reference methods are still based upon the use of acid hydrolysis and solvent extraction to obtain the total amount of fat. Examples are the Soxhlet [1, 2] and the Scmid-Bonzynsky-Ratslaff (SBR) [3] methods. These are examples of slow and laborious methods, which are dependent upon long experienced and properly trained personnel. Many of the routine methods used in food laboratories today are still performed by the use of hazardous solvents. Even if they are relatively closed systems, there always exists a risk of personnel being exposed to chemicals. An example of this category is the Foss-let fat analyser [4] that was introduced in 1973. The method uses a hazardous chlorinated solvent that is absorbed through the skin and requires highly skilled personal to make an accurate and safe analysis of fat in the meat. This instrument is no longer in production, but there certainly are many remaining in the food laboratories throughout the world.

One of the latest improvements to wet chemical methods was made by Büchi, the Büchi B-820 [5], which introduced a semi-automated dedicated gas chromatography(GC)-system to analyse the total amount of triglycerides. However, the system still makes use of some chemicals during the wet preparation of the samples before the fat analysis. Another important aspect is to really know what is quantified; raw fat or total fat? The accuracy of the result depends on the level of extractability by the specific solvent or combination of solvents used. With the LF-NMR, a specific proton signal is achieved relating to the absolute amount of hydrogen atoms proportional to the contained fat (lipids) in the sample. It is presumed that the phospholipid component that must be decomposed by acid hydrolysis using the extraction methods can be determined by use of LF-NMR. With the accuracy of the LF-NMR method for determination of the fat content in a sample, it can be used as a reference method for the calibration of other fast methods like NIR. The development of LF-NMR methods, that do not use chemicals, means great progress for the analytical chemist with respect to safety and simplicity of performing the fat quantification. The use of NMR at low magnetic field means no safety risk for the personnel operating the instrument. Training personnel to operate a LF-NMR and maintaining the instrument calibrated is a low cost task.

Smart Trac is an AOAC official method that applies a LF-NMR instrument and a microwave dryer to dry the sample prior to the NMR-measurement [6]. Another successful NMR method is the determination of solid fat content [7]. These NMR-methods are very simple methods to apply, but may not provide the total fat content due to the presence of high fractions of water and/or phospholipids. The first method only applies a single spin echo signal at a short observation time and does not discriminate between different types of fat, or whether there is any water present in the sample. The second method measures the FID only, and assigns the fast attenuating part to the solid fat, which just as well could have been a contribution from phospholipid signal [8]. Question then is whether the phospholipids should be defined as solid fat.

In the following sections, two different approaches for determining fat content in biological material will be discussed. These two methods have been developed due to the requirement for proper quality control of the fat content in various biological systems, in a user-friendly environment and without the availability of specialist NMR-skilled personnel.

9.1.1 The Oneshot Method

When assuming a Gaussian diffusion propagator and mono exponential attenuation of the NMR signal due to relaxation processes, the attenuation of the NMR signal is written [9]

$$I = I_0 e^{-\frac{t_1}{T_2}} e^{-\frac{t_2}{T_1}} e^{-\gamma^2 D \int_0^t \left(\int_0^{t'} g(t'')dt''\right)^2 dt'} \tag{9.1}$$

where

t_1	duration the NMR-signal is influenced by transverse relaxation processes
t_2	duration the NMR-signal is influenced by longitudinal relaxation processes
$g(t'')$	total magnetic field gradient, external and internal
D	diffusion coefficient
T_1	Characteristic longitudinal relaxation time
T_2	Characteristic transverse relaxation time
I_0	Initial intensity of the NMR signal

To resolve the fat signal from the water signal, a multi-pulsed magnetic field gradient spin echo experiment (m-PFGSE) is applied as shown in Fig. 9.1. As an extrapolation back to zero observation time will be conducted, it becomes unnecessary to perform an extra correction for longitudinal or transverse relaxation processes. In addition, the uncertainty due to eddy current transients is minimised as the same gradient strength is applied throughout the experiment. Otherwise, the eddy current transients that follow the onset and offset of the magnetic field gradient pulses, may cause unwanted dephasing of the NMR signal.

The echo attenuation for the m-PFGSE-sequence in Fig. 9.1 is written

$$I = I_0 e^{-n\bullet[\frac{2\tau}{T_2} - \frac{2\tau^3}{3}\gamma^2 G_i^2 D]} e^{-n\bullet[\gamma^2 g^2 D \delta^2 (\tau - \frac{\delta}{3})]} \tag{9.2}$$

G_i is the internal magnetic field gradient caused by changes in magnetic susceptibilities throughout the sample, g is the externally applied magnetic field gradient, δ is the gradient pulse length, τ is the time interval between 90° RF pulse and 180° RF pulse, γ is the gyromagnetic ratio, and n is the echo number (n = 1, 2, 3, etc.).

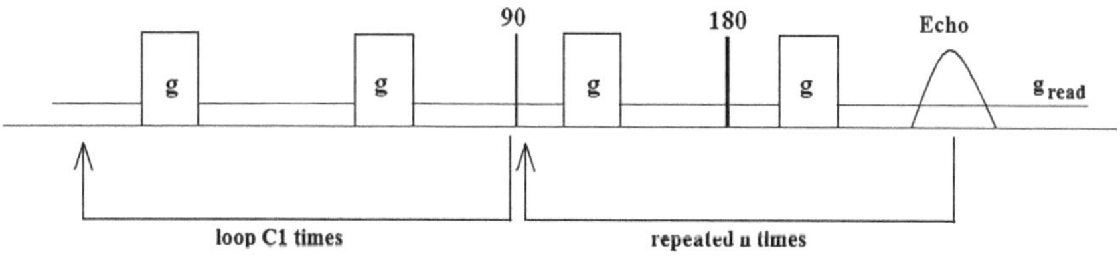

Fig. 9.1 The oneshot (m-PFGSE) sequence including preparatory gradients for saturation of the eddy current transients and a small read gradient throughout the experiment to focus the echo in the correct position

The contribution of the small read gradient, g_{read}, as shown in Fig. 9.1 is assumed to have negligible effect on the echo attenuation. By defining the unknown parameter

$$K = \frac{2\tau}{T_2} + \frac{2\tau^3}{3}\gamma^2 G_i^2 D + \gamma^2 g^2 D \delta^2 (\tau - \frac{\delta}{3}) \quad (9.3)$$

The attenuation simplifies to

$$I = I_0 e^{-n \bullet K} \quad (9.4)$$

Terms including relaxation and diffusion in the presence of internal and applied magnetic field gradients are then collected into one unknown, **K**.

To separate between the NMR signal from fat/oil and the other components such as water/protein, it is made use of the significant difference in mobility and transverse relaxation time. Fat has a significantly lower diffusivity than water, the order of 10^{-11} m²/s versus 10^{-9} m²/s at 25 °C. By fitting the applied field gradient pulse such that the water signal is suppressed during the first few echoes, the m-PFGSE experiment can be used to quantify the fat (oil) directly. Due to the short transverse relaxation times (≤ 1 ms [10]) of protein, its NMR signal will not contribute as the first measuring point ($n = 1$) in the m-PFGSE experiment is at approximately 5 ms or later. When applying strong enough gradients to suppress the water signal at $n = 2$, the attenuation is written as

$$I = I_{fat} e^{-n \bullet K_{fat}} \quad (9.5)$$

$\mathbf{I_{fat}}$ corresponds to the initial fat signal unaffected by relaxation and diffusion and $\mathbf{K_{fat}}$ includes the relaxation and diffusion terms of the fat component. A weighted linear fit [11] of the logarithm of (9.5) to the function

$$y = -ax + c \quad (9.6)$$

yields a value for c where

$$I_{fat} = e^c \quad (9.7)$$

By weighting the fit one takes into account that the model in (9.5) is not valid at all times, as the fat may exhibit a distribution of T_2 relaxation times. It may still be assumed to have a mono exponential decay at the shortest observation times, which are defined as the observation times that are shorter than the shortest T_2 value present in the T_2 distribution. As the observation time approaches the first measurement point at time equal to $2\tau(n \to 1)$, the validity of this assumption improves. The first measurement points are therefore given more weight than the last ones in the decay. Using this fitting procedure, corrections for diffusion and relaxation effects are made in the NMR signal, and the signal becomes a quantitative measurement of the content of fat or oil in the sample.

To measure the water content as well, the same experiment can be applied without pulsed field gradients, i.e. an ordinary CPMG-sequence [12]. Fitting the experimental data from the CPMG yields an initial NMR signal that contains the signal from fat and water, and a subtraction of the fat signal given by (9.7) results in a value proportional to the water content in the sample. However, it is important to note that this simple procedure for quantification of water does not apply when any molecule dissolved in the aqueous phase to enough concentration, as for example carbohydrates, has been added to the system. Carbohydrates dissolved in water will interfere with the water signal, as the diffusion and relaxation times will be of the same order of magnitude, and the initial NMR signal then contains signal from fat, water and carbohydrates dissolved in water.

9.1.1.1 Verification

The m-PFGSE method proposed in the previous section aims at measuring fat content in systems containing significant amounts of water. A typical result from the m-PFGSE experiment is shown in Fig. 9.2, where the gradient strength and duration of the gradient pulses were adjusted such that the water signal was suppressed to an insignificant level at the time of appearance of the first echo (n = 1). When extracting the intensities of the echo peaks, and fitting these as a function of echo number, the fitted initial intensity at zero observation time, corresponds to the total fat signal. Effects from relaxation, diffusion, and eddy current field transients are corrected for when employing this approach. The calibration of the method involves applying it to a sample of 100 % oil and of known weight. This provides a calibration value in arbitrary units, NMR-signal per gram of oil. When an unknown sample of known weight is measured, the NMR signal from this sample

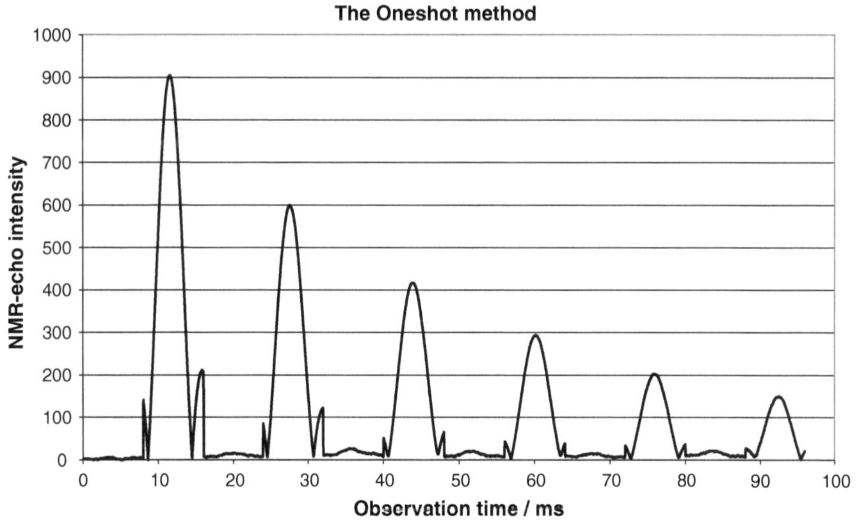

Fig. 9.2 The m-PFGSE performed on a sample containing fat. Only the recording of the echoes is shown in the graph (each echo is recorded during 0.25 ms), so the time axis only provides an approximate time when the echoes are being recorded

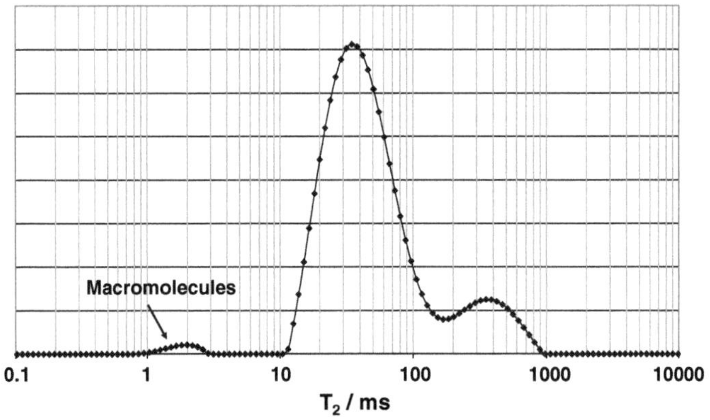

Fig. 9.3 T_2-distribution of a sample containing minced pork meat

is to be divided by the calibration value and the weight of the sample. The result is then the fraction of fat in the unknown sample.

42 samples of minced beef meat used in the verification consisted of deep-frozen samples that were measured by m-PFGSE and Foss-let methods [13] (results presented in Sect. 9.1.1.2). To check for any signal contribution from protein in the protein-water interface, a CPMG echo train was recorded and subjected to the 1D-ILT routine. The resulting T_2 distribution is shown in Fig. 9.3, where a small peak arising at approximately 1.5 ms is found. This signal cannot arise from free water as it is mobile in the biological material, which results in T_2 values significantly higher than 1 ms. The exchange between free and bound water is much faster than the duration of the τ-value in the CPMG-experiment [10], which supports the picture of a relatively long T_2 for the water. As long as the fat is in a liquid state, which is ensured by measuring at high temperatures (>40 °C), the peak at 1.5 ms cannot arise from fat either, leaving the protein or other macromolecules as most likely contributors to the peak signal. Thus, to quantify the fat and water content without any interference from other macromolecules as protein, the recording of the m-PFGSE or CPMG experiment should be started at an observation time of at least 5 ms. The signal arising from the macromolecules is then reduced to an insignificant amount, both when quantifying the fat content (m-PFGSE) and the water content (CPMG) (then assuming that the macromolecules are not water-soluble).

As the NMR method is a non-invasive method, the samples can be measured several times. It is therefore possible to adjust the experimental parameters until the desired accuracy is achieved. Figure 9.4 shows a recording of 100 measurements on a 100 % oil sample, where the standard deviation is 0.25, and during a period of 4 months, this calibration value did not change significantly. The standard deviation for 100 measurements was measured every second week, and all values for the standard deviation were within 0.35 %. Within this period of time, the experimental parameters were not changed, and the gradient probe was detached and attached to the NMR instrumentation several times. The standard deviation

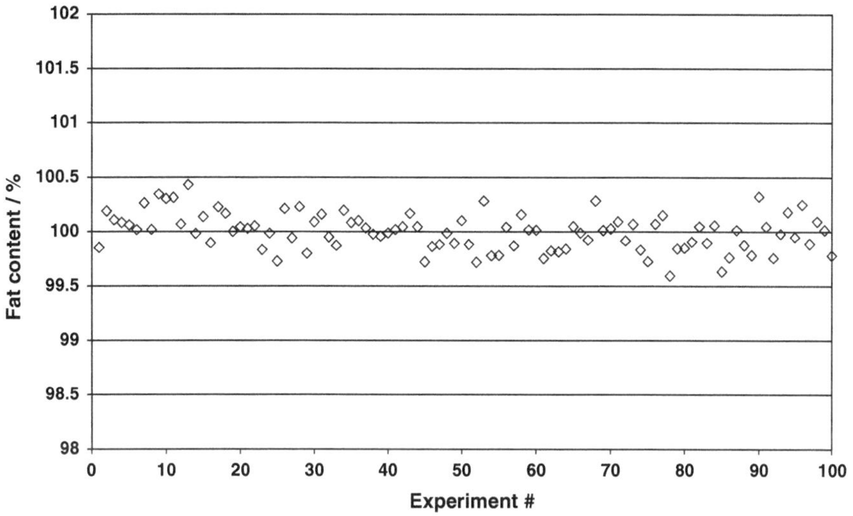

Fig. 9.4 100 measurements on a 100 % oil sample using the m-PFGSE method. The standard deviation found for the measurements is 0.26 %

does not show an increase over this period, and this reflects the stability of the m-PFGSE experiment; as long as the proton density of the calibration sample remains constant, the calibration value will also remain constant.

The calibration as indicated in Fig. 9.4 is required in order to measure the fat content of an unknown sample in the region 0.1–100.0 %. This method makes the basic assumption that the fat proton density does not change significantly from sample to sample. To investigate this effect on the calibration value the fat proton density of different types of fish oil extracted from wild salmon, bred salmon, herring, and mackerel have been measured. For the different fish oils, the proton density remained constant within a relative error of ± 1.0 %. The uncertainty for a sample of fat content of 10.0 % will then be ± 0.1 % due to a possible variation in proton density. However, as the proton density of fat from pork is found to be 8 % higher than of the fish oil, a systematic error would have been introduced if fish oil was used as a reference oil when measuring fat in samples of pork meat. It is therefore crucial to calibrate the NMR system with the same type of oil as is found in the samples that are to be measured.

As the m-PFGSE method corrects for systematic errors such as loss of signal due to relaxation, convection, and eddy current transients, probably the stability of the mains is the most critical factor for the performance of the method. Any instability on the mains may introduce extra and unwanted electrical currents looping through the gradient coils (ripple currents, see Chaps. 3 and 4). This will generate an unwanted and random dephasing of the NMR echo signal that increases as the duration of the NMR-experiment is increased. Through a sophisticated, yet simple, experimental setup, including preparatory and read gradients, the duration of the NMR experiment is reduced such that the degree of random dephasing of the NMR echo signal does not vary the measured fat content significantly. This is verified in Fig. 9.4, where 100 repeated measurements of 100 % fat yielded a standard deviation of ±0.26.

It should also be noted that the decays from the m-PFGSE method were examined with and without weighting on the linear fit to obtain the NMR signal arising from fat. Within the noise in the experiment, there was no difference between the two methods for fitting the experimental data. This confirms the validity of the assumptions of an approximately mono exponential decay at the shortest observation times indicated in Fig. 9.2 to be valid up to approximately 100 ms when the measurement temperature is 40 °C.

In order to verify that the m-PFGSE method for fat determination in water containing systems measures the fat content correctly, it was tested on a finely minced reference material made by Swedish Meats R&D (SMRD2000). The reference sample contained lean pork, water, potato flour, and nitrite salt, and the reference fat value is 14.3 % with ±0.4 % as expanded uncertainty. Reference methods used were Soxhlet [1, 2], Weibull Stoldt [14, 15], and SBR [3]. The reference value for water is 68.8 % ± 0.1 %, when drying the sample in an oven at 100–110 °C as the reference method. 4 samples containing 3.5 g of the meat paste were measured, and the fat content was found to be 14.3, 14.4, 14.6 and 14.7 % at a measurement temperature of 40 °C. The water content for one sample was found to be 68.7 %, which is in accordance with the reference value. This sample was dried for a period of time at 104 °C and the water content on the remaining sample was measured again by the m-PFGSE and CPMG methods. This procedure was repeated several times, and the result of measured water content as a function of loss of weight is shown in Fig. 9.5. Evidently, there is a linear relationship between the measured water content and the loss of weight. From the results achieved with the reference system SMRD2000, it can be concluded that the method for fat and water determination does separate the two components satisfactorily. When repeating the measurement 50 times as shown in Fig. 9.6, the standard deviation is found to be less than 0.1 %.

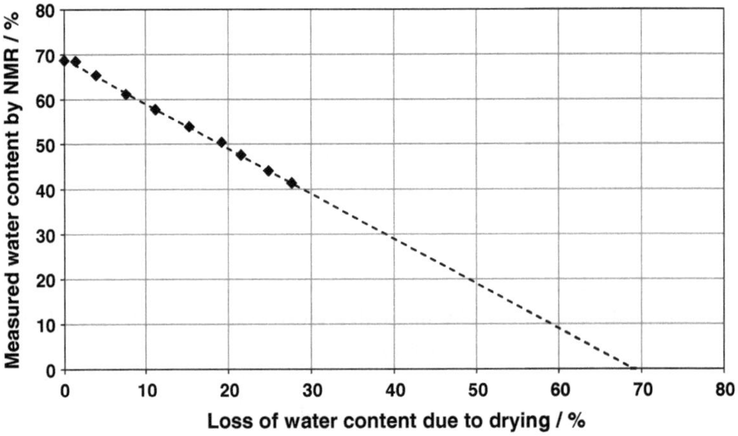

Fig. 9.5 Measurement of water in SMRD2000 as a function of loss of mass due to drying

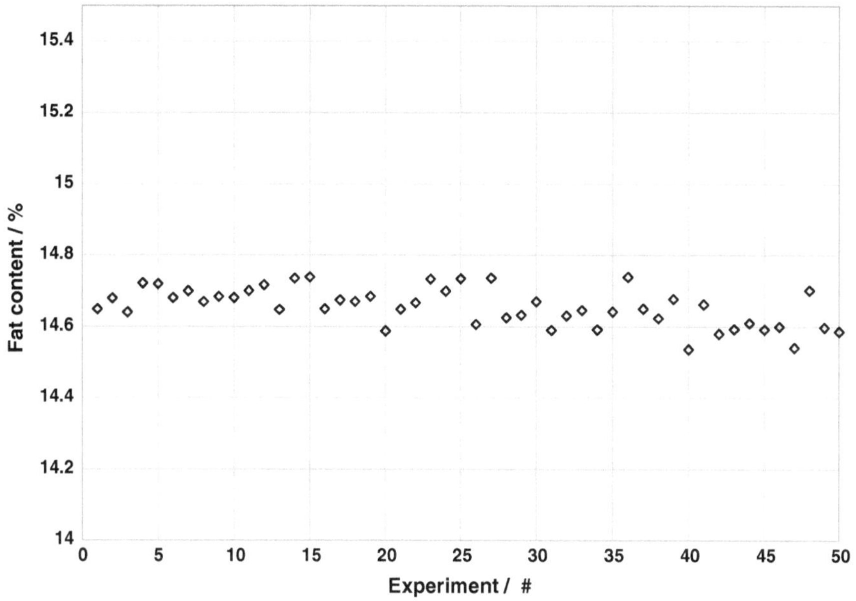

Fig. 9.6 50 repeated fat determinations in SMRD2000. The standard deviation is less than 0.1

9.1.1.2 The m-PFGSE Method Applied on Meat Samples

Figure 9.7 shows the fat content for the 42 samples of minced beef meat comparing two approaches, the non-invasive m-PFGSE method and the Foss-let chemical extraction method [13]. For the m-PFGSE method, the recording of the fat signal starts when the water signal is suppressed to an insignificant amount using the m-PFGSE sequence, and when the protein signal has decayed due to a short T_2 relaxation time. Comparing the m-PFGSE method against the Foss-let method provides a reasonably good correlation, as it is 0.975 when using Partial Least Squares Regression (PLS-1) [16]. When performing an ordinary linear regression without any constraints on where the fitted regression shall cross the axes, it do not cross at the centre, (0, 0 %). Instead, it indicates that there is a tendency for the m-PFGSE method to measure slightly higher fat content than the extraction method.

9.1.1.3 The m-PFGSE Method Applied on Fish Samples

As for the meat samples the mobility of fat found in fish is significantly lower than the mobility of the water. Thus the water signal may be suppressed to an insignificant amount while the initial fat signal can be found from fitting the intensity of the echo peaks containing fat signal only, to a single exponential decay as shown in (9.4). The fresh fish samples (wild salmon, bred salmon, mackerel and herring) were homogenised using a commercial food mincer before 3.5 g of each sample

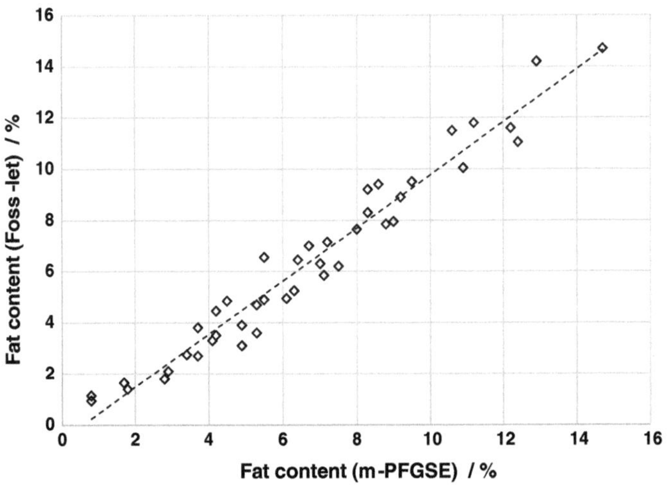

Fig. 9.7 Measured fat content in minced beef meat by NMR fresh, by NMR dried, and by Foss-let extraction

Table 9.1 Comparison between determined fat content from the m-PFGSE method and the ethyl acetate chemical extraction method

Fat content/%	m-PFGSE	Chemical extraction
Wild salmon	5.5	5.3
Bred salmon	11.1	10.9
Mackerel	30.0	29.8
Herring	17.2	16.9

was placed in the NMR tube. Numerous measurements from the same batch indicated that the sample preparation was sufficient to achieve proper homogenisation. To suppress the water signal 1.5 ms gradient pulses of 200 Gauss/cm were applied, and the first echo recorded was echo number 4. To avoid unwanted loss of echo signal due to convection caused by possible temperature gradients and/or vibration of the gradient coils when switching the gradients on and off, only even echoes were recorded and used when finding the initial fat signal [17, 18]. Table 9.1 shows results achieved on a set of different fish samples using the m-PFGSE method and compared against a standard chemical extraction method using Ethyl Acetate [19].

Even though the results are identical within the experimental noise, which is found to be ± 0.2 % in absolute units for both techniques, there seems to be a systematic difference between the results achieved with the m-PFGSE technique and by the use of chemical extraction. Such an offset is also indicated in the fat measurements of meat in the previous section, and could be explained by a small amount of fat remaining in the tissue from which the chemical solvents did not extract the fat completely. To confirm the existence of this offset, homogenised filet of salmon, from which the fat should have been extracted by the use

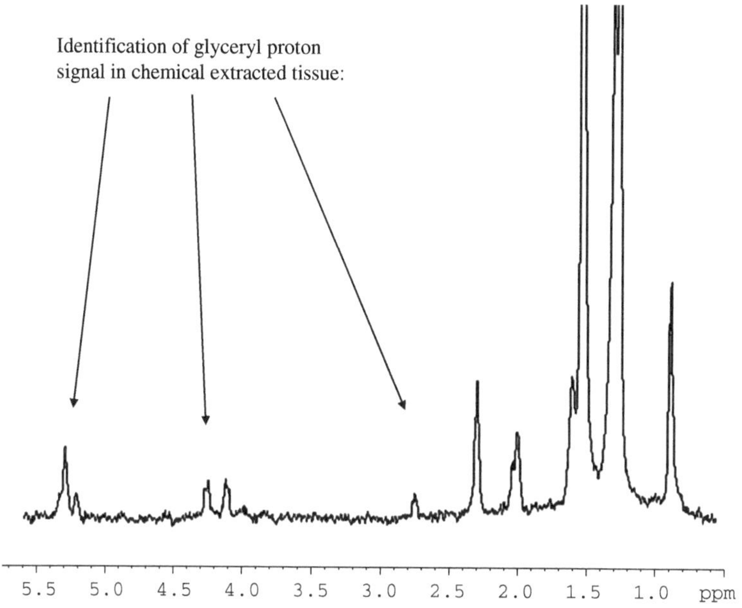

Fig. 9.8 Identification of fat in chemical extracted tissue from which the fat should have been removed

of chemical solvents was analysed in a 14.1 T high resolution NMR spectrometer. The result shown in Fig. 9.8, confirms that there are significant amounts of fat remaining. The high resolution NMR spectrum originates from the extracted salmon tissue dissolved in deuterated chloroform, and shows the characteristic chemical shifts of triglycerides. Typical signals are the glyceryl proton signals at 5.26 ppm (-CH_2-CH-CH_2), 4.29 ppm (-CH_2-CH-CH_2-) and 4.14 ppm (-CH_2-CH-CH_2) [20, 21]. Other typical fat signals occur at 5.34 ppm and in the region 2.90–0.80 ppm [20]. Using the CPMG method the amount of remaining fat within the tissue is found to be of the order of 0.1 % and may thus explain the discrepancy between the m-PFGSE method and chemical extraction methods.

These results show that the m-PFGSE compares well with solvent extraction techniques. There is also an indication that there is a tendency to measure slightly higher fat content using the NMR approach compared to the extraction methods. This is verified by high resolution NMR analysis on biological tissue from which the fat should have been extracted. While the extraction methods rely on the possibility of extracting the fat from the biological material, the NMR approach is non-invasive and measures the total fat. In a laboratory situation, the NMR-approach is the most user-friendly method, but the initial cost of investment in the NMR instrumentation has been and still is the major obstacle for having the NMR instrumentation to replace chemical extraction equipment. Nevertheless, at Skretting Aquaculture Research Centre in Stavanger, Norway, PFG NMR instrumentation has been operating since October 2002. It measures fat in fish, fish feed and its ingredients. Ståle Eilertsen, former scientific responsible chemist at Skretting ARC commented in 2004: "*The m-PFGSE returns high quality results all the way. At the least it seems to produce just as good results as for example Acid Hydrolysis. In addition it saves the environment by not using any form of solvents; during 100 measurements using the m-PFGSE, we save* 6 L *of solvent and* 10 L *of strong acid. This together with a less cumbersome measuring process reduces the cost per analysis significantly. By offering the fat determination from m-PFGSE to almost half the price as compared to Acid Hydrolysis, it will not only benefit us but our customers as well.*"

9.1.2 The NMR Dried Method

A widely used method for dried or partially dried samples is based on the fat being the only component with a relatively long transverse relaxation time [6]. The recording of one spin echo signal at an observation time of approximately 0.3 ms is then assumed to represent the fat signal from the sample. Any potential fat signal with short T_2 would not be completely accounted for. The m-PFGSE as described in the previous section, suppresses the water to an insignificant amount at the first echo while the attenuation of the fat signal due to diffusion and relaxation effects is recorded in a oneshot sequence. The attenuation of the fat signal is then fitted to an exponential decay resulting in a fitted initial fat signal where loss of signal due to diffusion and relaxation effects has been corrected for.

While both methods seem to work well on initially wet samples there is a major group of systems were both methods fail to determine the total fat content. On powdered samples where the water content is small (~10 %) and the phospholipid content is relatively high, a fat component with short T_2 is not completely accounted for by either of the methods. The spin echo method fails because significant amounts of the signal from the phospholipids have already decayed at the first measuring point. Consequently, the measured fat content will be underestimated. The m-PFGSE method fails because the first spin echo is recorded at approximately 10 ms to avoid interference from protein signal from the interface between protein and water [10], and as it takes a finite duration to have gradient pulses incorporated into the spin echo sequence. At an observation time of 10 ms the phospholipid from the powdered sample with small water content has decayed to an insignificant amount, and an extrapolation back to zero observation time will not recapture the initial amount of phospholipids.

9.1.2.1 Resolving Fat in Systems of Low Water Content

A proper quantification of the total fat signal is possible when the fat signal can be isolated from the components as protein, carbohydrates and water. Without chemical shift information using low resolution NMR equipment this can be achieved by making use of significant differences in the dynamic behaviour of the components, as relaxation times. To resolve the fat signal from the other components producing a proton NMR signal, the NMR spin echo technique can be used. In general, the attenuation in a CPMG echo train is written

$$I = \sum_k \sum_j I_{j,k} e^{-\frac{2n\tau}{T_2^{j,k}} - \frac{2n\tau^3}{3}\gamma^2 G_i^2 D_{j,k}} \tag{9.8}$$

Σ_k is the sum over the components present in the sample while Σ_j reflects the multi exponential behaviour within each component, $\boldsymbol{I_{j,k}}$ is the initial intensity, $\boldsymbol{D_{j,k}}$ is the diffusion coefficient, $\boldsymbol{T_2^{j,k}}$ is the transverse relaxation time, $\boldsymbol{G_i}$ is the strength of the internal magnetic field gradient, $\boldsymbol{\tau}$ is the time interval between 90° RF pulse and 180° RF pulse, and **n** is the echo number (**n** = 1, 2, 3, etc.).

In powdered samples there is typically 1–15 % water content present. With an overlap in transverse relaxation times (T_2) between water bound to the solid matrix and possible amounts of fat with short T_2's (bound fat), it is difficult to isolate the bound fat signal from the water. To isolate the total fat signal from any other component it is therefore important to dry out the approximately 10 % of water prior to the NMR measurements. When drying the water out of the sample the protein and carbohydrate signals becomes solid-like with a T_2 of the order of 10–20 μs. If the first echo in the CPMG echo train is recorded at more than 100 us, the signal from these components will then have decayed to an insignificant amount. However, as the bound fat also exhibits a relatively short T_2, it is important to start the recording of the CPMG echo train at the shortest observation time possible and record as many points as possible during the first few milliseconds. As will be verified in the following sections it is possible to resolve the fat components from the other components within a dried sample by applying a τ-value of 40 μs and start the recording of the CPMG train at echo number 2 which occurs at an observation time of 160 μs. With a 180° RF pulse duration of 9 μs, the protein and carbohydrate signal do not contribute to the echo signal at 160 μs even though there are $T_{1\rho}$ effects at such short τ-values [22].

Provided the protein and carbohydrate NMR signals have decayed and the water is dried out, the attenuation in (9.8) is simplified to

$$I = I_{bound\_fat} e^{-\frac{2n\tau}{T_2^{bound\_fat}}} + I_{free\_fat} e^{-\frac{2n\tau}{T_2^{free\_fat}}} \tag{9.9}$$

I_{bound_fat} corresponds to the fat fraction with relatively short T_2, $T_2^{bound\_fat}$, and I_{free_fat} corresponds to the fat fraction with relatively long T_2, $T_2^{free\_fat}$. The fat signal is then resolved and total fat content can be found from a two exponential fit of (9.9) as a function of echo number **n.** Even though the experiment in principle is straight forward, the successful resolution of fat signal is dependent on a careful setting of the experimental parameters to start the recording of the echo train as early as possible without any interference from other macromolecules exhibiting proton signal.

The proposed method is tested on various types of fish feed, and marine powders such as cod, salmon and crabs. The water was dried out from the samples prior to the NMR measurements; in an electric oven, operating at 104 °C.

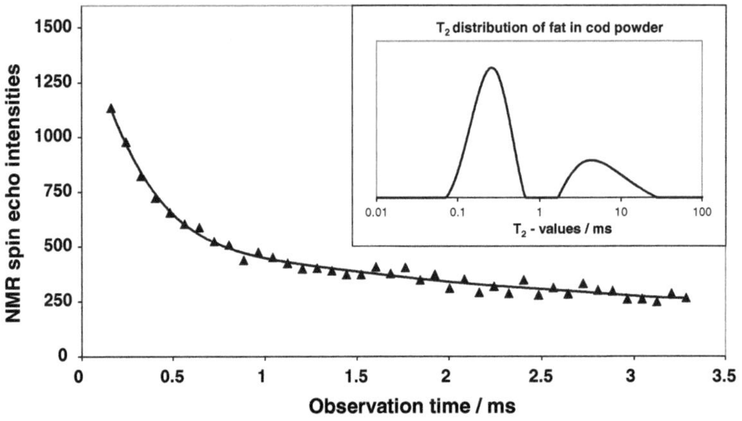

Fig. 9.9 The CPMG attenuation from dried cod powder and its corresponding T_2 distribution

The determined fat contents by NMR were compared in the following sections against results achieved from approved standard chemical extraction techniques as Soxhlet [1, 2], Ethyl Acetate [19], and Acid Hydrolysis [23].

9.1.2.2 Fat Determination in Marine Powders and Fish Feed

When the water is dried out, the mobility of the part of the fat, which is coupled to the solid matrix, is significantly reduced. As most of phospholipids are found in the cell membranes, it leads to a reduction in their transverse relaxation times when the tissue is subjected to drying. In Fig. 9.9, the initial CPMG attenuation from cod powder is shown together with the resulting T_2 distribution. The acquisition of the CPMG attenuation in this particular setup starts at 180 μs to avoid any interference of signal from protein and/or other macromolecules. The signal consist of two main components, which are easily resolved using the inverse Laplace transform analysis on the CPMG attenuation, and shown in the upper right corner of Fig. 9.9. For the cod powder, the amount of mobile fat ($T_2 > 1$ ms) is less than the bound fat ($T_2 < 1$ ms), i.e. the phospholipids are the main contributor to the fat signal in the cod powder.

This approach for determining the total fat content has been applied on 38 different samples of various marine products (powdered coalfish, crabs, mussels, salmon and a selection of powdered white fish). Together with the optimised CPMG method, Acid Hydrolysis was applied on the powdered systems for comparison. This was a part of a verification process of the NMR dried method, and the target was to prove the NMR approach to be as reliable as the extraction method. If successful, NMR would replace chemical extraction in the production facility, as the preferred method for quality control of fat content in production of marine powders. In Fig. 9.10, the NMR results are plotted against the Acid Hydrolysis results for powdered systems with fat content ranging from 3 to 30 %.

Out of 38 samples, there was only one sample showing a significant discrepancy between the NMR approach and the approved solvent extraction method.

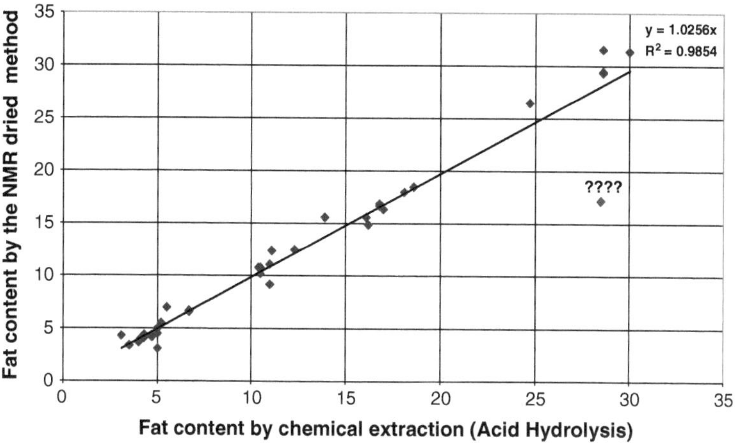

Fig. 9.10 Comparison between the NMR dried method and chemical extraction by acid hydrolysis. The fat contents were unknown prior to the NMR measurements

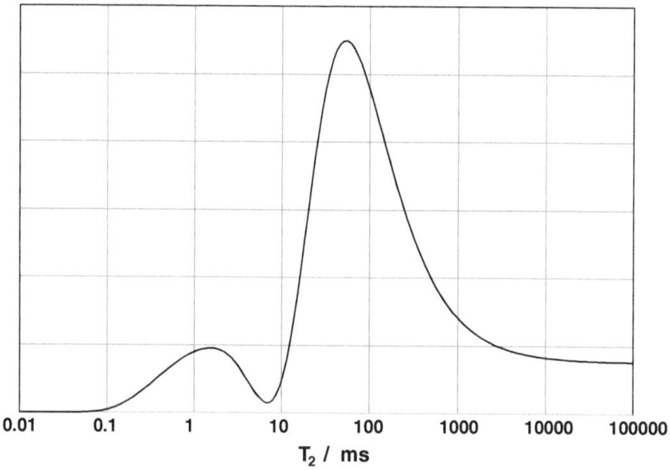

Fig. 9.11 The T_2 distribution arising from the sample where the NMR approach returns a fat content of 29.5 % while acid hydrolysis returns a fat content of 17.2 %

Several samples were measured on this powder, but the NMR approach returned 29.2 ± 0.3 % while the Acid Hydrolysis returned a value of 17.2 %. As the NMR approach was the method being evaluated, this method was questioned and not the extraction method.

Figure 9.11 shows the T_2 distribution arising from the sample and confirms that the fat consists of mainly less bound fat. Consequently, it is not a satisfactorily explanation to assign the larger fat content as measured by NMR to a macromolecule that has been misinterpreted as bound fat (left peak in Fig. 9.11). As there were no indications that the NMR dried method measures anything other than fat, it was decided to perform a pre-extraction procedure (adding ether to the sample). The measured

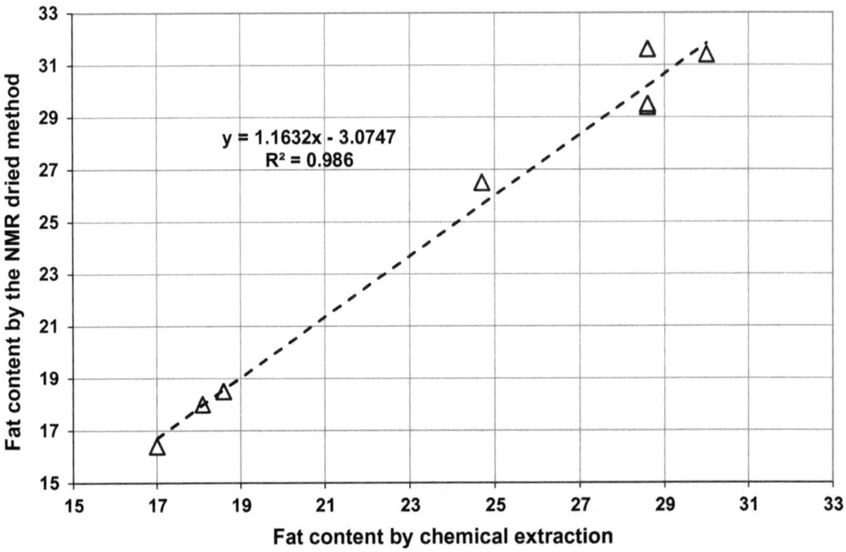

Fig. 9.12 Comparison between the NMR dried method and chemical extraction by acid hydrolysis

fat content using chemical extraction then increased to 28.5 %, i.e. 10.3 % higher than achieved using ordinary Acid Hydrolysis. For the marine powder investigated, the conclusion is that the NMR approach seems to be more reliable than the ordinary Acid Hydrolysis. The NMR dried method then replaced the solvent extraction method in this factory, and has been running satisfactorily since 2002.

Also seen in Fig. 9.10 is a slight discrepancy between NMR and Acid Hydrolysis for the powders with higher fat contents. In Fig. 9.12, the data are plotted for one type of marine powders, and the linear regression curve do not cross the junction of the axes (0 %, 0 %). The most likely cause of this discrepancy is that the NMR method measures the total fat content while the Acid Hydrolysis method does not extract the fat from the samples with higher fat content as well as from the samples with lower fat content. This is confirmed when a double chemical extraction is performed on the marine powders, as the measured fat content by chemical extraction then increases and there is no longer a significant discrepancy between those and the results achieved with the NMR dried method.

9.1.2.3 Fat Determination in Animal Feed, Bird Feed and Its Ingredients

In Fig. 9.13 a plot is shown where a variety of animal feed, bird feed and their ingredients have been measured by the NMR dried method and compared against chemical solvent extraction (Weibull Stoldt [14, 15]). The same calibration is applied for all samples, i.e. a constant hydrogen index is assumed for the fat

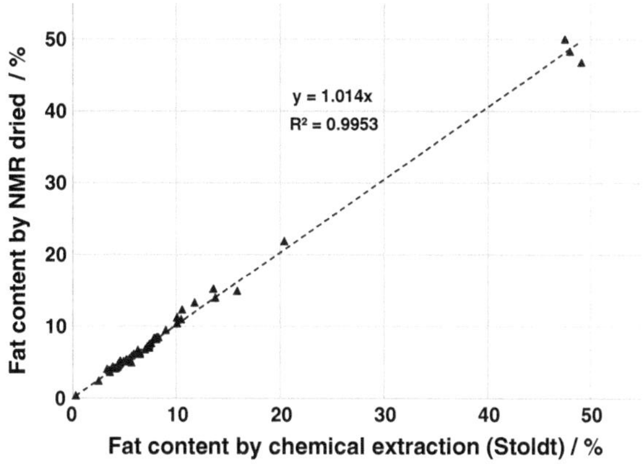

Fig. 9.13 Comparison between the NMR dried method and chemical extraction (Weibull Stoldt) on 57 samples from various types of animal feed, bird feed and their ingredients

present in all the samples. Taking into account the variety of the samples measured on, the fit is considered as quite good, and clearly demonstrates the performance of the NMR dried method.

9.1.2.4 Resolving Phospholipids Using the NMR Dried Method

As shown in Figs. 9.9 and 9.11 the transverse relaxation times can be divided into two clearly separated regions for the fat in the marine powders and other dried systems. The most obvious explanation to this is the different behaviour of phospholipids and other fat as the triglycerides when there are small or no amounts of water present in the samples. The mobility of the phospholipids is strongly reduced because they are attached to a much more rigid structure compared to a wet system where the water makes the texture less rigid. The other fat, the triglycerides, will be less affected by the low water content, as they still will be in a liquid state.

In Fig. 9.14 the high resolution spectra from cod (upper spectrum) and salmon powder (lower spectrum) are shown. The high-resolution CPMG spectra were recorded on a 14.1 T Bruker Avance DRX system. In the CPMG pulse sequence a τ value of 2 ms was employed and data were collected at 16 different echo times ranging from 8 ms to 4 s. The data were processed using the standard Bruker processing tool, xwinnmr. No zero filling and an exponential line broadening of 5 Hz was employed. The marine powders investigated at high resolution were dissolved in deuterated chloroform and analysed directly. TMS (Tetra Methyl Silane) was used as an internal chemical shift reference.

It is evident that the dominant fat in cod powder are the phospholipids. This is particularly seen from the phospholipid head group signals at 4.42 ppm (-O-CH_2-) and 3.37 ppm (-CH_3)^3-N+) and from the glycerol signals at 5.25 ppm (-CH_2-CH-CH_2) and 4.44–3.78 (-CH_2-CH-CH_2) [21, 24]. However, that the major fat content of cod powders are triglycerides is apparent from the glyceryl signals at 5.26 ppm (-CH_2-CH-CH_2), 4.29 ppm (-CH_2-CH-CH_2) and 4.14 (-CH_2-CH-CH_2).

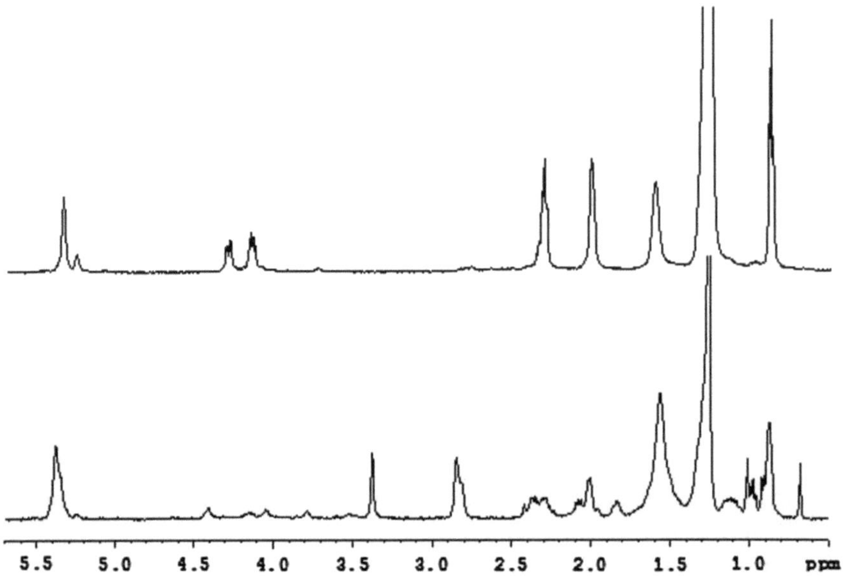

Fig. 9.14 High-resolution NMR spectra arising from cod powder (*lower spectrum*) and Salmon powder (*upper spectrum*)

From the cod powder spectrum, both unsaturated and saturated fatty acids can be seen to be bound to the phospholipids. Unsaturated fatty acids gives the following resonances: 5.44–5.28 ppm (-HC=CH-), 2.89–2.76 ppm (=CH-CH_2-CH=) and 2.07 ppm (=CH-CH_2-) [21, 25]. The signal at 0.97 ppm, in addition to the usual -CH_3 signal at 0.87 ppm, indicates a n3 unsaturated fatty acid [25–27]. n-3 unsaturated fatty acids, like DHA (docosahexaenoic acid; C22:6(n-3)) is a well-known component of fish oils and is mostly attached to the *sn*-2 position of phospholipids [28]. When comparing spectra from salmon with spectra from cod the very same phospholipid peak is seen at 1.6 ppm, but the cod fat contains a relatively larger amount of the phospholipids. This is confirmed in the T_2 distributions achieved from the low-resolution NMR instrumentation.

The high-resolution CPMG spectra on the dissolved powders show that the peaks arising from phospholipids exhibit a much shorter T_2 than the peaks arising from the other fat. Again, this is in accordance with the results achieved on the low field NMR instrumentation. It is found that the relative fraction of fat with short T_2 is 0.75 of the total fat content in the cod powder while it is only 0.05 in the salmon powder [8].

In order to verify that the components with relatively short T_2 (bound fat) correlate to the amount of phospholipids while the components with relatively long T_2 (free fat) arise from other fat, the total amount of fat (bound + free) was compared against the Acid Hydrolysis extraction method and the amount of free fat against the Soxhlet extraction method. While the Acid Hydrolysis method is assumed to extract the total fat content, the Soxhlet method has shown to extract only the part of the fat not as strongly attached to the solid matrix of the sample. The extraction

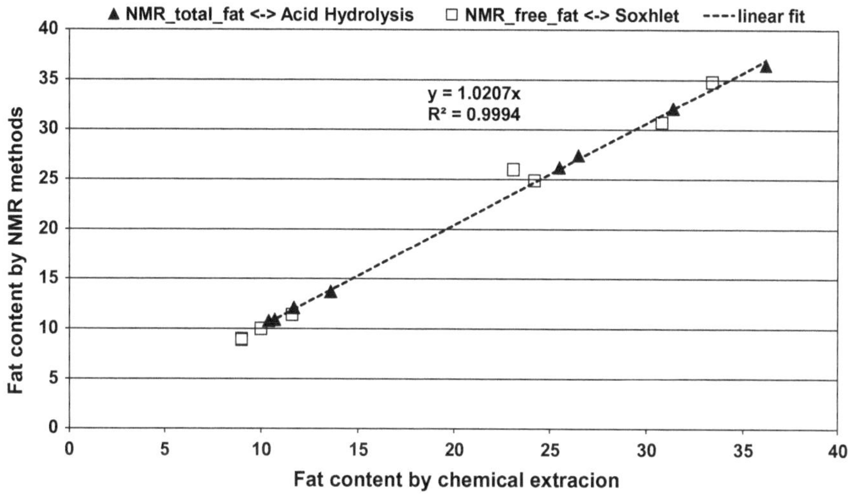

Fig. 9.15 Comparison between NMR total fat (NMR dried) and acid hydrolysis and between NMR liquid fat (m-PFGSE) and Soxhlet

results were mean values achieved from the results reported by more than ten independent laboratories participating in a round robin [29]. Figure 9.15 shows that the free fat content, as measured by the NMR dried method (only the fat component with long T_2's), correlates to the extraction method that does not extract the phospholipids (Soxhlet), while the total fat content, as measured by the NMR dried method correlates to the extraction method that do extract the phospholipids as well (Acid Hydrolysis). This strongly supports the picture of a low-resolution NMR method that is capable of measuring the correct total fat content and in resolving the amount of phospholipids from the other fat.

9.2 Two and Three Dimensional Applications on Biological Materials

The use of combined Diffusion, T_1, or T_2 measurement provides data that after processing may be used to resolve the response from different components, as fat and water, better than the ordinary one-dimensional experiments. This is a consequence of the correlation between Diffusion, T_1, or T_2 in the two dimensional data which is not apparent in the one-dimensional data. In biological material, the water and fat signals are likely to partially overlap, and appear in the T_2 range 50–500 ms. Then the resulting T_2 distribution from a CPMG-experiment will not provide any resolved information from the different components, as is demonstrated in Fig. 9.16. In a diffusion experiment, it is much easier to resolve the contribution from the fat and water, as their molecular mobility differs by at least

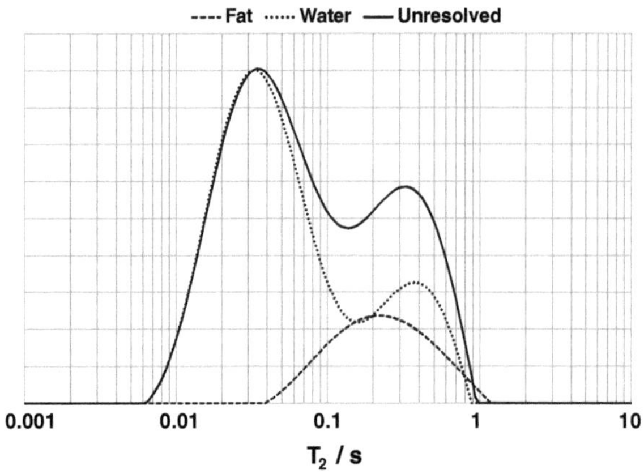

Fig. 9.16 The resolved T_2-distributions from fat and water together with the unresolved T_2-distribution arising from salmon A at 35 °C

one order of magnitude at 25 °C. When combining the PFGSE- and the CPMG-sequences, as shown in Chap. 1, a constraint is imposed on the CPMG-echo attenuations by weighting them due to the differences in the mobility of the components. Initially, at weak applied gradient strength, both components are present in the attenuation, but as the applied gradient strength is increased, increasingly more weight is on the slow diffusing component, the fat. If the gradient applied are strong enough, the water signal is suppressed to an insignificant amount, due to its high mobility, and the recording of the CPMG-echo attenuation from fat only may be conducted. The corresponding T_2 distribution is shown in Fig. 9.16. By comparing the two T_2 distributions, one with water suppression and one without, the signal from the water can be identified. To produce a resolved T_2 distribution from the water alone, one shall subtract a rescaled version of the fat-signal from the PFGSE-CPMG acquired with weak gradients [30]. Using such a procedure, a T_2 distribution of water may be found, and is also shown in Fig. 9.16.

In the following section, the method for resolving the fat and water prior to being subjected to the 2D-inverse Laplace transform will be applied and the separated 2D-ILT distributions will be compared against the 2D-ILT on the unresolved data sets. D-T_2 and T_1-T_2 processed data will be presented on various cheese samples, which will motivate the introduction of a three dimensional method that combines Diffusion, T_1 and T_2 in a combined spoiler recovery – oneshot method.

9.2.1 Separation of Fat and Water Signals Prior to Analysis

The method for separation of fat and water signal prior to the 2D-ILT analysis, initially developed for crude oil and brine located simultaneously in rock core plugs [30, 31], depends on a successful resolution of the oil and water signal. This is easier to achieve in biological systems, as the fat is well defined compared to a crude oil. The fat consist mainly of triglycerides for which the mobility is in the lower range

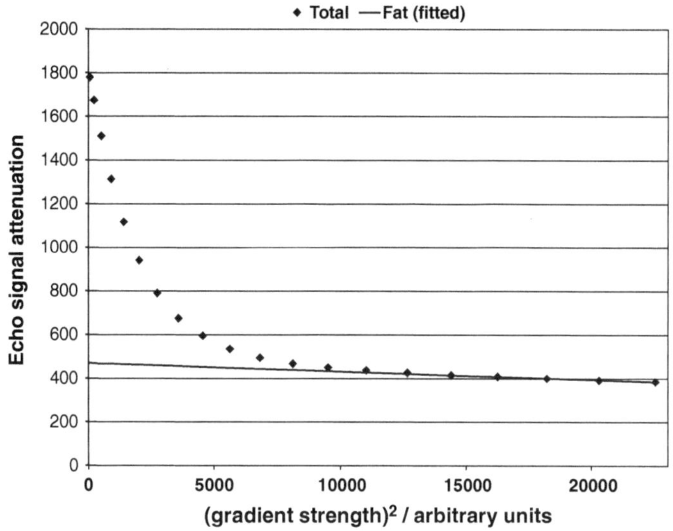

Fig. 9.17 The attenuation of the intial CPMG signal from a combined PFGSE-CPMG sequence (Chap. 1) on salmon A. The solid line represents the fitted attenuation of the fat signal

of 10^{-11} m^2/s at room temperature when the water is in the range of 10^{-9} m^2/s. A typical attenuation from the PFGSE part is shown in Fig. 9.17, where the solid line indicates the fitted value of the oil signal. The procedure for separation of the oil from the water signal is then to apply the fitted intensities for the fat signal at each gradient strength, and produce a set of rescaled CPMG-echo attenuations from oil. This is done for each gradient strength by using a CPMG-echo attenuation where the water signal is suppressed, i.e. a CPMG-echo attenuation from oil only, and rescale it to fit a CPMG-echo attenuation from oil only for the appropriate gradient strength. The scaling value for each appropriate gradient strength is found along the solid line in Fig. 9.16, and this set of CPMG-echo attenuations from oil can then be subtracted from the total CPMG-echo attenuations, to produce CPMG-echo attenuations from the water component only at the appropriate gradient strengths [31]. Finally, the separated datasets may be analysed using the 2D-ILT routine, and compared against the 2D-ILT routine performed on the unresolved data set.

In Figs. 9.18 and 9.19 the resolved and unresolved two dimensional distributions are shown for the salmon sample used to produce Figs. 9.16 and 9.17. While the water signal, which adds up to approximately 75 % of the total signal, does not change when comparing the two approaches for producing the 2D-ILT distributions, the fat signal does change significantly. In the unresolved distribution (Fig. 9.19), the fat signal is less in intensity and the T_2 values are shifted towards longer relaxation timcs. The fitted diffusion coefficients remains unchanged. When performing the same procedure on a salmon sample of less fat content, the diffusion coefficients were also shifted when comparing the two approaches. While the diffusion coefficients were found in the same region as shown in Fig. 9.18 for the resolved

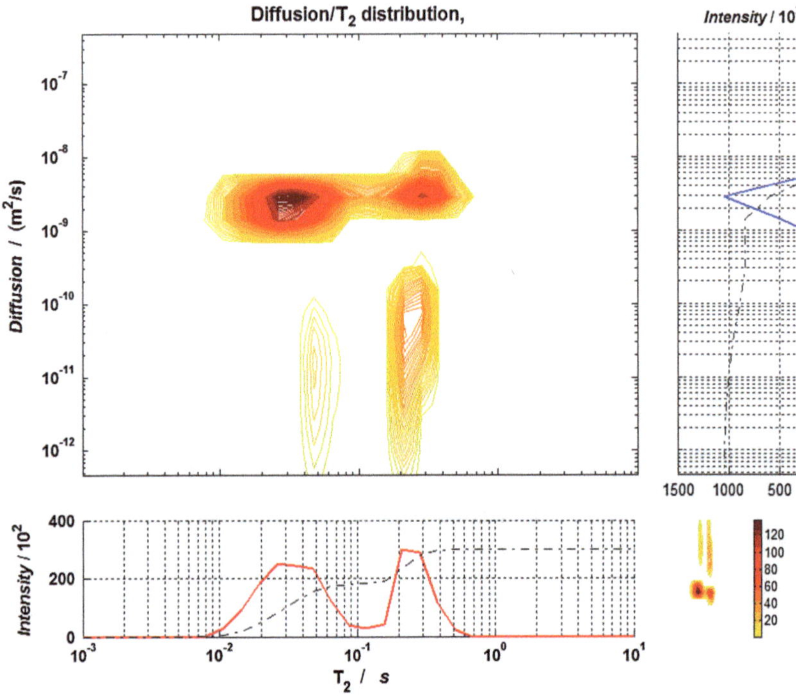

Fig. 9.18 The two-dimensional Diffusion-T_2 distributions arising from separated water and fat signals from salmon A. The distributions were combined after being subjected to 2D-ILT analysis

distributions, the diffusion coefficients from the unresolved approach were shifted to the edge of the distribution, ~10^{-12} m^2/s. In order to evaluate which method provides the most correct picture of the two dimensional distribution from the fat signal, the fat content was determined by the oneshot method [32], and the average T_2 of fat was found from the T_2-distribution shown in Fig. 9.16. The Anahess routine described in Chap. 5 was also applied on the unresolved data set for comparison, and all the results are summarized in Table 9.2 for the two salmon samples investigated.

The average T_2 value found from 1D-ILT arises from one-dimensional raw data, and provides the most accurate value for the average T_2 of fat while the oneshot method provides the most accurate value for the initial fat signal. When the fat and water signal are not separated prior to the 2D-ILT analysis ("Unresolved" in Table 9.2), the average T_2 values are not far away from the values found from 1D-ILT. However, the fitted fat signal is significantly less than the values found using the oneshot method. When the fat signal is resolved from the water signal prior to the 2D-ILT analysis ("Resolved" in Table 9.2), the fitted fat signal is much closer to the values achieved using the oneshot method. However, the T_2 for the sample with the lowest fat content is shifted to a much shorter value than the one achieved from 1D-ILT. The average values found from Anahess do fit reasonably well for the average T_2-values, or at least as well as any of the 2D-ILT results. In contradiction to the 2D-ILT results

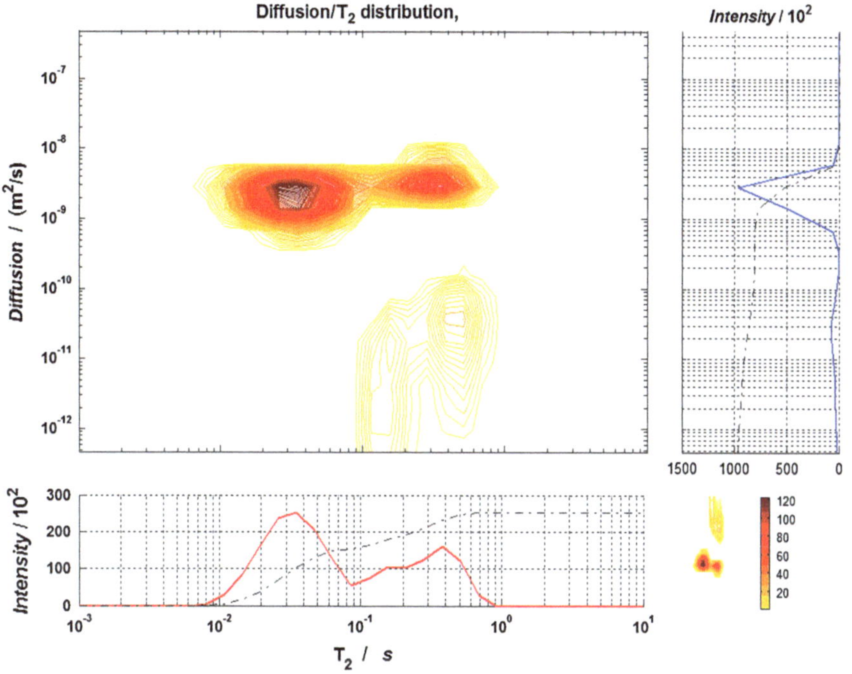

Fig. 9.19 The two-dimensional diffusion-T_2 distribution arising from Salmon A without any separation prior to the 2D-ILT analysis

Table 9.2 Summary of the average T_2 values and intensities for the fat component found from different approaches for analysing the raw data

	Unresolved	Resolved	Anahess	1-DILT	Oneshot
Salmon_A					
Average T_2/ms	320.0	250.1	264.3	**285.1**	–
Fat signal/arbitrary units	307	401	496	–	**501**
Salmon_B					
Average T_2/ms	277.4	121.8	215.8	**249.0**	–
Fat signal/arbitrary units	39.6	91.9	138.4	–	**136**

The 1-DILT and oneshot only produce the average T_2 and the initial fat signal respectively

Anahess fits as well to the initial fat signal as the oneshot method. The diffusion coefficients found from Anahess are also in the region 10^{-11} m^2/s for both samples.

It is evident from Table 9.2 that the 2D-ILT routine encounters difficulties providing correct values for the component with the lowest signal intensity, even if this component adds up to 25 % of the total signal. The results from using the Anahess routine provides a good fit of the fat signal, even if the fat content is down to 8 % (Salmon B). The 2D-ILT provides a qualitatively reasonable plot of the Diffusion and T_2 values, but fails in quantifying the different components properly.

9.2.2 Characterisation of Cheese by D-T_2 and T_1-T_2

In Figs. 9.20 and 9.21 the D-T_2 and T_1-T_2 are shown for two types of cheese, where the water is located to the right of the fat signal in the T_2 dimension (Cheese A) or to the left (Cheese B). The measurement temperature was 35 °C, and the spoiler recovery sequence was used in order to reduce the acquisition time and measure the cheese before it melted. With the spoiler recovery sequence the total experimental time for acquiring both the D-T_2 and T_1-T_2 attenuations was less than 20 min. Based on the analysis done in the previous section, it is assumed that the D-T_2 and T_1-T_2 provides reasonably good description of the typical D, T_1, and T_2 found in the systems. From the D-T_2 plot it can be seen that Cheese A has a water component with rather long T_2, around 200 ms, and an apparent diffusion coefficient 2×10^{-9} m^2/s. The fat signal, which is less intense, is found at a T_2 of 100 ms and a diffusion coefficient of 1×10^{-11} m^2/s. From the T_1-T_2 both components are close to being along the diagonal ($T_1 = T_2$). For Cheese B, the fat signal is more intense, but located in the same region as for Cheese A. The water signal is now located at a T_2

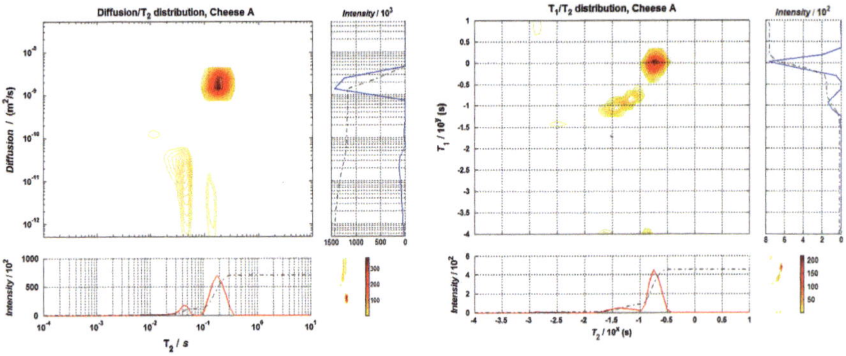

Fig. 9.20 The D-T_2 (*left*) and T_1-T_2 (*right*) distributions arising from cheese A

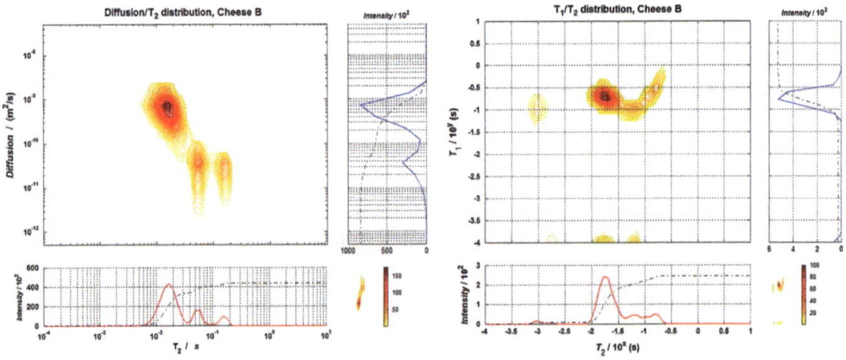

Fig. 9.21 The D-T_2 (*left*) and T_1-T_2 (*right*) distributions arising from cheese B

of approximately 20 ms, the diffusion coefficient is 7×10^{-10} m^2/s, and the signal is located away from the diagonal in the T_1-T_2. This shows that water in cheese B is located in smaller cavities than in cheese A, where the water is allowed to move more freely. Apparently, the fat content is higher in Cheese B relative to the water content. This may be a consequence of a significant loss of water signal in Cheese B signal due to the finite duration of the PFGSE-part of the PFG NMR experiment. During the PFGSE part of the sequence, which lasts 10 ms, significant amounts of water signal are lost due to short T_2's of 20 ms or less. For cheese A there is hardly any loss of water signal, as its T_2 is around 200 ms.

Knowing the observation time in the PFGSE part of the sequence, the apparent diffusion coefficient of the water may be used to estimate the size of the cavities in the cheese (through a fit to the surface to volume ratio (S/V) using the short observation time expansion given in Chap. 2). However, as the 2D-ILT distributions only provide reasonable values, it would be more accurate to use the 1D-projection as shown in Fig. 9.17, and perform a two-component fit to find the apparent diffusion coefficients of fat and water.

9.2.3 A Three Dimensional D-T_1-T_2 Method Applied on a Salmon Tissue

In the previous sections, it is shown that the D-T_2 and T_1-T_2 provide qualitative information on the different components located in the sample. However, the T_1-T_2 suffers from less resolution of the fat from the water, as their T_1 and T_2 values may be much more similar than the apparent diffusion. By placing applied gradients in the τ-intervals of the CPMG, one may alter the apparent T_2 that is measured depending on the molecular mobility of the components, as in the one-shot method. The combined spoiler recovery PFG-NMR sequence shown in Fig. 9.22 provides such a tool where a set of T_1-T_2^*-values can be found for each applied gradient strength. The SRD provides the resolution in the T_1 dimension while the repeated n-loop provides the resolution in the T_2^* dimension.

The equation for the sequence shown in Fig. 9.22 is written for each component as

$$I = I_0(1 - e^{-\frac{SRD}{T_1}})e^{-2n\tau\cdot\left[\frac{1}{T_2}+\gamma^2 g^2 \delta^2 D\left(\frac{3}{4}-\frac{\delta}{6\tau}\right)\right]} = I_0(1 - e^{-\frac{SRD}{T_1}})e^{-\frac{2n\tau}{T_2^*}} \quad (9.10)$$

which defines T_2^*. As there is a term involving diffusion in T_2^*, a different response is expected to take place for fat and water. When there is a significant drift in T_2^* for the water signal as a function of applied gradient strength, there is only a small and hardly noticeable change in T_2^* for the fat component. Then one may, in principle, alter the values in the T_1-T_2 plot such that the fat and water components become satisfactorily resolved.

In Fig. 9.23 the SRD is fixed and the oneshot is recorded on a sample containing fat and water with similar T_2-values (~200 ms), using weak (0.5 Gauss/cm) and stronger applied gradient (50 Gauss/cm). When applying the weak gradient,

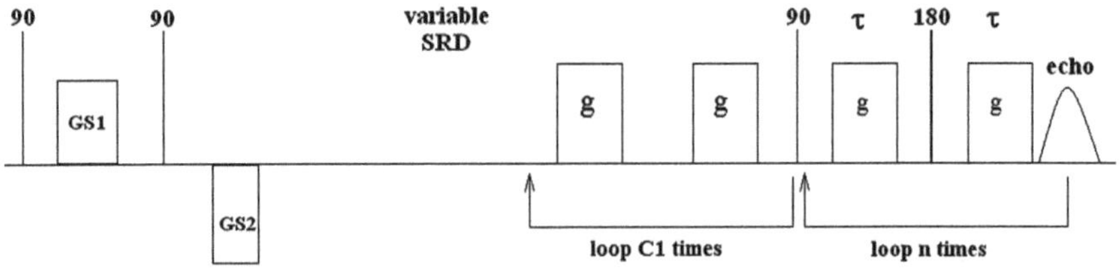

Fig. 9.22 The combined spoiler recover oneshot sequence

Fig. 9.23 The oneshot acquisition of 32 echoes, using an applied gradient g of 0.5 Gauss/cm (~mono exponential decay, *grey plot*) and 50 Gauss/cm (bimodal decay, *black plot*)

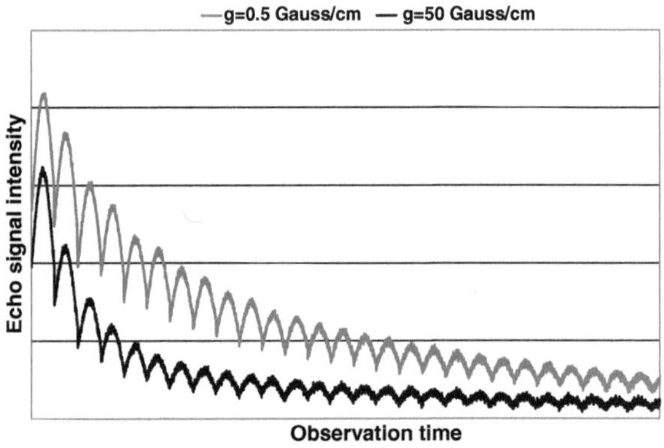

the attenuation of the echoes is very much mono exponential with a time constant corresponding to approximately 200 ms. As the strongest gradient is applied, the attenuation becomes bimodal, with hardly any change in T_2^* for the fat while the T_2^* for the water is now approximately 20 ms. The fat signal is now resolved from the water signal. To produce a three-dimensional data set, 32 values of the SRD was applied for each gradient strength, which varied from 0.5 Gauss/cm up to ~ 50 Gauss/cm. As processing tool for the two-dimensional data set, Anahess is chosen because it has shown to be the most accurate and sensitive tool when analysing systems containing components of relatively low concentration (see Chap. 5).

Figure 9.24 shows the output from the Anahess routine on data sets arising from the combined spoiler recovery—oneshot sequence applied on a pure oil sample and a pure water sample. For each applied gradient strength a set of (T_1, T_2^*)-values are found, the water sample fits to one pair of (T_1, T) whilst the oil sample fits to two pairs (T_1, T_2^*), i.e. a two component fit. The results arising from the water sample show no significant dependency on T_1, as expected, but a considerable drift towards shorter T_2^*. For the oil component there is a small drift towards shorter (T_1, T_2^*), which may be caused by a higher weighting of the fraction in the oil with the lowest mobility, as the oil with higher mobility will tend to move towards shorter T_2^*'s and contribute less in the echo attenuation. Apparently, shorter T_1- and T_2-values will then be fitted to the two components representing the oil phase. However, compared to the considerable drift of the water component in T_2^*, this change in (T_1, T_2^*) can be regarded as negligible.

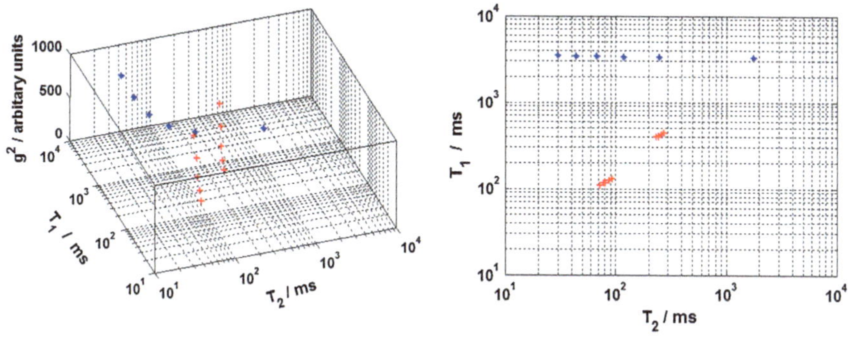

Fig. 9.24 The processed results using the Anahess algorithm on data sets arising from a sample of fat ('+') and from a sample of water ('+'). The figure to the *left* shows a 3D-visualisation of (T_1, T_2^*) as a function of the square of the gradient strength, while the *right* figure shows the projection of all the pairs of (T_1, T_2^*) on to the T_1-T_2 plane. The intensities of the components are not shown, but they do not vary as a function of applied gradient strength for the individual pairs of (T_1, T_2^*)

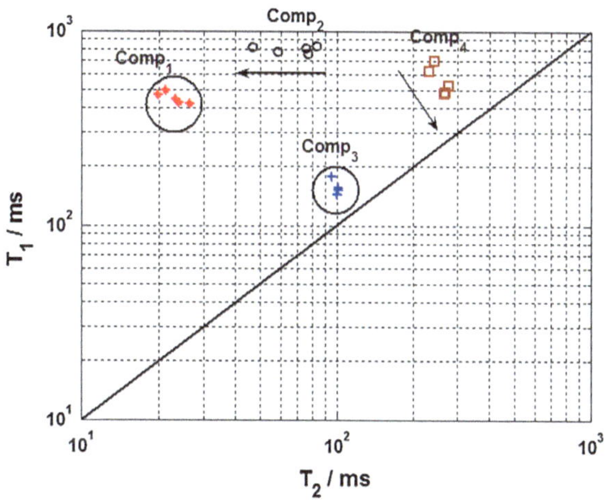

Fig. 9.25 The Anahess results from a 4 component fit of the pairs of (T_1, T_2^*) from a salmon sample at various applied gradient strengths. The *arrows* indicate the direction of the drift of the (T_1, T_2^*) as the gradient strength is increased in the oneshot part of the combined spoiler recovery-oneshot sequence

In Fig. 9.25, the results from a 4-compnent fit using the Anahess routine on a salmon sample are shown for various applied gradient strengths. At the lowest gradient strength component 1 originates from water only, component 3 originates from oil only, while the two other components originates from both oil and water. As the gradient strength is increased component 2 and 4 drift in very different directions, component 2 towards shorter T_2's (as the bulk water sample) and component 4 towards the diagonal ($T_1 = T_2$). Simultaneously the intensity of component 2 increases while component 4 decreases, as shown in Fig. 9.26. As the value of the square of the applied gradient strength exceed 200 in arbitrary units, the intensities of components 2 and 4 does not change, and the water signal is now resolved and found in components 1 and 2 whilst the oil signals are found

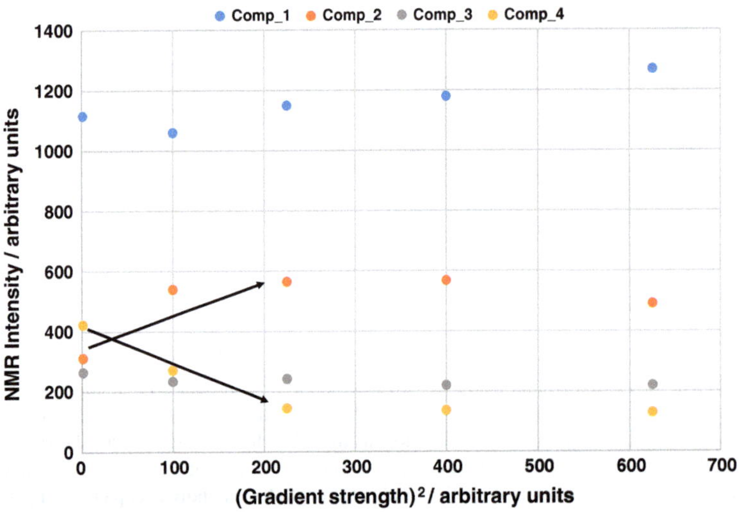

Fig. 9.26 The fitted intensities of the 4 components using the Anahess routine. The *arrows* indicate the change in intensities of components 2 and 4 going in the opposite direction, i.e. a transfer of signal from component 4 to component 2

in components 3 and 4. Even though there is no significant drift in component 1, it arises from water. As the true T_2 is short for this component, it requires a substantial gradient strength to affect the drift of this component. It will only start to drift when component 1 merge with component 2.

The method for resolving fat and water has potential of measuring fat and water in biological tissue as a quality control. By choosing a sufficiently strong gradient strength and number of SRD values, a two dimensional data set is acquired within just a few minutes. Thereafter the Anahess routine may fit the data to the components as shown in Fig. 9.11, and fat and water signals within the sample are quantified. Effects from T_2, Diffusion and T_1 are corrected for using the Anahess routine. Finally, the intensities are calibrated against 100 % oil and water samples of known weight, to produce the absolute fat and water content of the unknown sample.

The visualisation of the dynamic change for the pairs of (T_1, T_2^*) with increasing applied gradient strength as shown in Fig. 9.25, provides more information on the sample than performing 2D-ILT on ordinary T_1-T_2 or T_2-D data sets. The sensitivity of the Anahess makes it possible to produce plots as shown in Fig. 9.25, and will provide new and valuable insight to the systems being investigated. However, as the functionality in the gradient dimension is known from (9.10), there is the potential of performing a three-dimensional fit of the data set.

9.3 Characterisation of Feed

In the production of fish feed, there are at least two stages involved that are crucial to the efficiency of the pellet as a feed; first is the production of the porous pellet, and second is to fill the pellet with oil, a main ingredients in the feed. Some important questions are then raised about the functionality of the pellet. How much oil can be filled into the pores of the pellet, i.e. determination of porosity required? The more oil you can fill into the pellet, the more efficient is the pellet as feed to the salmon. Once the oil is filled into the cavities of the pellet, will there be a significant oil leakage from the pellet? When the feed leaves the factory, and is transported to the sites where the bred salmon is located, the feed may experience significant variations in temperature. This will change the viscosity of the oil, and if the temperature for example exceeds the one in the factory, significant amounts of oil may leak out of the pellet. Then the oil become waste instead for food for the fish, neither very beneficial for the breeders nor for the environment.

In the following section PFG NMR methods for determining porosity of the pellet, pore size distributions, and oil leakage probing are proposed. The most important challenge to solve when characterising the feed is to have a liquid medium that may probe the cavities of the pellet during the lifetime of the NMR signal, i.e. the order of a few seconds. The oil that is used for filling of the pores of the pellet has triglycerides as its main component, and thus the relaxation times are just a few hundreds of milliseconds and the diffusion coefficient is ~10^{-11} m^2/s. During the lifetime of the NMR signal from this oil, the root of the mean squared displacement does not exceed ~10 μm. As will be shown in Sect. 9.3.2 this is much less than the typical/average pore sizes found in the pellet, and consequently it is not a good probe for the heterogeneity of the pellet. As the dried pellet is very hydrophilic, it is not a good solution to fill the pores with water, as the water then will penetrate into the solid matrix of the pellet and make it swell.

The solution proposed is to make use of a light oil as heptane, and fill the pores with the oil that should not affect the matrix of the pellet. The diffusion of bulk heptane is found to be 3.25×10^{-9} m^2/s at a measuring temperature of 30 °C, and thus its root of mean squared displacement exceeds 100 μm during the NMR lifetime. When filling the pores of the pellet with heptane the dried pellet was exposed to heptane under vacuum (~0.1 bar) for an hour. After an hour of exposure, air bubbles were still arising the pellet, but the vacuum was removed, as there was enough heptane inside to test the proposed methods for determining pore sizes and probing fat leakage. Four different pellet types, shown in Fig. 9.27, were filled with heptane and subjected to the PFG NMR measurements as documented in the following.

Fig. 9.27 The pellets to be exposed to PFG NMR

9.3.1 Determining Porosity

If the volume of the pellet is known, V_{pellet}, it is straightforward to measure porosity by calibration of the heptane signal (NMR signal of bulk heptane per unit volume, $NMR_{heptane}$). Once the signal from the heptane inside the pellet, NMR_{pellet}, is measured, the porosity is written

$$\text{Porosity} = 100 \cdot \left(\frac{NMR_{pellet}/NMR_{heptane}}{V_{pellet}} \right)$$

However, this method does not apply for the pellet in general, as they take on various shapes and dimensions, which makes it difficult to measure the accurate volume of the pellet(s). The other and preferred method is then to apply a profile experiment as for the brine profile experiment described in Chap. 7. In order to avoid the fat signal from the solid matrix of the pellet, delays of 0.5 and 1 s were applied before acquiring the profile, where the two delays were used to correct for relaxation, i.e. to produce the initial profile unaffected by relaxation (Fig. 9.28).

A procedure for determining the porosity of the pellet is as follows:

- Define a region of interest in which the pellet is located, perform a profile experiment on bulk heptane only, and calculate the area within the region; $\mathbf{A_{100\,\%}}$.
- Perform a profile experiment on the pellet filled with heptane, and calculate the profile area within the region; $\mathbf{A_{pellet}}$.
- Fill the tube with heptane, perform an identical profile experiment, and calculate the profile area within the region; $\mathbf{A_{pellet+bulk\ heptane}}$.

The porosity of the pellet is then readily found as

$$\text{Porosity} = 100 \cdot \left(\frac{A_{pellet}}{A_{100\%} - (A_{pellet+heptane} - A_{pellet})} \right)$$

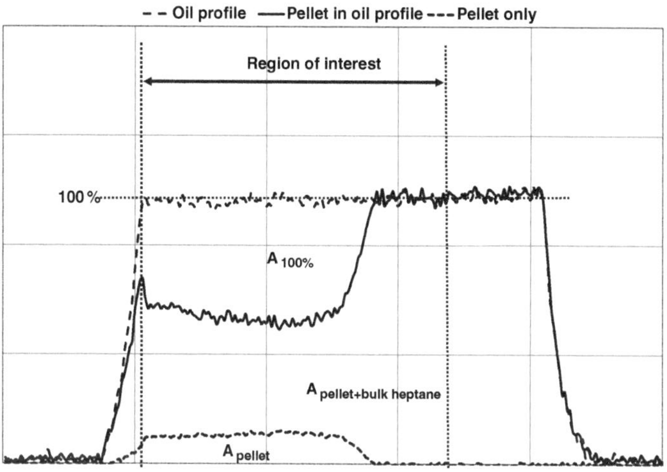

Fig. 9.28 The acquisition of profiles for determination of porosity of the pellet

Table 9.3 Apparent porosity of the different pellets

Sample	A	B	C	D
Porosity/%	25.0	19.4	32.9	14.8

The results from the procedure of measuring the porosity is summarized in Table 9.3, but the porosity must be denoted an apparent porosity as only a fraction of the pellet is filled with heptane. As the pellets are not completely filled, this may also reflect into which sample the heptane is most easily filled.

9.3.2 Pore Size Distributions and 2D-Experiments

The pore size distributions are determined by combining the diffusion measurements at short observation times with the CPMG sequence for the recording of the decay due to T_2 relaxation, as shown in Fig. 9.29. However, as the solid matrix contains significant amounts fat (~5 %), a z-storage delay must be applied within the combined PFG CPMG sequence, also shown in Fig. 9.29. The z-storage delay applied was 1 s to ensure that all signals arising from the solid matrix had decayed to an insignificant amount. As the T_1 of the heptane is of the order of seconds, there was still enough signal from the heptane present in the pores of the pellet to characterise the pore space.

Figure 9.30 shows the normalized pore size distributions for the four different pellets. While the shape of the distribution is very much Gaussian on a log scale, the average "spherical" pore diameter vary from 50 to 100 μm. C and A have the largest pores while B is the intermediate and D the one with the smallest pores. According to the apparent porosities and pore size distributions, the amount of oil that can be filled into the pores will vary for the different pellet types.

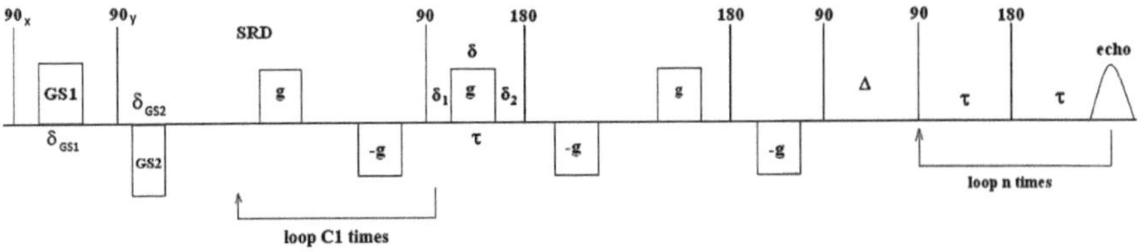

Fig. 9.29 The PFG-NMR sequence applied for determining pore size distributions in fish feed. It consists of the SR approach to avoid long recycle delays, a bipolar 11 interval PFGSE (with preparatory pulses) to measure the apparent diffusion at shorter observation times, a z-storage delay, Δ, to suppress signal from the solid matrix, and at the end the CPMG to record the attenuation due to T_2 relaxation

Fig. 9.30 The achieved pore size distributions from the four pellet systems

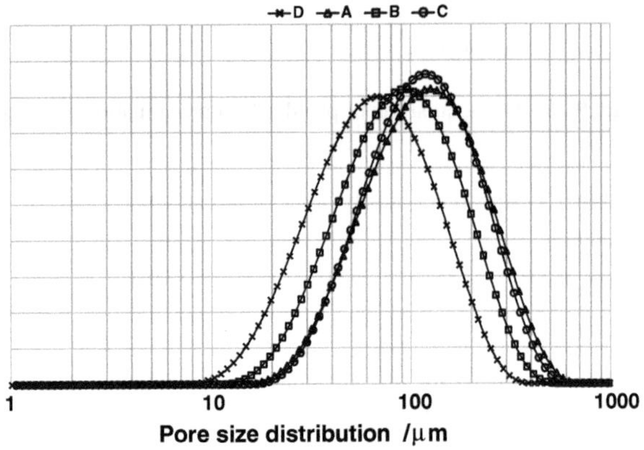

Figure 9.31 is a representative picture of the T_1/T_2 correlated distributions arising from a pellet filled with heptane. The main peak found along the diagonal ($T_1 = T_2$), is the heptane signal, while the signal at a T_2 of ~100 ms slightly off the diagonal arises from the signal within the solid matrix of the pellet. This is mainly from the triglycerides, and should not be interpreted as a contributor to a bimodal pore size distribution. It is only the heptane signal that can probe the cavities due to its long NMR lifetime and high mobility.

To say something about the connectivity of the pores, diffusion measurements at various observation times were conducted. This will provide an indication on how easy the oil may move through the pellet, and out of it, i.e. a measure of the pellet permeability.

9.3.3 Probing Oil Leakage

As for the determination of pore size distribution, the contribution from the fat signal from the solid matrix can be removed by adding a z-storage delay prior to the

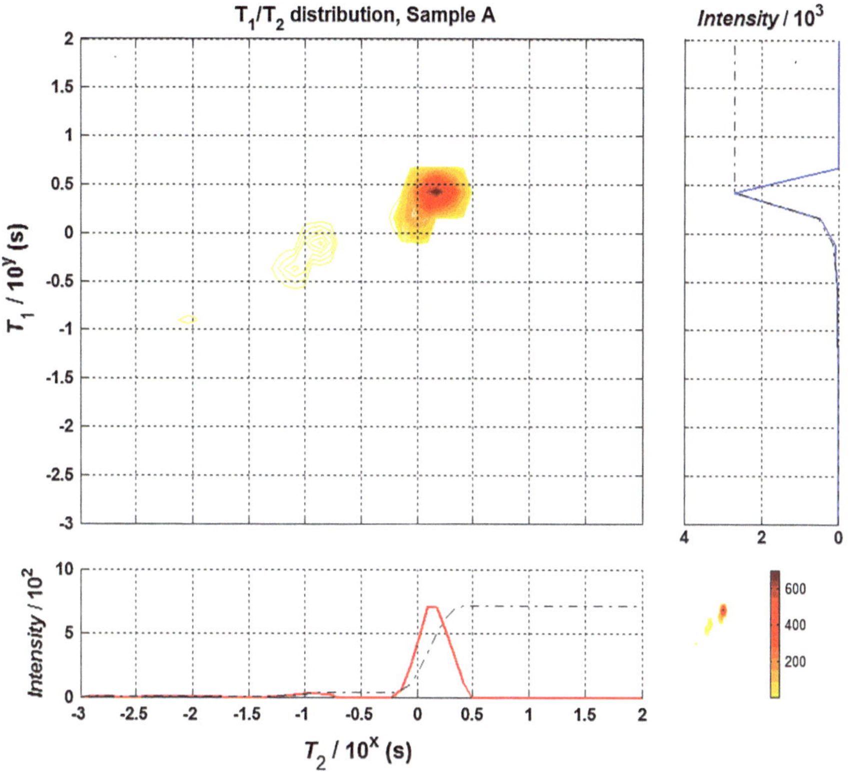

Fig. 9.31 The T_1/T_2 correlated distribution for sample A

time dependent diffusion measurements (Fig. 9.32). The procedure of acquiring fast diffusion measurements, as described in Chap. 6, was applied to ensure that as little heptane as possible evaporated during the measurements.

Figure 9.33 shows the apparent diffusion measurements normalized against the bulk diffusion of heptane, and as a function of the normalized root of the mean squared displacement. The normalization factor along the horizontal axis is the inverse of the measured surface to volume ratio $(S/V)^{-1}$. This represents the typical pore size found in the pellet, and to probe any connectivity, the value along the horizontal axis must exceed 1 by a reasonable amount. When increasing the observation time from 10 to 2,000 ms, the heptane probes pore space equivalent to between 5–8 times the $(S/V)^{-1}$. Within this range, a significant difference response is found for the four different pellets. For sample C, the attenuation (D/D_0) is only down to 0.8 during the 2 s of observation times. As the $(S/V)^{-1}$ is the largest for this system, it only probes up to 5 times $(S/V)^{-1}$. At the other extreme, sample D is found, where the attenuation is down to 0.5 and it probes up to 8 times the $(S/V)^{-1}$. In the intermediate range samples A and B are found. However, they differ as A, with the largest value on $(S/V)^{-1}$ probes only 5 times $(S/V)^{-1}$, while B probes 7 times $(S/V)^{-1}$ during the observation time of 2 s.

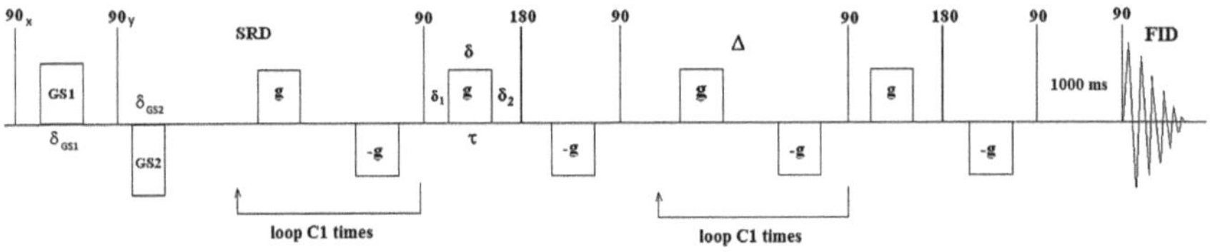

Fig. 9.32 The PFG-NMR sequence for performing observation time dependent apparent diffusion measurements. It consist of the SRD to avoid long recycle delays, a bipolar 13 interval PFGSE (with preparatory pulses) to measure the apparent diffusion at various observation times, Δ, and a z-storage delay of 1,000 ms to suppress signal from the solid matrix in the diffusion measurements.

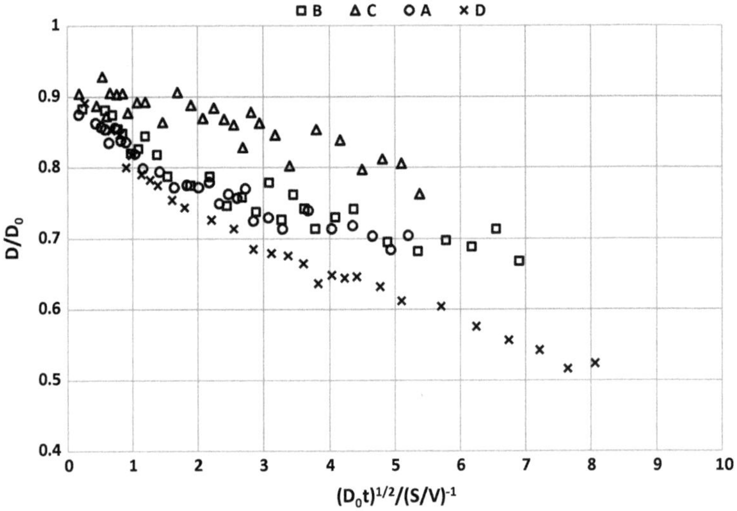

Fig. 9.33 The apparent diffusion measurements normalized against the bulk diffusion of heptane, and as a function of the normalized root of the mean squared displacement

Table 9.4 shows the approximate values of the connectivity at 5 times $(S/V)^{-1}$. As the horizontal axis is scaled against the individual pore sizes of the different pellet systems, any difference at a given value (along the horizontal axis) will origin from differences in connectivity. This confirms that even though the shape of the pore size distributions is very similar, there is a significant difference in the connectivity between the pores for the various pellet systems. Whether this will have any impact on oil leakage will depend on the other parameters as typical pore size and porosity. If the pores are too big the channels connecting the pores must be small in order to prevent leakage, but then again it becomes more difficult to fill the whole pore space of the pellet with the oil. If the pores are small and well-connected the porosity may be low and the amount of oil that can be filled will be less due to the lower porosity.

Finally, a direct measure of fat leakage can be obtained as shown in Fig. 9.34. Here, the profiles of the heptane in pellet C (right) and D (left) are shown as a function of time and measured at a temperature of 70 °C. The labels in the plots indicate the time

Table 9.4 Connectivity at 5 times $(S/V)^{-1}$

Sample	A	B	C	D
D/D_0	0.7	0.7	0.8	0.5

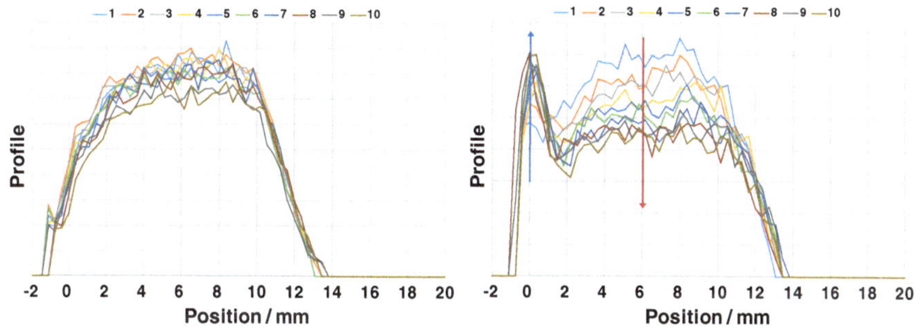

Fig. 9.34 The acquired profiles of pellets C (*right*) and D (*left*) every minute after being exposed to 70 °C

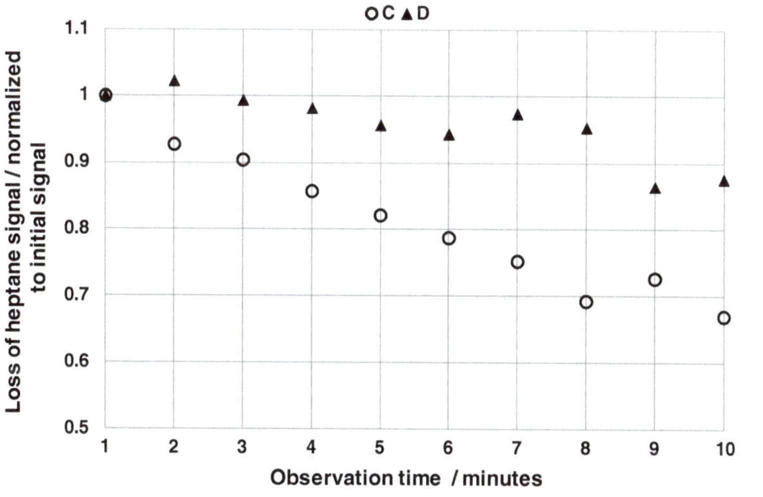

Fig. 9.35 The loss of heptane signal from pellets C and D as a function of time

of profile measurement after the pellets are exposed to 70 °C. The heptane starts immediately to leak out of the pellet and the profile intensity is decreasing with time, except at the bottom of the sample tube (position = 0 mm). Here the difference between the two pellets is readily seen, as the intensity of the profile increases the most for sample C. This confirms that the heptane more easily leaks out of the pellet in C than in D.

Figure 9.35 shows the loss of signal from pellets C and D as a function of time, which has a close resemblance to the connectivity measurements shown in Fig. 9.33. Instead of conducting the diffusion measurements, it may be sufficient to apply the procedure leading to Figs. 9.34 and 9.35, which is a much more direct way of probing the oil leakage. Through the measurements on heptane saturated pellets, PFG NMR has shown to be a tool that provides porosity-values, pore size distributions, degree of connectivity between pores in the pellet, and probes oil leakage. These factors are important for quality control during production of pellets filled with oil.

9.4 Characterisation of Emulsions in Margarine

The determination of droplet size distributions in a biological system, as margarine, is one of the most successful low field NMR applications currently in use. It applies the method proposed by Packer et al. in 1972 [33]: A log normal distribution of the droplet size distribution is assumed and this is fitted to a dataset describing restricted diffusion of the water trapped inside the droplets. The oil signal is assumed to be removed by applying an inversion sequence prior to the diffusion experiment, where the inversion recovery delay, t_{rd}, is set to

$$t_{rd} = \ln(2) \cdot T_1^{oil} \tag{9.11}$$

There are some drawbacks using this method for determining the droplet size distribution in a water in oil emulsion. The assumption of a log normal distribution is a good assumption for many system, but the method will not be sensitive to bimodal distributions or any other distribution that does not fit to the log-normal one. Instead of using a single log normal distribution, one may fit the data to a distribution containing two log normal distributions. However, there is still a need for defining a shape of the distribution instead of letting the experimental data dictate the form of the distribution. A successful removal/suppression of the oil signal depends on the span of T_1 values found in the oil. The T_1 of vegetable oils and animal fat do span a region of T_1 values, and this will affect the removal of oil signal (as demonstrated in Chap. 7). The recording of the restricted diffusion attenuation is lengthy, at least compared against the method presented in the next section. With 8 gradient strengths, 9 scans for each gradient strength (one dummy scan), and a recycle relay of 5 times 1.5 s, the total experimental time will be approximately 10 min.

In the following section, a methodology presented in Chap. 7 has been evaluated for finding a fast and accurate method for obtaining its droplet size distributions in margarine.

9.4.1 *Choice of Method*

The margarine contains a large fraction of oil and a smaller water fraction that are present as droplets. The first parameters to find are the typical relaxation times for the oil and water. As the margarine is solid like at 5 °C, it is likely that suppression due to significant difference in relaxation times will apply. In Fig. 9.36, the achieved

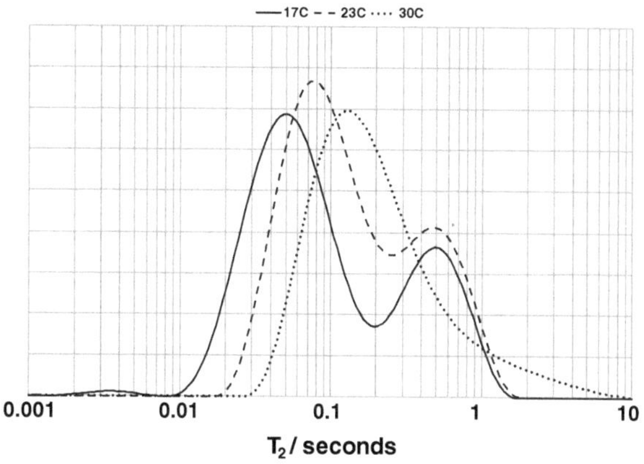

Fig. 9.36 The T_2 distributions from margarine at various temperatures

T_2 distributions at three different temperatures are shown. At 17 °C, the distributions arising from the fat (left peak) and water (right peak) are only partially overlapping, and it will be possible to resolve the water from the fat signal by their differences in relaxation times. As the temperature increases the degree over overlap increases and at 30 °C there is only one peak with a broad shoulder to the right. At this temperature, the emulsion is not stable as the fat melts, forms a bulk fraction at the top of the sample, and the water droplets coalesce, and forms a bulk phase at the bottom.

In Fig. 9.37, the 13 interval PFGSTE is applied on the margarine when it is present as an emulsion, and when the same sample has melted at 30 °C. The observation time is 107 ms, i.e. a root of mean squared displacement of ~40 μm for bulk diffusion of water. This is significantly larger than the droplet sizes that are expected to be found in the margarine (<10μm [34]). The attenuation exhibits a bimodal pattern, where initially there is oil and water present, but as the gradient increases, the fat component is suppressed, and one is left with the emulsified water. This is confirmed by melting the sample and performing a repeat of the 13 interval PFGSTE using the identical experimental settings. The component with the apparently lowest diffusion coefficient has disappeared, i.e. the emulsified water is no longer present, and an approximately mono-exponential decay is observed. The bulk water that is produced from the coalesced emulsion, does not contribute in the attenuation, as the gradient strength and pulse duration are chosen to discriminate between emulsified water and fat. Any signal from bulk water is suppressed to an insignificant amount already following the first gradient strength applied in the attenuation.

Even though the differences in relaxation times are sufficient to resolve the water from the oil signal, and the diffusion measurements at short observation times can be applied to extract the droplet size, it is more accurate to extract the droplet size from the asymptotic approach (Chap. 7). The reason is that the short observation time approach requires a value of the bulk diffusion of water inside the droplets, and this is unknown as there are most likely other water-soluble components within the droplets that may alter the bulk diffusion value from the one

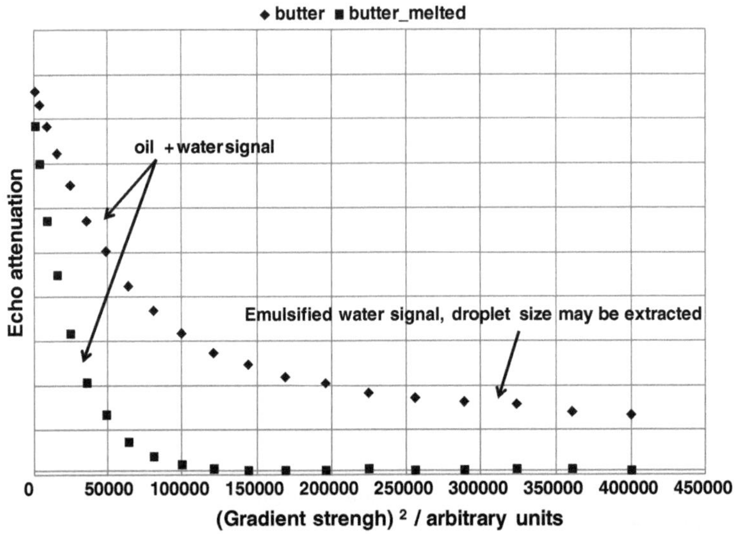

Fig. 9.37 The 13-interval PFGSTE sequence applied on margarine, in its initial and melted states

found for distilled water. In the asymptotic limit, there is no requirement to know the unrestricted diffusion of the water phase, and this simplifies the experimental set-up and makes it much faster.

9.4.2 Droplet Size Distribution of Water in Margarine

The chosen PFG NMR sequence for determining droplet size distributions is shown in Fig. 9.38. The asymptotic limit for the attenuation from restricted diffusion within a sphere is written (Chap. 7)

$$\left(\frac{I}{I_0}\right) \approx 1 - \frac{1}{5}\gamma^2\delta^2 g^2 \bar{R}^2 \tag{9.12}$$

A measure for the average droplet radius is found by measuring the attenuation at two different gradient strengths, where the fat signal is suppressed to an insignificant value. The attenuation due to T_2 relaxation, which is recorded simultaneously, may then be transformed to a droplet size distribution as prescribed in Chap. 7. In Fig. 9.39 the achieved droplet size distribution is shown, and the shape of the distribution is close to being, but not identical to, log normal with an average value of the droplet diameter around 2.3 μm.

There are a few advantages of using the proposed method for measuring the droplet size distribution of water in margarine. Primarily because the acquisition time is quite short. Under the assumption that four scans are sufficient (+one

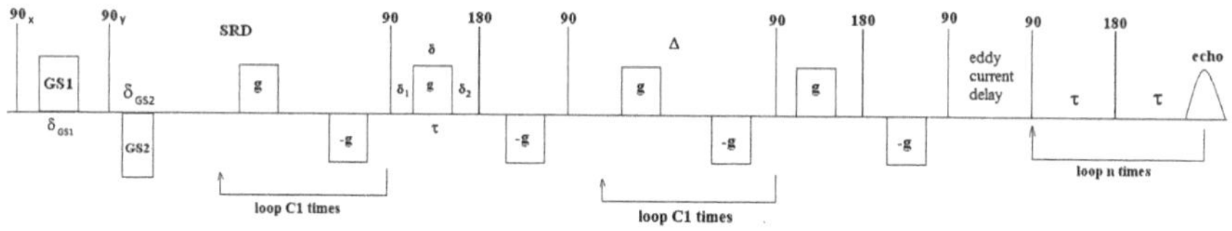

Fig. 9.38 The PFG-NMR sequence applied for determining droplet size distributions in margarine. It consist of the SR approach to avoid long recycle delays, a bipolar 13 interval PFGSE (with preparatory pulses) to measure the average droplet radius at the asymptotic limit, an eddy current delay, and at the end the CPMG to record the attenuation due to T_2 relaxation

Fig. 9.39 A droplet size distribution of water in margarine

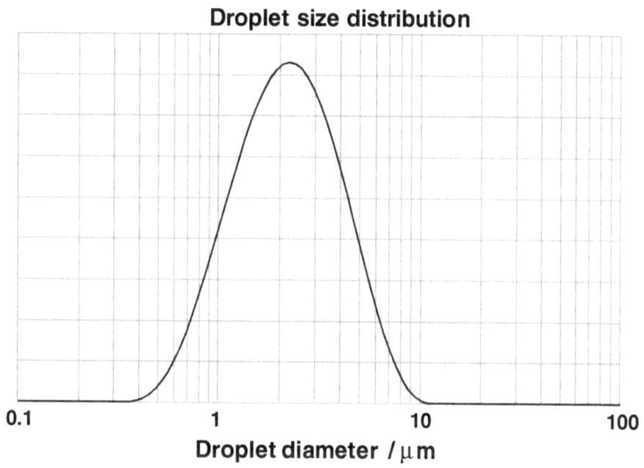

dummy scan) for each applied gradient strength, the following delays are included for each scan; a spoiler recovery delay of 2 s, a recycle delay of practically zero, a z-storage delay of 200 ms in the diffusion sequence, and a duration of the CPMG-decay of 2.5 s. The total acquisition time using two gradient strengths is therefore approximately 50 s. Due to the short measuring time and a model (9.12) that does not include the bulk diffusion, the sample may be taken directly from the fridge and measured instantly in a magnet tempered at 30 °C. During the 50 s measurement period there is not enough time to heat up the sample to a level where it affects the margarine. To validate this assumption, the T_2 distribution was measured before and immediately after the recording of the droplet size distribution, and there was hardly any change in the shape and position of the distributions.

The CPMG-decay produces a T_2 distribution through the inverse Laplace transform routine. This is transformed into a droplet size distribution as described in Chap. 7. Consequently, the droplet size distribution results from thousands of measuring points in the CPMG-decay, and not from a limited number of measuring points from a restricted diffusion experiment as described by Packer et al. [33]. The diffusion experiment is used to measure the average droplet size, and the accuracy of this two-point determination of the slope of the PFG-attenuation of the emulsified water is within a few percent in relative units, as documented in Chap. 6.

Finally, the distribution may take on any arbitrary shape and is not restricted to a predefined shape as the log normal distribution. As the distribution arises from an inverse Laplace transform, it is important to keep the smoothing parameter to a fixed value. Otherwise, it will not be possible to compare droplet size distributions against each other, as the broadening depends on the smoothing parameter. The latter is probably the most important pit-fall to avoid in the proposed method for determining the droplet size distribution by PFG NMR.

9.5 A Medical Application

In the field of orthopedic surgery, the degeneration of tendon is one of the main reasons for pain in the extremities as well as loss of functionality. Good examples are degenerative ruptures, tendinosis and symptoms from the achilles tendon in the ankle, quadriceps and patellar tendon in the knee, and "tennis and golfers elbow". Where there is a tendon, there may be a tendon problem, acute or degenerative.

In the shoulder, degeneration and rupture of the rotator cuff tendon is an important reason for long-term sick leave, pain and loss of ability to perform labor. The development of surgical arthroscopic procedures and equipment, have given the shoulder surgeons the ability to repair the ruptures of the tendon and restore the function of the joint. This surgery can be challenging, and not all patients can be helped. Magnetic resonance imaging (MRI) is an important diagnostic tool for the orthopedic surgeon to make a preoperative assessment and plan the surgery. The MRI provides images of the joint and tendon, and the size and shape of the rupture is visualized, as well as retraction of the tendon and muscle-tendon-junction from its insertion/footprint. This may provide whether an indication that the rupture is reparable or not.

The MRI instrumentation for medical purposes does not provide very specific information of the quality or dynamics of the tendons. From the images, information of fat infiltration in the muscle and tendon is however extractable. This infiltration is a sign of atrophy and poor quality of the muscle-tendon-motor unit. The more fatty infiltration of the muscle on the images, the poorer quality, and when there is more than 50 % of fat in the muscle, the risk is high that the tendon injury is not reparable [35, 36].

Dynamic PFG NMR is a potential tool for probing the quality of the tendons, as it is sensitive to the structure of the material measured on. For example, when measuring apparent diffusion of the water at longer observation times, the water connectivity in the tissue may be probed. Two-dimensional experiments as T_1-T_2 and/or Diffusion-T_2 may reveal information on the homogeneity of the tissue, which can be related to the degree of degeneration and loss of structure in the tendons. In the following section, focus will be on a tendon present in the shoulder, the caput longum biceps tendon. A part of this tendon is removed in a standardized surgical procedure, and can be analyzed without loss of functionality of the shoulder. This tendon is analyzed to see whether any difference in age and pathology (trauma and degeneration) can be probed by the means of PFG

NMR. Both diffusion and two-dimensional measurements are performed on the tendons by applying a low field (0.5 T) PFG NMR spectrometer operating at a temperature of 37.5 °C.

9.5.1 Anatomy and Sample Preparation

The anatomy of the shoulder is complex. In the development of the species, the primates have developed the shoulder into a joint with an extensive range of movement and flexibility. The price the human species pay for this development is the instability of the shoulder joint, as the stability of the human shoulder relies entirely on the stabilizing soft tissue; capsule, labrum and muscle/tendons. Consequently, the tendons are in constant wear and tear. Many of the problems in the late adult patient group are due to degeneration or mechanical wear of the stabilizing tendons, where the supraspinatus tendon and the biceps caput longum tendon are often mostly affected. Both of these tendons are important in the shoulder function and stability of the primates. However, the biceps caput longum tendon in humans is thin and not powerful, and it can rupture without any significant loss of functionality of the shoulder. Consequently, the contribution of the caput longum muscle/tendon is of less importance in the normal human range of movements/activity, as seen for patients with spontaneous proximal ruptures of the caput longum tendon.

Because the biceps caput longum tendon in human is thin and less powerful than in the other primates, it is more exposed to wear and tear in the relatively unstable shoulder joint. It is an important pain generator, and can also subluxate in the entrance of the sulcus and affect the other tendons in the shoulder, especially the supraspinatus and supscapularis tendons. Many different surgical procedures address the degenerated biceps caput longum, where one is to perform a so-called subpectoral bicepstenodesis. This involves suturing the musculotendinous junction of the biceps caput logum tendon to the lower border of the pectoralis major tendon. The proximal part of the tendon, proximal to the tenodesis, is routinely cut and ends up as excised tissue. Orthopedic- and shoulder surgeon Kristian Sørland Humborstad at Haraldsplass Deaconal Hospital, Bergen (Norway), performed the surgeries that produced the excised tissue, and this tissue is what provides an opportunity for applying PFG NMR experiments on tendon tissue.

In Fig. 9.40, the human right shoulder is shown, and the position of the caput longum tendon is found within the red circle. To the right an image taken during surgery is shown, from which the proximal part is cut and used as a part of a PFG NMR analysis. To the lower right the tendon is cut in two, one part representing the intra-articular part (i) and the other the sulcus part (s). These two parts were measured separately by various PFG NMR experiments as shown in the following sections. As there are fibres in the tendon, an anisotropy of the diffusion coefficient may be expected. To prevent any discrepancies due to anisotropic diffusion, the tendons were located in the same direction in the sample tube.

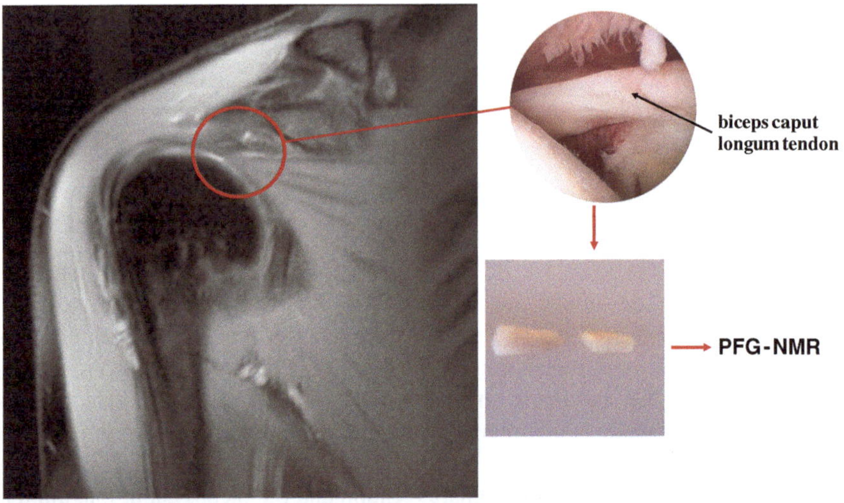

Fig. 9.40 An MRI of the human right shoulder, high lightning the caput longum tendon (within the *red circle*)

9.5.2 Diffusion Measurements

The diffusion measurements were performed on two tendons, where one of them did not show any signs of being damaged by preoperative assessment (B), while the other clearly was degenerated (A). The tendon was divided into two sample tissues, one from the sulcus region (s) and one from the intra-articular region (i) (see Fig. 9.40). By visual inspection, it is noted that the sulcus part appears to be denser as the diameter of the tendon decreases when going into the sulcus region.

In Fig. 9.41 the normalized natural logarithm of the attenuation as a function of the square of the applied gradients strength from a bipolar 13 interval PFGSTE is shown for all four samples and a bulk water sample as comparison. The z-storage delay was set to 50 ms, and then one should expect a departure from the behaviour of bulk water due to the restricting geometries. The protein signal does not contribute to the attenuation due to its short T_2 (the total duration of phase encoding and decoding intervals was 8 ms and T_2 of the protein is less than 1 ms [10]).

There is a clear difference in the results arising from the two regions of the tendon. Initially, all four samples have a slightly less steeper slope than the one from the bulk water, but the two samples from the sulcus attenuates less than the two from the intra-articular region. As the gradient strength is increased it appears that the two samples from the sulcus region produces a bimodal attenuation, which is a typical attenuation found for a sample containing fat and water [32]. For the intra-articular region there is hardly any fat signal present, as the signal attenuates down to 0.7 % of

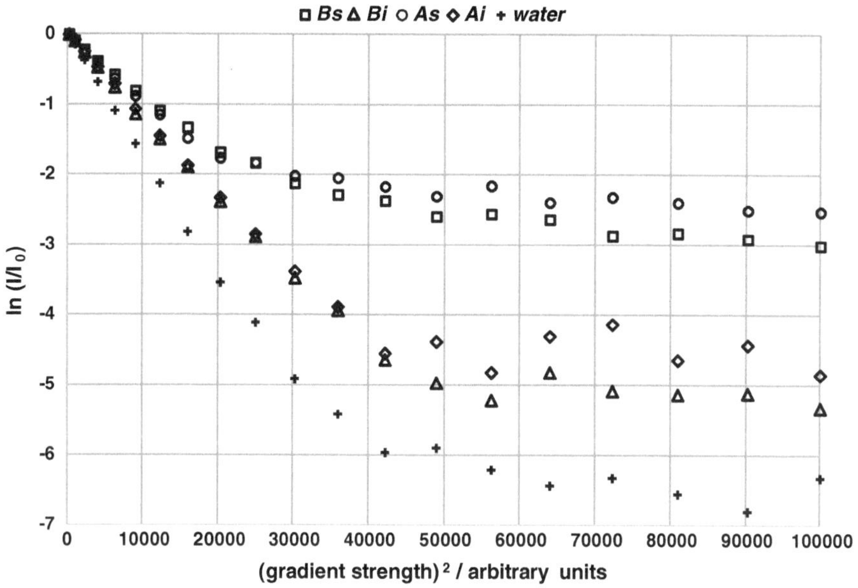

Fig. 9.41 The bipolar 13-interval PFGSTE sequence applied on 4 tendon samples arising from two tendons (*A* and *B*). The observation time was 50 ms. *Ai* and *As* represents the sulcus and intra-articular regions respectively

the initial signal ($\ln(I/I_0) = -5$). Knowing the weight of the samples and the NMR signal per gram of fat, the fat content was found to be approximately 4.5 % for *As* and 2.9 % for *Bs*. As fatty infiltration indicates poorer quality of the tendon, a diffusion measurement confirms the existence and higher fat content in *As* and consequently confirms the degeneration of this part of tendon A. Subtracting the fat signals from the attenuations, the apparent diffusion coefficients for the four samples varied slightly. In the intra-articular region both samples returned a diffusion coefficient of 2.1×10^{-9} m^2/s, while it was found to be 1.7×10^{-9} m^2/s for *Bs* and 1.8×10^{-9} m^2/s for *As*. The mobility seems to be more restricted in the sulcus region, which is expected as the diameter of the tendon decreases as it enters the sulcus.

In Fig. 9.42 the apparent diffusion measurements are shown for the tendons arising from the sulcus region, and the method applied is the one proposed in Chap. 6. The apparent diffusion coefficient for sample *Bs* starts around 1.7×10^{-9} m^2/s, drops gradually below 1.5×10^{-9} m^2/s and thereafter rises to a plateau value around 1.5×10^{-9} m^2/s. The apparent diffusion coefficient for sample *As* starts around 1.8×10^{-9} m^2/s, drops gradually down to 1.6×10^{-9} m^2/s and rises to a plateau value around 1.9×10^{-9} m^2/s. The drop in the apparent diffusion coefficient followed by an increase is due to the small but significant fat signal that contributes most at the shorter observation times. As the observation time is increased, the fat

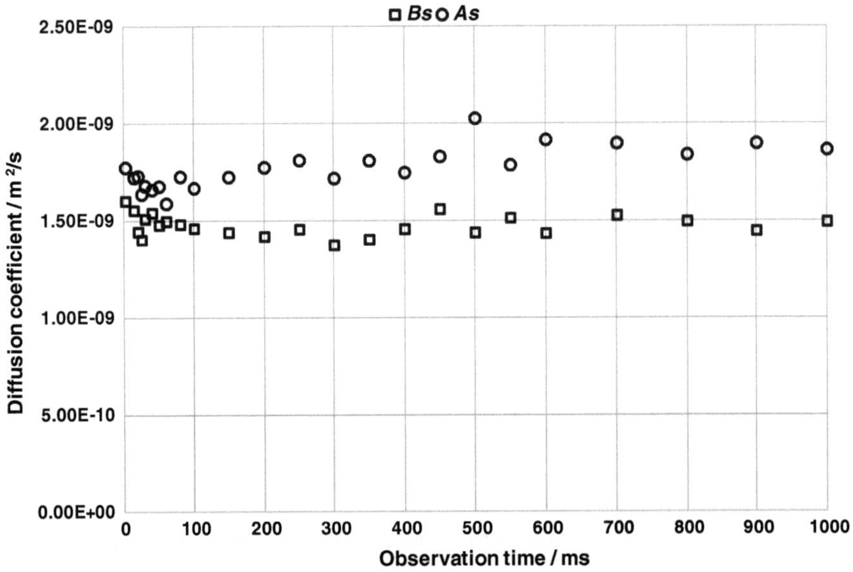

Fig. 9.42 Measured apparent diffusion coefficients as a function of observation time within the sulcus region

signal contributes less and less due to its shorter relaxation time, and after 400 ms the apparent diffusion coefficient originates from water signal only. The drop and rise in apparent diffusion coefficient is most readily seen for *As,* which contains the largest fraction of fat. Observation time dependent diffusion measurements were conducted on the intra-articular regions as well, but it only produced a small sample variation, ranging from 2.1×10^{-9} to 2.0×10^{-9} m^2/s.

In addition to the determined higher fat content in the degenerated tendon, a significant difference in tortuosity is found when comparing the regular and degenerated tendons taken from the sulcus region. The interpretation could be that in the degenerated tissue, the mobility of the water is less restricted, and the muscle fibres are not as rigid as for the normal tendon. It could also be that there are some ruptured region on a microscopic scale, which allows the water to move more freely in the degenerated tendon. The diffusion measurements shown in Fig. 9.3 probe length scales up to ~100 μm.

It is shown that PFG NMR may reveal differences that can be related to the functionality of the tendon, but no information on the heterogeneity of the tissue is provided, in which the water is moving. However, the application of two-dimensional experiments will provide more insight to the heterogeneity of the tissue, as will be shown in the next section.

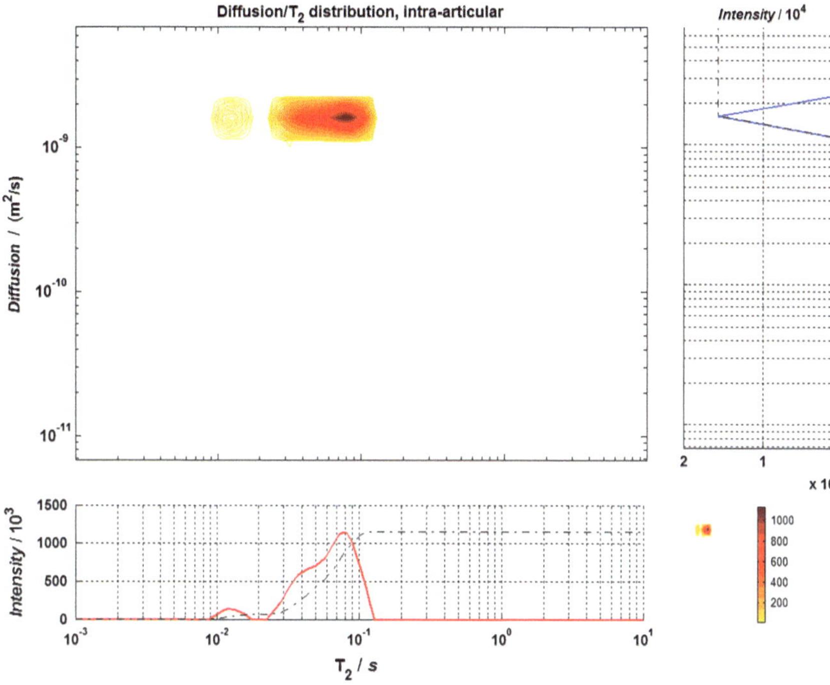

Fig. 9.43 The D-T_2 distribution arising from intra-articular region of the tendon

9.5.3 Diffusion-T_2 and T_1-T_2 and Measurements

In Figs. 9.43 and 9.44 the distributions arising from the 2D-ILT on Diffusion-T_2 and T_1-T_2 correlated experiments are shown for the intra-arcticular region of the tendon, while Figs. 9.45 and 9.46 provides the distributions arising from the identical experimental procedure, but now from the sulcus region.

From Figs. 9.43, 9.44, 9.45 and 9.46, it is seen that the tendon outside sulcus is much better defined, especially with much less scatter in the T_1-T_2 distribution. In the sulcus region the distributions are broadened over larger regions, and the presence of fat is detected as a component with a diffusion coefficient in the region 10^{-11} m^2/s. Another finding is the significant reduction in T_2 when going from the intra-articular to the sulcus region, while the T_1 seems to be unaltered. In order to provide quantitative information on the average T_1 and T_2, 5 tendons (divided into intra-articular and sulcus), were measured using the spoiler recovery T_1-T_2 method, and data were processed using the Anahess algorithm. The results for the average values of T_1 and T_2 are plotted in pairs in Fig. 9.47. While there seems to

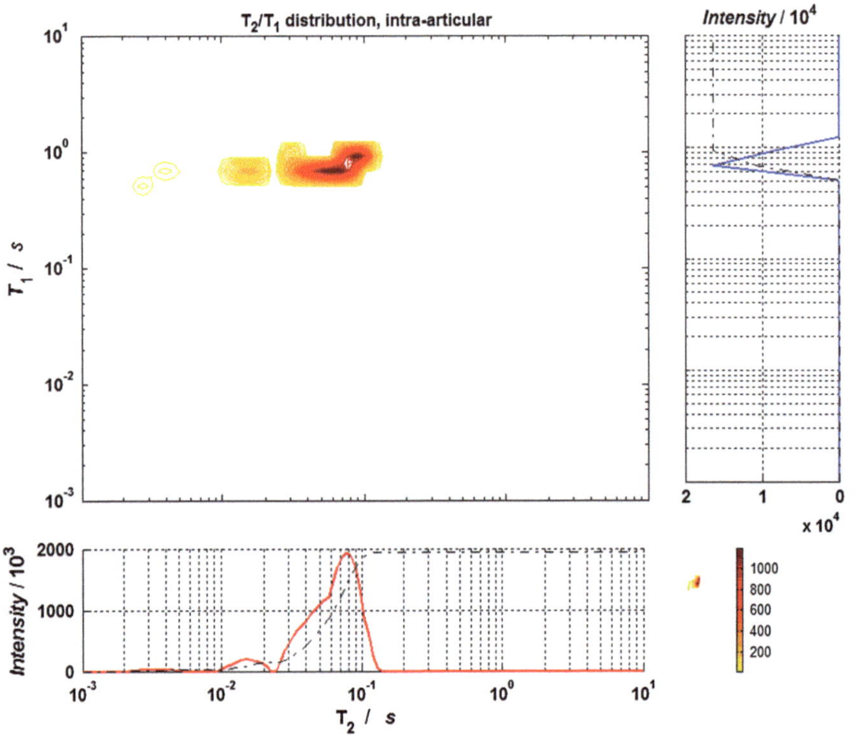

Fig. 9.44 The T_1-T_2 distribution arising from intra-articular region of the tendon

be a slope dependency on the pair of (T_1, T_2)'s for the intra-articular region, the sulcus region produces pair of (T_1, T_2)'s where T_1 is located around 500 ms and T_2 around 50 ms (with an outlier at 110 ms). The degree of degeneration of the tendons is unknown, except for the ones that were used for producing the diffusion measurements in the previous section, and these values are provided in Table 9.5. A striking effect when comparing A and B is that the T_1 and T_2 does not change much for sample A when going from the intra-articular to the sulcus region, while there is a significant decrease in the values for sample B. This supports the picture of a ruptured tendon in the sulcus region for sample A.

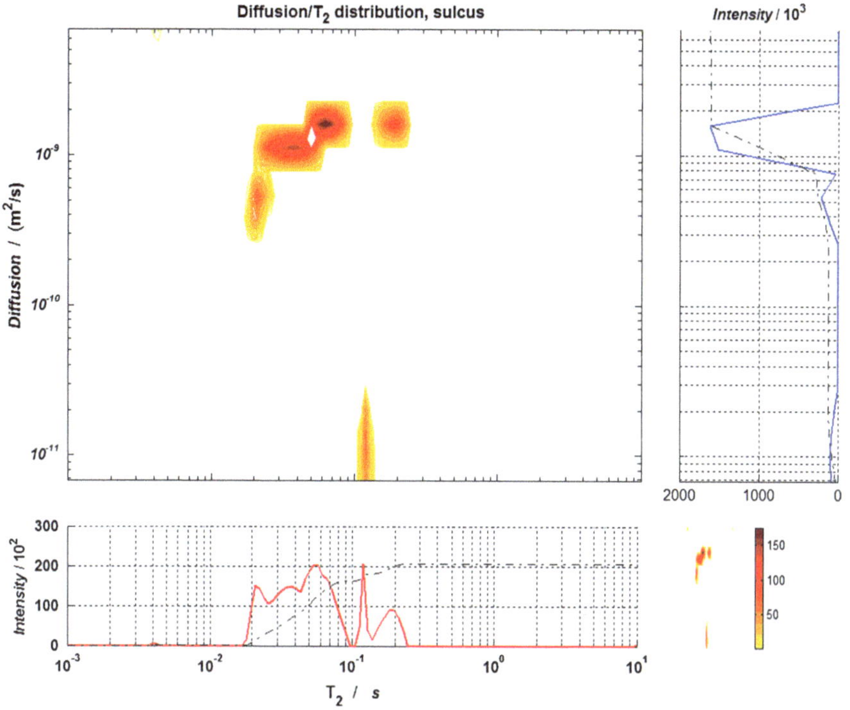

Fig. 9.45 The D-T_2 distribution arising from sulcus region of the tendon

9.5.4 A Potential Non-invasive Method?

The motivation behind applying PFG NMR on the tendons, is to see whether there are any parameters that can be used for probing the functionality of the tendons non-invasively, i.e. using the medical MRI available. It is shown that the sulcus region is a good probe for the functionality of the tendon. When this part degenerates, its degree of degeneration may be quantified from diffusion weighted images, where the 13 interval sequence is implemented with an observation time of 500 ms (to suppress the contribution from the fat signal). Diffusion measurements may

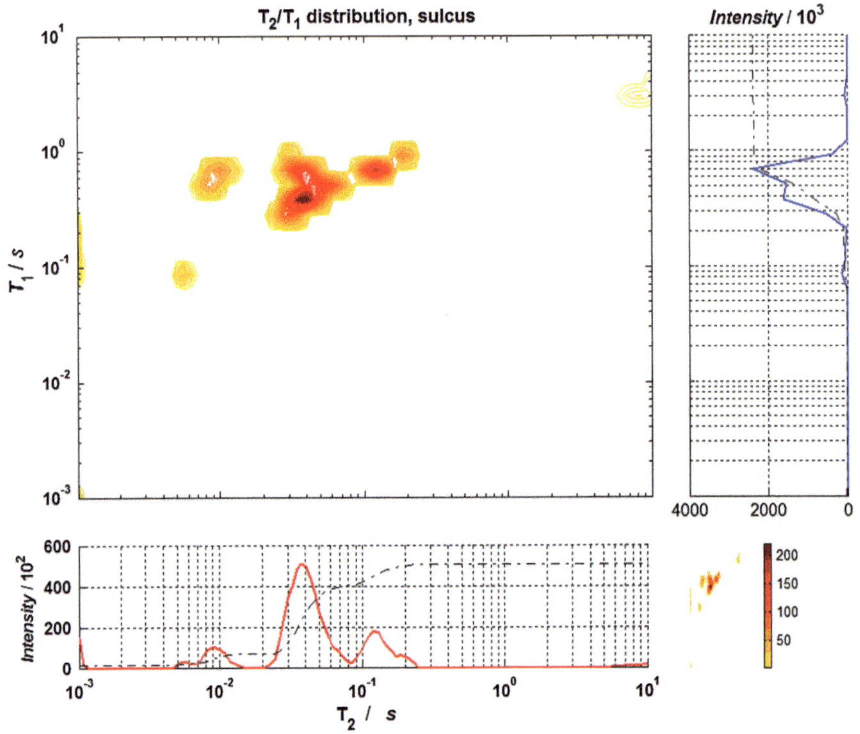

Fig. 9.46 The T_1-T_2 distribution arising from sulcus region of the tendon

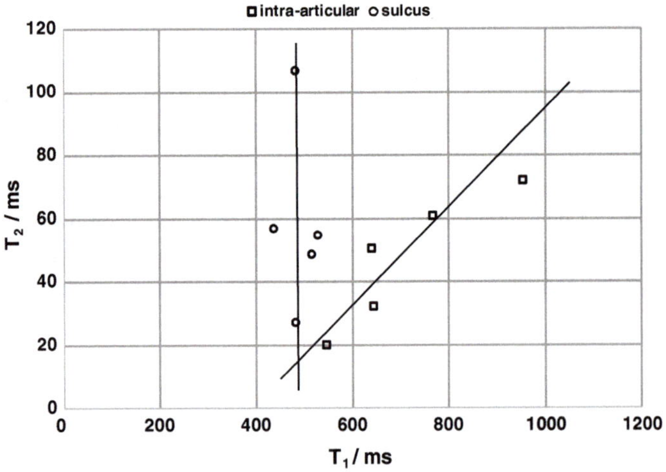

Fig. 9.47 Average pair of T_1- and T_2-values for tendon tissue arising from the intra-articular ('□') and sulcus ('o') region

Table 9.5 Average T_1- and T_2-values for tendon A and B in milliseconds

	T_1 (intra-articular)	T_2 (intra-articular)	T_1 (sulcus)	T_2 (sulcus)
A	638.5	50.7	513.8	48.9
B	953.0	72.2	480.1	27.4

then probe the degree of degeneration, as the PFG NMR method probes the connectivity within the tissue. The fraction of fat content can be also be determined from a 11 interval PFGSE diffusion measurements as shown in Fig. 9.2. Another approach could be to quantify T_2^s/T_2^i ratio from T_2-weigthed images, as Table 9.1 indicates some interesting features when comparing a normal tendon intra-articular/sulcus with a degenerated tendon.

References

1. W.B. Jensen, The origin of the Soxhlet extractor. J. Chem. Educ. **84**(12), 1913 (2007)
2. H. Greenfield et al., *Food Composition Data: Production, Management, and Use* (Food and Agriculture Organization of the United Nations, Rome, 2003)
3. S. Willart, R. Nilsson, Y. Rasmusson, The influence of lipolytic activity on the determination of the fat content in milk and dairy products with the Schmid-Bondzynski-Ratzlaff method. Alnarp, (1959). Print
4. A.S.f. Testing, and Materials, *Standard Test Method for Fat in Meat and Meat Products by the Foss-Let Fat Analyzer Technique* (American Society for Testing and Materials, Pennsylvania, 1976)
5. R. Pendl et al., Determination of total fat in foods and feeds by the caviezel method, based on a gas chromatographic technique. J. AOAC Int. **81**(4), 907–917 (1998)
6. T.P. Leffler et al., Determination of moisture and fat in meats by microwave and nuclear magnetic resonance analysis: collaborative study. J. AOAC Int. **91**(4), 802–810 (2008)
7. A. Dieffenbacher et al., *Standard Methods for the Analysis of Oils, Fats and Derivatives*, 1st supplement to the 7th edn. (Wiley, Oxford, 1992)
8. G.H. Sorland et al., in *On the Use of Low-Field NMR Methods for the Determination of Total Lipid Content in Marine Products*. Magnetic Resonance in Food Science: The Multivariate Challenge. The Royal Society of Chemistry, Cambridge, 2005), pp. 20–27
9. R.F. Karlicek Jr., I.J. Lowe, *A modified pulsed gradient technique for measuring diffusion in the presence of large background gradients*. J. Magn. Reson. **37**(1), 75–91 [(1969) 1980]
10. P.A. Bottomley et al., A review of 1H nuclear magnetic resonance relaxation in pathology: are T1 and T2 diagnostic? Med. Phys. **14**(1), 1–37 (1987)
11. W.H. Press, *Numerical Recipes in C++: The Art of Scientific Computing* (Cambridge University Press, New York, 2002)
12. S. Meiboom, D. Gill, Modified spin-echo method for measuring nuclear relaxation times. Rev. Sci. Instrum. **29**(8), 688–691 (1958)
13. C.D. Usher, C.J. Green, C.A. Smith, The rapid estimation of fat in various foods using the Foss-Let density apparatus. Int. J. Food Sci. Technol. **8**(4), 429–437 (1973)
14. K. Kolar et al., An intercomparison of methods for the determination of total fat in a meat reference material. Fresenius J. Anal. Chem. **347**(10–11), 393–395 (1993)
15. L.M.L. Nollet, F. Toldra, *Handbook of Processed Meats and Poultry Analysis* (Taylor & Francis, Boca Raton, 2010)
16. V.E. Vinzi et al., *Handbook of Partial Least Squares: Concepts, Methods and Applications* (Springer, Berlin, 2010)

17. P.T. Callaghan, Y. Xia, Velocity and diffusion imaging in dynamic NMR microscopy. J. Magn. Reson. [(1969), 1991]. **91**(2), 326–352
18. G.H. Sørland, D. Aksnes, Artefacts and pitfalls in diffusion measurements by NMR. Magn. Reson. Chem. **40**(13), S139–S146 (2002)
19. N.S. Radin, *Lipid Extraction* (Humana Press, Clifton, 1989), pp. 1–61
20. G.G. Haraldsson, B.Ö. Gudmundsson, Ö. Almarsson, The synthesis of homogeneous triglycerides of eicosapentaenoic acid and docosahexaenoic acid by lipase. Tetrahedron **51**(3), 941–952 (1995)
21. M. Kriat et al., Analysis of plasma lipids by NMR spectroscopy: application to modifications induced by malignant tumors. J. Lipid Res. **34**(6), 1009–1019 (1993)
22. A. Abragam, *The Priciples of Nuclear Magnetism* (Oxford Universiy Press, New York, 1983), p. 618
23. Y.H. Hui, W.K. Nip, R. Rogers, *Meat Science and Applications* (Taylor & Francis, Boca Raton, 2001)
24. J.G. Lammers et al., Syntheses and spectroscopic properties of α- and β-phosphatidylcholines and phosphatidylethanolamines. Chem. Phys. Lipids **22**(4), 293–305 (1978)
25. M. Aursand, J.R. Rainuzzo, H. Grasdalen, Quantitative high-resolution 13C and 1H nuclear magnetic resonance of ω3 fatty acids from white muscle of atlantic salmon (Salmo salar). J. Am. Oil Chem. Soc. **70**(10), 971–981 (1993)
26. F.D. Gunstone, High resolution NMR studies of fish oils. Chem. Phys. Lipids **59**(1), 83–89 (1991)
27. B.W.K. Diehl, High resolution NMR spectroscopy. Eur. J. Lipid Sci. Technol. **103**(12), 830–834 (2001)
28. L.L. Holte, F.J.G.M. van Kuijk, E.A. Dratz, Preparative high-performance liquid chromatography purification of polyunsaturated phospholipids and characterization using ultraviolet derivative spectroscopy. Anal. Biochem. **188**(1), 136–141 (1990)
29. Nutreco_ARC, *Round Robin* (2002)
30. G.H. Sørland et al., Exploring the separate NMR responses from crude oil and water in rock cores. Appl. Magn. Reson. **26**(3), 417–425 (2004)
31. J.G. Seland et al., Combining PFG and CPMG NMR measurements for separate characterization of oil and water simultaneously present in a heterogeneous system. Appl. Magn. Reson. **24**(1), 41–53 (2003)
32. G.H. Sorland et al., Determination of total fat and moisture content in meat using low field NMR. Meat Sci. **66**(3), 543–550 (2004)
33. K.J. Packer, C. Rees, Pulsed NMR studies of restricted diffusion. I. Droplet size distributions in emulsions. J. Colloid Interface Sci. **40**(2), 206–218 (1972)
34. J.P.M. van Duynhoven et al., Scope of droplet size measurements in food emulsions by pulsed field gradient NMR at low field. Magn. Reson. Chem. **40**(13), S51–S59 (2002)
35. D. Goutallier et al., Fatty muscle degeneration in cuff ruptures. Pre- and postoperative evaluation by CT scan. Clin. Orthop. Relat. Res. **304**, 78–83 (1994)
36. J. Bernageau et al., Magnetic resonance imaging findings in shoulders of hemodialyzed patients. Clin. Orthop. Relat. Res. **304**, 91–96 (1994)

Index

G.H. Sørland, *Dynamic Pulsed-Field-Gradient NMR*,
Springer Series in Chemical Physics 110, DOI 10.1007/978-3-662-44500-6

MIX
Papier aus verantwortungsvollen Quellen
Paper from responsible sources
FSC® C105338

If you have any concerns about our products,
you can contact us on
ProductSafety@springernature.com

In case Publisher is established outside the EU,
the EU authorized representative is:
Springer Nature Customer Service Center GmbH
Europaplatz 3, 69115 Heidelberg, Germany

Printed by Libri Plureos GmbH
in Hamburg, Germany